FROM ELECTROCATALYSIS TO FUEL CELLS

From Electrocatalysis

to Fuel Cells

Edited by G. SANDSTEDE

Published for BATTELLE SEATTLE RESEARCH CENTER

by the UNIVERSITY OF WASHINGTON PRESS

SEATTLE *and* LONDON

CHEMISTRY

Library of Congress Cataloging in Publication Data
Main entry under Title:

From electrocatalysis to fuel cells.

   Papers read at a seminar at the Battelle Seattle
Research Center.

   Includes bibliographical references.
   1. Fuel cells--Congresses. 2. Catalysts--
Congresses. I. Sandstede, Gerd, 1929-   ed.
II. Battelle Memorial Institute, Columbus, Ohio.
Seattle Research Center.
TK2931.F76   621.35'9   79-38116
ISBN 0-295-95178-8

PREFACE

The title of the Proceedings "From Electrocatalysis to Fuel Cells" indicates that they cover by and large the research field of fuel cells, with special emphasis being given to results on electrocatalysts. They contain the papers which were read during a three-day seminar at the Battelle Seattle Research Center under the same title. The program of the seminar had been so devised that it comprised both reports on new theories and results of electrocatalysis as well as discussions on the state of the art in fuel cell technology for a multitude of systems. In addition, papers on new secondary battery systems were presented. On this basis it was possible to compare the properties and applications of the different electrochemical energy conversion devices in the light of the latest findings and to discuss in what way and to what extent fuel cells and secondary batteries can supplement each other.

The seminar on fuel cells was held in a phase of development in which some uncertainty existed about the potential applications of fuel cells. After the spectacular successes achieved in space technology, fuel cells were thought to be immediately applicable in traction and to have other applications in the not too distant future. However, they proved to be too expensive or not sufficiently developed. Thus, after a stimulating period of development there have been years of doubt and reflection. Meanwhile substantial progress has been made in the field of engineering, and at present new electrode structures and new electrocatalysts are being developed, which might make it possible to obtain cheaper and more efficient electrodes and less complex accessories.

Considering the advances made in the field of electrocatalysis one must take into account that ten to fifteen years ago we had only just obtained some knowledge on the hydrogen and oxygen electrode; little research, however, had been done on the mechanism of organic reactions, and the possibility of a direct anodic oxidation of the hydrocarbons themselves was not yet known. Although extensive knowledge existed on the electrolytic double layer at mercury, next to nothing was known about the double layer at solid electrodes, so that there was hardly any information available on the surface of the electrocatalyst. Together with the introduction of the term electrocatalysis a closer investigation of the effect of potential on electrosorption and the mechanism of electrooxidation has been started. Even today, theories about electrocatalysis are still in their infancy, so that much progress can be expected as soon as the results of interdisciplinary research, *i.e.*, the input of *e.g.*, quantum chemistry, solid state physics, surface physics, electrodics, will become available.

The more than thirty papers contained in this volume deal in particular with the development, trends and prospects of electrocatalysis and fuel cells, including such interesting fields as electrotraction and implantable fuel cells. Special emphasis has been placed on a detailed discussion of the new inorganic and organic electrocatalysts. As to the outcome of the seminar the reader is referred to the concluding remarks at the end of the Proceedings.

If this seminar encourages further investigations into the development of fuel cells and if these Proceedings stimulate new research, this will be thanks to the participants who, with their papers and

PREFACE

contributions to the discussions, gave a review of the results of fuel cell research and an outline of those problems still confronting us. Everybody has profited from the expertise of the participants and having been the chairman of the meeting I would like to thank the speakers, discussion chairmen and discussion contributors very much for their participation in the seminar and for preparing the papers for publication in these Proceedings.

Many sincere thanks are also due to the Battelle Organization for arranging the meeting and to the Battelle Seattle Research Center staff for the preparation of the seminar and of the Proceedings.

Frankfurt/Main, April, 1971
Germany

Gerd Sandstede

ORGANIZATION

Dr. Ron S. Paul	-	Director, Battelle Memorial Institute, Pacific Northwest Division
Dr. Tommy W. Ambrose	-	Executive Director, Battelle Seattle Research Center
Louis M. Bonnefond	-	Conference Coordinator, Battelle Seattle Research Center
A. W. Roecker	-	Librarian, Battelle Seattle Research Center
Dr. Gerd Sandstede	-	Battelle Institut e.V., Frankfurt, Chairman of the Seminar

LIST OF ATTENDEES

Dr. Otto J. Adlhart
Research and Development of
  Engelhard Industries
497 Delancy Street
Newark, New Jersey 07105

Dr. John S. Batzold
Esso Research & Engineering Co.
Government Research Laboratory
P.O. Box 8
Linden, New Jersey 07036

Dr. Werner Baukal, Head
Solid State Electrochem. Group
Battelle-Institut e.V.
6 Frankfurt/Main 90
Römerhof/Wiesbadener Strasse
Germany

Dr. Klaus D. Beccu, Head
Division of Electrochemistry
Institut Battelle
7, route de Drize
1227 Carouge-Genève
Switzerland

Dr. Ted R. Beck
Materials Science Laboratory
Boeing Scientific Res. Labs.
P. O. Box 3981
Seattle, Washington 98124

Dr. Horst Binder, Asst. Head,
Interfacial Res. & Electro-
  chemistry Division
Battelle-Institut e.V.
6 Frankfurt/Main 90
Römerhof/Wiesbadener Strasse
Germany

Dr. Carlo V. Bocciarelli, Director
Pennsylvania Res. Assoc., Inc.
3401 Market Street
Philadelphia, Pennsylvania 19104ˈ

Professor John O'M. Bockris
Electrochemistry Laboratory
University of Pennsylvania
Philadelphia, Pennsylvania 19104

Dr. Elton J. Cairns
Chemical Engineering Division
Argonne National Laboratory
9700 South Cass Avenue
Argon, Illinois 60439

Dr. Michael A. Callahan
U.S. Army Mobility Equipment
  Research & Development Ctr.
Attn: SMEFB-EA
Ft. Belvoir, Virginia 22060

Dr. Paul D. Cohn
Chemistry & Metallurgy Div.
Battelle Memorial Institute
Pacific Northwest Laboratories
Battelle Boulevard
Richland, Washington 99352

Dr. Denis Doniat, Head
Electrical Processes Section
Institut Battelle
7, route de Drize
1227 Carouge-Genève
Switzerland

Dr. Duane Faletti
Applied Physics Laboratory
University of Washington
Seattle, Washington 98105

Dr. Sol Gilman
Dept. of the Army Headquarters
U.S. Army Electronics Command
Electronic Components Lab.
Ft. Monmouth, New Jersey 07703

Dr. José Giner, Asst. Director
Corporate Research Division
Tyco Corporate Technology Ctr.
16 Hickory Drive
Waltham, Massachusetts 02154

Dr. Derek P. Gregory, Asst. Dir.,
  Engineering Research,
Institute of Gas Technology
3424 South State Street
Chicago, Illinois 60616

LIST OF ATTENDEES

Dr. Sidney Gross
Boeing Company
M.S. 88-06
P.O. Box 3999
Seattle, Washington 98124

Dr. Manfred A. Gutjahr, Head
Sect., Energy Storage & Conversion
Institute Battelle
7, route de Drize
1227 Carouge-Genève
Switzerland

Dr. Carl E. Heath, Project Mgr.
Jersey Enterprises, Inc.
30 Rockefeller Plaza
New York, New York

Dr. James R. Huff
Commanding Officer
U.S. Army Mobility Equipment
   Res. & Dev. Center
Attn: SMEFB-EE
Ft. Belvoir, Virginia 22060

Dr. Karl V. Kordesch
Union Carbide Corporation
Consumer Products Div.
P.O. Box 6116
Cleveland, Ohio 44101

Dr. Wolfgang H. Kuhn
Electrochemistry Group
Battelle-Institut e.V.
6 Frankfurt/Main 90
Römerhof/Wiesbadener Strasse
Germany

Mr. John McHardy
Electrochemistry Laboratory
University of Pennsylvania
Philadelphia, Pennsylvania 19104

Dr. Larry G. Morgan
Chemistry & Metallurgy Div.
Battelle Memorial Institute
Pacific Northwest Laboratories
Battelle Boulevard
Richland, Washington 99352

Mr. William T. Reid, Sr. Fellow
Mechanical Engineering Dept.
Battelle Memorial Institute
Columbus Laboratories
505 King Avenue
Columbus, Ohio 43201

Dr. Gerd Sandstede, Head
Chemistry Department
Battelle-Institut e.V.
6 Frankfurt/Main 90
Römerhof/Wiesbadener Strasse
Germany

Mr. Jan Spalek
Mechanical Engineering Dept.
Battelle Memorial Institute
Pacific Northwest Laboratories
Battelle Boulevard
Richland, Washington 99352

Mr. Bill Smyrl
Boeing Scientific Res. Labs.
P.O. Box 3981
Seattle, Washington 98124

Dr. Edward F. Sverdrup
Manager, Fuel Cell Research
Westinghouse Electric Corp.
Research & Development Ctr.
Beulah Road
Pittsburgh, Pennsylvania 15235

Dr. Helmut Tannenberger, Head
Physical Chemistry Group
Institut Battelle
7, route de Drize
1227 Carouge-Genève
Switzerland

Dr. Michel Voinov
Physical Chemistry Group
Institut Battelle
7, route de Drize
1227 Carouge-Genève
Switzerland

Dr. Sidney K. Wolfson, Jr., M.D.
Director, Surgical Res. Lab.
Michael Reese Hospital & Med. Ctr.
   904 Dreyfus Building
29th Street & Ellis Avenue
Chicago, Illinois 60616

Dr. S. J. Yao
Department of Surgery
Michael Reese Hospital & Med. Ctr.
      Present Address
Department of Surgery
School of Medicine
University of Pittsburgh
Pittsburgh, Pennsylvania 15213

CONTENTS: PAPERS AND SPEAKERS

CONTENTS: PAPERS AND SPEAKERS

CONTENTS: PAPERS AND SPEAKERS

*Not included.

**Paper refers also to Session VII

xi

WELCOME

T. W. AMBROSE

Battelle Seattle Research Center
Seattle, Washington

I extend a hearty welcome to each of you from Battelle Memorial Institute. Fundamentally, Battelle is a not-for-profit research organization. It was created from the Will of Gordon Battelle in 1923. Today there are four Battelle research laboratories throughout the world, two in this country and two in Europe. One, at Columbus, Ohio, which is where Battelle was started, has between 2700 and 2800 people. The other in the United States is at Richland, Washington, with 1400 to 1500 people. And in Europe there are two, the Laboratory at Frankfurt, Germany of about 800 people; and the one at Geneva, Switzerland, with 600 to 700 people.

At the Battelle Research Center here in Seattle there is no laboratory equipment and our research is done primarily with pencil and paper. Our function at Seattle is aimed at another part of the Battelle Will which goes something like, "Contribute to knowledge, the communication of knowledge for the good of mankind." One way we do this is by routinely holding conferences, such as this one. Secondly, we have a collection of visiting research people that come and spend some time with us; all the way from a day or two up to a year, working on papers, writing books, and some for retreading and a chance for study. Some of the conferences that we hold here are for Battelle only and some are for Battelle and the scientific community and then some for community services only. The type of conferences that we are having today really pertains to Battelle's interest in this field as well as in the intense interest of the scientific community.

INTRODUCTION: TYPES OF FUEL CELLS

G. SANDSTEDE

Battelle-Institut e.V., Frankfurt

SUMMARY

Fuel cells may be classified according to the following characteristics:

1) Type of electrolyte
2) Type of electrode
3) Type of fuel
4) Temperature
5) Type of catalyst

The electrolytes have to be distinguished by the following four types of ions conducting the electric current:

$$O^{2-}, CO_3^{2-}, OH^-, H^+$$

The most important types of electrodes are the following:

Electrode with a conductive covering layer with fine pores (double porosity electrodes)

Electrode with a non-conductive covering layer with fine pores

Electrode-electrolyte vehicle combination

Electrode ion exchange membrane electrolyte combination

Hydrophobic electrode

Flow-through electrode

Immersion electrode

In this short systematic survey the different types of fuel cells are distinguished by temperature and the following fuels:

Biological fuel - Carbon - Hydrocarbons - Methanol - Formic Acid - Formaldehyde - Hydrazine - Hydrogen - Ammonia - Metals

The survey also includes the different types of converters. A classification of the fuel cells by catalysts is not expedient as these have not yet been systematized according to theoretical criteria. In addition to various types of fuel cells, the biological oxidation in the respiratory chain is discussed, because it may yield information on fuel cell catalysis.

INTRODUCTION

This survey covers the electrochemical oxidation, including the biological oxidation, and methods for the classification of the different types of fuel cells. While the different types of cells are discussed, no reference will be made to their historical development. Several books and many reviews on fuel cells were published in the last few years [1-21], in which quite a number of fuel cells and their history are described in detail. Applications, too, will not be discussed in this context (since they will be dealt with in the following papers). The principles of the different cells, however, will be described as they are the basis for the different applications.

As an introduction the principle of the respiratory chain (Fig. 1) is discussed, because the mechanism of electrochemical oxidation taking place in nature may yield information on fuel cell processes. For millions of years electrochemical reactions have been taking place in the biological organism. The free energy of food is converted into electric, mechanical or chemical energy in another substance (ATP). This conversion of carbonaceous and hydrogen-containing substances involves the biological oxidation by oxygen, which is — at least partly — an electrochemical reaction. In this respect the respiratory chain is a type of fuel cell. It is supplemented by the citrate cycle to the effect that the biological oxidation to water and carbon dioxide yields carbon dioxide in the citrate cycle, whereas hydrogen is converted in the respiratory chain. The reaction of hydrogen with oxygen does not take place in one stage, but requires a large number of intermediate steps (in the Figure from bottom to top).

Looking at the different steps in the respiratory chain one may speculate upon gaining knowledge that could be used in fuel cell catalyst research. If immediately applicable findings about the mechanism of catalysis can not be obtained, there may at least be a stimulation of research directions. Some evidence to this effect has already been obtained.

In general the enzyme catalysts consist of the prosthetic (active) groups and a large mass of protein, which acts not only as a carrier (support) but has a promoting effect and is responsible for the specificity. Thus, particular linking groups, for example nitrogen or sulfur, may also have a favorable effect when they are introduced into the surface of a fuel cell catalysts support (first starting points are discussed in [22] and [23]).

In the reduction of oxygen, intermediate formation of a chemisorption bond to a cytochrome is necessary. The bond is generated by coordinative bonding to the central iron ion of a porphyrin derivative. This knowledge is now being used in designing oxygen catalysts for fuel cell cathodes [23].

Other metal chelates, again mainly iron porphyrin derivatives, serve in the respiratory chain as redox systems for electron transport, whereas hydrogen ions are responsible for ionic conduction. At present we have no useful information about the electrochemical oxidation of the hydrogen. It is true that some of the active species involved are known, but the potential-determining step of hydrogen-ion formation has not yet been definitely formulated. The most important reaction seems to be the oxidation of ubihydroquinone by cytochrome b_2Fe^{III}, but the first step involving nicotine amide adenosine dinucleotide may also play a role.

Quinones are not only used in redox couples in the respiratory chain but also in other biological systems. For this reason they are excellent candidates for cathode materials of fuel cells and batteries as well [24].

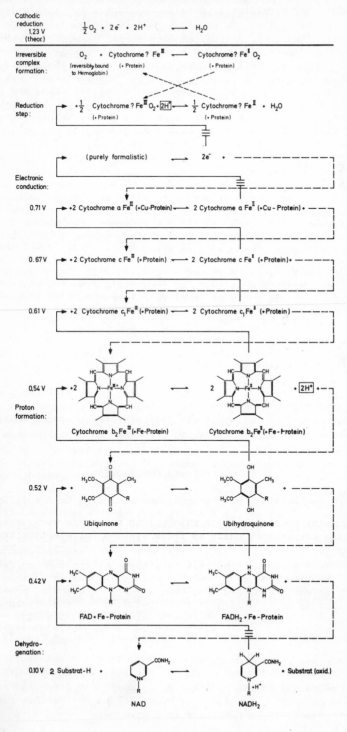

FIG. 1 Biological oxidation of combined hydrogen by oxygen in the respiratory chain (including standard potentials of various redox reactions).

There is ample information on the basic structure of the respiratory chain. Fig. 1, for example, shows the redox potentials (hydrogen potential equals zero in the same electrolyte) of the individual redox systems [25]. The order of the reaction steps, however, as shown in the Figure may change because the concentration of the substances and the type of protein have a yet unknown effect on the potential. As the Figure suggests, there are certainly further intermediate steps and fairly little is known about the numerous individual steps and their catalysts so that the principle of bionics does not yet substantially benefit fuel cell research. However, as soon as we obtain more detailed knowledge about the biological events, bionics may well become useful, as may be inferred from the results achieved by systematic investigation of porphyrin derivatives already mentioned.

CLASSIFICATION ATTEMPT

If the technological fuel cell is compared with the biological equivalent, one might assume at first glance that in technology the subtility of nature has been replaced by the simple concept of anode-electrolyte-cathode. At a closer look, however, it is noted that electrochemical reactions at technical electrodes frequently involve many steps. However, the reaction mechanisms will not be discussed here, because the different types of fuel cells can adequately be described in terms of gross electode reactions. Only if electrocatalysis is discussed it is necessary to specify the reactions.

Fuel cells may be classified by the following criteria:

1) Type of electrolytes
2) Type of electrodes
3) Type of fuel
4) Temperature
5) Type of catalysts

The following discussion of the different types of fuel cells* is based on this classification, making reference to some findings of earlier studies [5,12].

ELECTROLYTES AND ELECTRODES

The electrolytes can be distinguished according to the following four ions, which conduct the electric current: Oxygen ions, carbonate ions, hydroxide ions, hydrogen ions. The types of cells derived from these electrolytes are shown in Fig. 2. The two cells on the left are high-temperature cells provided either with solid electrolytes, e.g., zirconium oxide doped yttrium oxide, or fused salts; the two cells on the right are low-temperature cells. The cells need no further explanation, it may be noted, however, that water as a reaction product is formed at the anode if the current is conducted via the oxygen-containing ions, and at the cathode if an acid electrolyte is used.

The most important types of electrodes are shown in Fig. 3:

*The references quoted here are by no means complete; the reader is referred to the books listed in the bibliography for further references and in particular, if no reference is indicated.

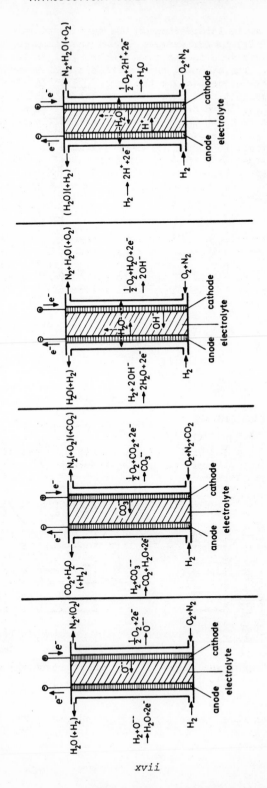

FIG. 2 Basic types of fuel cells according to the electrolytes used.

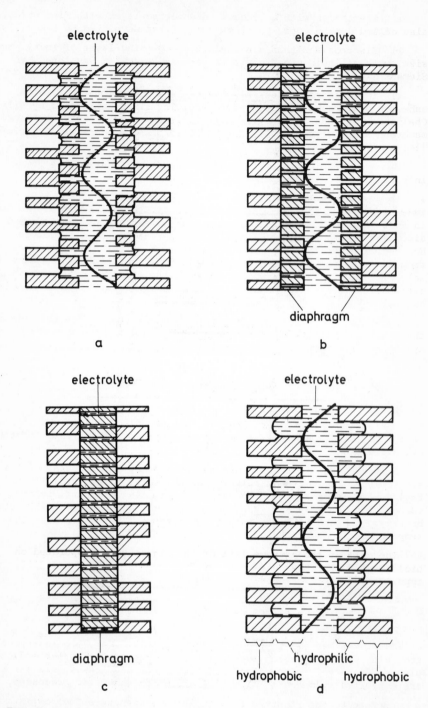

FIG. 3 *Basic types of fuel cells according to the structure of electrodes.*

a) Electrodes with a conductive covering layer with fine pores, also called double porosity electrodes, first developed by Bacon [19].

b) Electrodes with a non-conductive covering layer of small pore size, with a liquid electrolyte in between, a system developed by Siemens [16].

c) The electrode-electrolyte vehicle combination, *e.g.*, with asbestos diaphragms as electrolyte vehicle, first introduced by Allis Chalmers [17]. An alternative version is the electrode-ion exchange membrane-electrolyte combination, as used first by General Electric [14].

d) Hydrophobic electrodes, which operate without excess pressure, in particular those developed by Kordesch [15,26].

There are also flow-through electrodes — *e.g.*, by Varta [20] — and immersion electrodes [3,10]. The latter are shown in Fig. 4. Anodes in fuel cells, which are fed with dissolvable fuel, need not have a gas side, and can therefore be completely immersed in the electrolyte. However, they can also be supplied with fuel on the side turned away from the electrolyte [19]. At the left the Figure shows a methanol cell with an alkaline electrolyte and in the middle a cell with an acid electrolyte. These cells can also be fed with formic acid or aldehydes as fuel.

Hydrazine as a fuel is used in the cell on the right-hand side of the Figure. Unlike the alkaline methanol cell, the electrolyte of the hydrazine cell remains invariant.

EXPLANATION OF THE CELL TYPES

After the brief description of electrolytes and electrodes, fuel cells classified by temperature and fuel are discussed in the following on the basis of a diagram, in which both parameters are used for the classification.

In Fig. 5 the fuel cells are represented by the formulas of the electrolyte. In addition to temperature, fuel and electrolyte, this figure also shows types of converters, including crackers, reformers, carbon dioxide removers, *etc*. The different types of fuel cells are briefly described by moving along the abscissa of the figure. The temperature is indicated on the ordinate.

The cell at the extreme left of the Figure can be operated on biological fuel either directly or after conversion. Three different types can be distinguished:

1) Biological oxidation in the respiratory chain.
2) Implantable fuel cells
3) Biochemical fuel cells

The first has already been described above. For (2) electrodes are being developed which convert in the organism the oxygen and the glucose dissolved in the blood. In the biochemical fuel cell (3) fuels are converted by means of microbial and other enzymatic processes.

Further to the right we find the high-temperature fuel cells. The high-temperature fuel cell with solid electrolyte may be operated either with gasified carbon or reformed hydrocarbons. The operating temperature

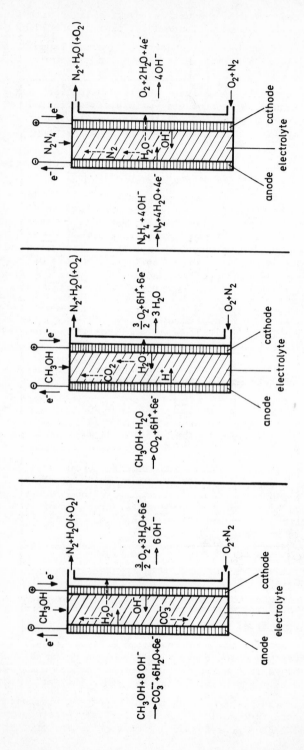

FIG. 4 *Types of fuel cells with immersion electrodes for electrolyte dissolvable fuels.*

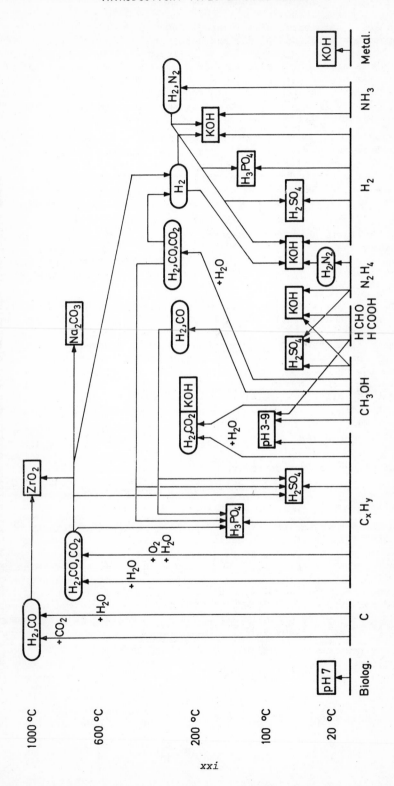

FIG. 5 Types of fuel cells classified by temperature and fuel as well as electrolyte (including various types of converters and purifiers).

of the fused carbonate cells is about 600°C to 700°C. Owing to the decomposition into carbon they are unable to convert the hydrocarbons directly, but must be operated with an integrated reformer. This temperature gives optimum efficiency for a cell provided with a reformer. At higher temperatures the thermodynamic efficiency decreases substantially. In connection with low-temperature cells, on the other hand, the heat loss of the reformer is considerable owing to the endothermal reaction and the dissipation of heat involved.

Nevertheless, most of the research effort is concentrated on low-temperature cells, because their prospects are the most promising from the technical point of view. Theoretically, the phosphoric acid cell is the most suitable because it supplies the heat required for the vaporization of water and rejects the carbon dioxide. It should also convert the carbon monoxide or should at least not be adversely affected by it.

The sulfuric acid cell operating at temperatures below 100°C does not supply the heat necessary for vaporization. It still has good prospects at present, because highly active nonplatinum metal catalysts have been found, $e.g.$, [23,27]. The activity and the lifetime of the oxygen electrode, however, still need improving.

The best cells developed so far are the alkaline hydrogen cells. But if carbonaceous fuels are to be used, they are theoretically the most unfavorable cells, because the carbon monoxide has to be shifted to carbon dioxide, and carbon dioxide has to be removed. After the purification steps — marked in the figure by H_2 — the hydrogen can be used in two types of cells with alkaline electrolyte. One arrow of the picture points to the potential application of the intermediate-temperature cell, the other to the low-temperature cell. These two types of cells are the most advanced cells, also operated with non-platinum metal catalysts.

The direct conversion of hydrocarbons indicated by three arrows has been closely investigated in the cells with phosphoric acid and sulfuric acid or with dissolved carbonate and other solutions (pH 3-9) [14]. The arrows in the Figure start in each case from the hydrocarbon C_xH_y. Most of the investigations have been postponed because the amounts of platinum metal required as catalyst are too large.

An attempt to use an internal reforming cell failed because of the insufficient lifetime of the palladium membrane. As indicated by the two arrows, either a light hydrocarbon or methanol can be used in the internal reforming cell [17]. Reforming with water vapor is also possible with methanol; this can be effected at moderate temperatures of about 300°C [12,28]. The gas mixture can then be used in the way described for hydrocarbons (see same Figure). Thermal cracking of methanol to hydrogen and carbon monoxide at a temperature of 220°C has been demonstrated recently [27].

The direct methanol cells [10,17,29,30], indicated by the arrows from the bottom of Fig. 5 to H_2SO_4 — or KOH-cell have already been mentioned. So far, the use of nonplatinum metal catalysts has not become known.

Formic acid-air cells with acid electrolyte have not yet been manufactured, but can now be developed without difficulty [22].

Formate-air cells are available [19]. Formaldehyde electrodes and cells with nonplatinum metal catalysts are in their development stage [27].

Hydrazine is used either directly or after decomposition. In view of the high reaction rate of hydrazine in alkaline electrolytes nearly all companies conducting fuel cell research have tried to develop hydrazine cells. However, at present the price for the hydrazine prohibits the use of such cells for non-military purposes.

Next, Fig. 5 shows four types of hydrogen cells: low- and medium-temperature cells both with alkaline and acid electrolyte. The development of all four types is being continued. The catalyst required for the phosphoric acid cell is platinum; for the other three types batteries with nonplatinum metal catalysts also are being developed.

In all types of hydrogen cells cracked ammonia gas may be used as fuel. For the direct oxidation of ammonia in alkaline electrolytes too much platinum metal (Ir) is required [17,21].

Finally, Fig. 5 shows an alkaline low-temperature fuel cell which operates on metallic fuels. Cells of this type include sodium amalgam/air and zinc/air cells. With zinc/air cells the zinc can be regenerated by electrolysis of the electrolyte. A zinc/air cell and other metal/air cells can also be designed as secondary batteries.

Regenerative fuel cells for practical applications have not yet been developed although many systems have been investigated. They will not be discussed here as they would exceed the scope of this survey.

CONCLUSIONS

Classification of fuel cells by catalysts used would be an interesting task. However, to this end the catalysts would have to be classified first. Lately, several new electrocatalysts have been discovered; in addition, theoretical studies have been undertaken which led to a better understanding of the mode of action of some types of catalysts. A classification of electrocatalysts on the basis of theoretical viewpoints would be desirable. Besides the catalyst problem also the materials problem is of great significance to the fuel cells. In view of the three parameters fuel, electrolyte and temperature the variety of potential fuel cells is vast. Practical application of these cells will depend on how the specific requirements and the economic feasibility of the different types can be harmonized.

REFERENCES

[1]   E. Justi, M. Pilkuhn, W. Scheibe and A. Winsel, *Hochbelastbare Wasserstoff-Diffusions-Electrode*, Franz Steiner Verlag, Wiesbaden (1959).

[2]   G. J. Young (Ed.), *Fuel Cells I and II*, Reinhold Publ. Corp., New York, I (1960), II (1963).

[3]   E. Justi and A. Winsel, *Kalte Verbrennung — Fuel Cells*, Franz Steiner Verlag, Wiesbaden (1962).

[4]   W. Mitchell, Jr. (Ed.), *Fuel Cells*, Academic Press, New York (1963).

[5]   G. Sandstede, "Galvanische Brennstoffzellen" in *Dechema-Monographien*, <u>49</u>, Verlag Chemie, Weinheim/Bergstr. (1964).

[6]   *Fuel Cell Systems I and II*, Advances in Chemistry Series Nos. 47 and 90, American Chemical Society, Washington, D. C. I (1965), II (1969).

[7]   W. Vielstich, *Brennstoffelemente*, Verlag Chemie, Weinheim/Bergstr. (1965).

[8]   B. S. Baker (Ed.), *Hydrocarbon Fuel Cell Technology*, Academic Press, New York (1965).

[9]   *Les Piles à Combustible*, Institut Français du Pétrole, Editions Technip, Paris (1965).

[10]   K. R. Williams (Ed.), *An Introduction to Fuel Cells*, Elsevier Publ. Co., Amsterdam (1966).

[11]   M. Barak, "Fuel Cells — Present Position and Outstanding Problems," *Advanced Energy Conversion*, Vol. 6, Pergamon Press, Oxford (1966), p. 29.

[12]   G. Sandstede, "Elektrochemische Brennstoffzellen," in *Fortschr. Chem. Forsch.*, <u>8</u>, Springer Verlag, Berlin, Heidelberg, New York (1967).

[13]   A. B. Hart and G. J. Womack, *Fuel Cells: Theory and Application*, Chapman and Hall, London (1967).

[14]   H. A. Liebhafsky and E. J. Cairns, *Fuel Cells and Fuel Batteries*, John Wiley & Sons, New York (1968).

[15]   C. Berger (Ed.), *Handbook of Fuel Cell Technology*, Prentice Hall, Englewood Cliffs, New York (1968).

[16]   F. v. Sturm, *Elektrochemische Stromerzeugung*, Verlag Chemie, Weinheim/Bergstr. (1969).

[17]   J. O'M. Bockris and S. Srinivasan, *Fuel Cells: Their Electrochemistry*, McGraw-Hill Book Co., Inc., New York (1969).

[18]   M. W. Breiter, *Electrochemical Processes in Fuel Cells*, Springer-Verlag, Berlin, Heidelberg, New York (1969).

[19]   W. Vielstich, *Fuel Cells*, Wiley-Interscience, New York (1970).

[20]   H. H. v. Döhren and K. J. Euler, *Brennstoffelemente*, (Varta Fachbuch Band 6) VDI Verlag, Düsseldorf (1970).

[21]   J. O'M. Bockris and A.K.N. Reddy, *Modern Electrochemistry*, Plenum Press, New York (1970).

[22]   H. Binder, A. Köhling and G. Sandstede, "Platinum Catalysts Modified by Adsorption or Mixing with Inorganic Substances," *these Proceedings*, p. 59.

[23]   H. Alt, H. Binder, W. Lindner and G. Sandstede, *these Proceedings*, p. 113.

[24]   H. Alt, H. Binder, A. Köhling and G. Sandstede, *these Proceedings*, p. 335.

[25]   E. Buddecke, *Grundriss der Biochemie*, Walter de Gruyter & Co., Berlin (1970).

[26] K. V. Kordesch, *Proceedings of the 21st Power Sources Conference,*
      PSC Publications Committee, Red Bank, New Jersey (1967), p. 14.

[27] K. v. Benda, H. Binder, A. Köhling and G. Sandstede, *these
      Proceedings,* p. 87.

[28] O. Bloch, C. Dezael et M. Prigent, *Comptes Rend. Troisièmes
      Journées Internat. d'Etude des Piles à Combustibles,* Presses
      Académiques Européennes, Brussels (1969) p. 180.

[29] H. Binder, A. Köhling, W. H. Kuhn, W. Lindner and G. Sandstede,
      these Proceedings, p. 131.

[30] H. Binder, A. Köhling and G. Sandstede, "Effect of Alloying
      Components on the Catalytic Activity of Platinum in the Case
      of Carbonaceous Fuels," *these Proceedings,* p. 43.

[31] E. L. Simons, D. W. McKee and E. J. Cairns, *Comptes Rend.
      Troisièmes Journées Internat. d'Etude des Piles à Combustibles,*
      Presses Académiques Européennes, Brussels (1969) p. 131.

SESSION I

METAL CATALYSTS FOR FUEL CELLS

Chairman

J. O'M. Bockris

MECHANISMS AND CATALYSIS OF HYDROCARBON FUELS ON PLATINUM

J. R. Huff

U.S. Army Mobility Equipment Research & Development Center
Fort Belvoir, Virginia

ABSTRACT

The mechanisms and catalysis of direct hydrocarbon oxidation on fuel cell electrodes has long been of interest since it offers the key to a simple, reliable source of electric power with many other desirable qualities. Over the past few years, a number of modes of attack have been employed in attempts to understand both the mechanism of oxidation and the catalytic factors which might increase the rate of reaction. Multipulse potentiodynamic techniques and other electrochemical methods in combination with isotopic exchange and gas chromatography have given some insight into the major problems and also some hope in solving the chief mechanistic and catalytic roadblocks to a direct hydrocarbon fuel cell.

The mechanism and catalysis of hydrocarbon oxidation on platinum has long been of interest to investigators of fuel cell phenomena. Proper utilization of direct oxidation offers great gains in efficiency and reliability. A number of modes of approach have been exploited in attempts to understand both the mechanism of oxidation and the catalytic factors which might increase the rate of reaction. Definition of an overall reaction mechanism was accomplished by Brummer [1] at Tyco Laboratories. Further use of the knowledge gained in these studies has led to some interesting catalytic effects. Other lines of attack have included isotopic exchange and gas chromatography combined with electrochemical techniques. These combined efforts are directed towards achieving direct oxidation of hydrocarbon fuels at a fuel cell electrode.

Brummer [1] has shown that the oxidation of a hydrocarbon to carbon dioxide in phosphoric acid involves adsorption of the molecule on the electrocatalyst followed by rapid formation first of an intermediate designated CH-α then further rapid conversion to a species identified as O-type. The subsequent oxidation of O-type to CO_2 is the slow step in the reaction. A third species, CH-β, may also be formed under certain conditions, but it is not of concern in the direct oxidation path. The scheme, as shown in Fig. 1, holds for potentials less than 0.35V *vs.* RHE. At higher potentials (*e.g.*, 0.50V), the slow step becomes the initial adsorption process itself and the adsorbate reacts rapidly to form CO_2.

It is clear that there is much more that needs to be known about the reaction for a complete understanding of the mechanism. However, the reaction is sufficiently well understood that the present state of knowledge can be used as a guide in catalyst development.

The use of acid electrolyte is advantageous from the point of view of CO_2 rejection, but causes severe corrosion problems. Because of the

3

$$C_3H_8 + 6H_2O \xrightarrow{Pt} 3CO_2 + 20H^+ + 20e^-$$

<u>at low potentials</u> ($\leq 0.45V$ *vs.* R.H.E.)

$$C_3H_8 \xrightarrow{-5e^-} \text{CH-CH-CH} \xrightarrow[+H_2O]{-6e^-} 3\text{C-OH} \xrightarrow[+H_2O]{-9e^-} 3CO_2$$

$$\text{fuel} \xrightarrow[\text{fast}]{} (CH-\alpha) \xrightarrow[\text{fast}]{} (O\text{-type}) \xrightarrow[\text{slow}]{} CO_2$$

<u>at high potentials</u>

$$\text{fuel} \xrightarrow[\text{slow}]{} (CH-\alpha) \xrightarrow[\text{fast}]{} \begin{array}{c}\text{intermediates}\\ \text{unknown}\end{array} \xrightarrow[\text{fast}]{} CO_2$$

FIG. 1 *Mechanism of hydrocarbon oxidation. (All Figures courtesy of Tyco Laboratories, Waltham, Massachusetts.)*

corrosive environment, the choice of catalytic materials is limited. Brummer's approach has been to seek some method of stabilizing non-noble metals so that they can be used. Adsorption on an inert substrate is one possibility. The essence of the method is that in sub-monolayer amounts the reversible potential for metal dissolution is shifted in the noble direction (Fig. 2). Therefore, the stability of the non-noble metal against corrosion is enhanced.

$$M^{+z} + ze^- \rightleftharpoons M$$

$$E = E_0 + \frac{RT}{zF} \ln \frac{a_{M^{+z}}}{a_M}$$

$$E = E_0' - \frac{RT}{zF} \ln a_M$$

$$a_M \ll 1 \text{ for adsorbed sub-monolayer}$$

FIG. 2 *Principle of adsorption stabilization.*

Experiments have been carried out using Pt as the substrate in 80% H_3PO_4 at 130°C. The effect of Cu, Ni, Al and Na on the oxidation rates of CH_4 and C_3H_8 was studied as a function of electrode potential and additive concentration. Cu was found to form a stable sub-monolayer, unfortunately it inhibited the reaction (Fig. 3). The other three metals do not adsorb on Pt in the potential region of interest, but they do produce a significant enhancement of the C_3H_8 oxidation rate when present as ions in solution. The concentrations used ranged from 20mM to 200mM for Ni^{+2}, to 400mM for Al^{+3}, and to 4M for Na^+. The source of Ni^{+2} was Baker Analyzed Reagent $NiCO_3$ dissolved in 80% H_3PO_4. Al^{+3} was added by dissolving Fisher Al_2O_3 (80-200 mesh) in 80% H_3PO_4. Na^+ was introduced in the form of Baker and Adamson Reagent $NaH_2PO_4 \cdot H_2O$ dissolved in 80% H_3PO_4. It was necessary to preelectrolyze the electrolytes to obtain "clean" systems, particularly in the case of Ni^{+2} where the impurity present inhibited the reaction to the extent that no enhancement was observed (Fig. 4).

The effects on methane and propane oxidation as a function of Ni^{+2} concentration are shown in Figs. 5 and 6. The enhancement increases with the ion concentration increase to a level of saturation. Maximum enhancement was achieved with Ni^{+2} ranging up to a factor of 10 depending upon the time of operation. One of the more interesting results is shown in Fig. 7. Here, the long term behavior of propane oxidation with

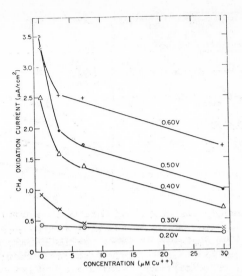

FIG. 3 *Oxidation rate of CH₄ on platinized Pt as a function of*
Cu⁺⁺ concentration.

and without Ni^{+2} in the electrolyte is depicted. It can be concluded
from this result that Ni^{+2} not only enhances the overall oxidation
rate, as seen from the initial current values, but also greatly improves the long term performance. The results of a different test to
examine the long term performance are shown in Fig. 8. The potential
was held at the same level as Fig. 7, 0.50V, but the concentration of
Ni^{+2} was varied. The duration of the test as well as the current levels sustained demonstrate that the Ni^{+2} does indeed enhance both the
performance level and the life of an electrode oxidizing propane.

Inasmuch as these additives would eventually be used in an actual
fuel cell, it was necessary to determine what effect, if any, they
might have on cathode performance. Preliminary experiments with Na^+
at concentrations of 2M and lower have shown an enhancement of the O_2
reduction current. Thus, in this case, the performance of both fuel
cell electrodes is improved.

Based upon these results, further efforts will concentrate on du-
plicating these performance levels in an actual fuel cell.

In order to obtain more information about the actual structure of
the intermediates comprising CH-α, Grubb [2], Niedrach and Tochner [3],
and Barger and Savitz [4] studied the reaction of propane on platinum
fuel cell electrodes. The CH-α moieties are desorbable if a cathodic
pulse is applied to the electrode, and each investigator detected
methane, ethane and propane when this procedure was followed. The relative amounts of methane and ethane obtained were found to be dependent on the adsorption potential [4]. These results gave a clear picture of the influence of potential on the course of the reaction and
also indicated the need for information about the bonding of the intermediates to the catalyst. Therefore, a series of isotopic exchange
experiments were conducted in order to gain some insight into bond
making and breaking on the catalyst surface. Deuterophosphoric acid
in D_2O was used as the electrolyte. The basic assumption in these
studies was that C-D bonds would be formed on cathodic pulse where

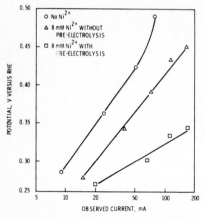

FIG. 4 C_3H_8 *oxidation currents on a partially immersed fuel cell electrode as a function of potential of adsorption and Ni^{++} concentration.*

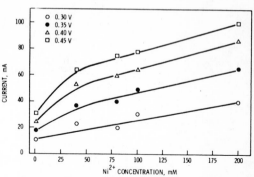

FIG. 5 CH_4 *oxidation currents on a partially immersed fuel cell electrode as a function of Ni^{++} concentration and potential of adsorption.*

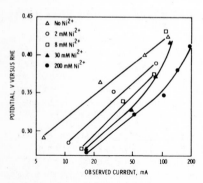

FIG. 6 C_3H_8 *oxidation currents on a partially immersed fuel cell electrode as a function of potential of adsorption and Ni^{++} concentration.*

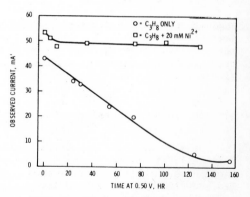

FIG. 7 *Time effect on the oxidation of C_3H_8 with and without Ni^{++}.*

C-catalyst bonds had been formed. To date, propane [5] and benzene [6] have been used as the adsorbents under a variety of conditions and over a potential range from 0.20V to 0.50V versus the dynamic hydrogen electrode [7]. The reactant was passed over a fuel cell electrode which had been subjected to a pretreatment regime to prepare a reproducible catalyst surface. The electrode was held at the potential of interest while adsorption took place, then it was cathodically pulsed and the desorbed products were collected and analyzed with a mass spectrometer. In the case of propane, at all potentials, the desorption products, methane, ethane and propane, were found to be either completely deuterated or to contain only one hydrogen. There was a trend towards increased amounts of species containing only one hydrogen as the adsorption potential was made more anodic. Adsorption times were varied from about 40 minutes to 1 minute without any appreciable effect on the

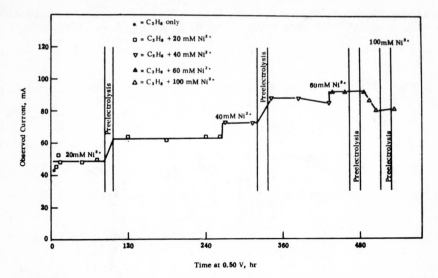

FIG. 8 *Time effect on the* C_3H_8 *oxidation current at nominal 0.5V, with variation of* Ni^{++} *concentration.*

extent of deuteration. These results indicate that exchange occurs much faster in this potential range than later steps in the reaction, *i.e.* C-H, C-D, and C-catalyst bonds are made and broken much faster than conversion to some other product closer to CO_2 on the reaction path takes place.

The behavior of benzene is more complex. The exchange occurs so rapidly that the process becomes diffusion controlled. The isotopic distribution in the products indicates that there may be two types of reactions occurring concurrently. One reaction appears to involve step-wise exchange and the other multiple exchange. It can be postulated that a different type of reaction site is required for each reaction. Further efforts are being made to clarify dependence of these reactions on potential.

Another approach aimed at elucidating the complex interaction between catalyst and hydrocarbon is also being pursued. This method is based on the assumption that by proper choice of hydrocarbons one can gain some insight as to the mode of adsorption on the electrocatalyst [8]. For example, dimethylcycloolefins will yield both *cis-* and *trans-* dimethylcycloalkanes upon cathodic pulse and the cycloalkane isomer ratio allows speculation as to the adsorption geometry. The olefins 1,2-dimethylcyclohexene, 2,3-dimethylcyclohexene, 1,2-dimethylcyclopentene and 2,3-dimethylcyclopentene were chosen for this study for several reasons: this type or quite similar types of molecules occur in substantial concentrations in logistic hydrocarbon fuels, the molecules have been extensively studied under normal heterogeneous catalysis conditions [9] so the effect of electrocatalyst potential, if any, could be determined, and the reduction products, *cis-* and *trans-* 1,2-dimethylcycloalkanes, are readily separable and identifiable by gas chromatography.

The working electrode in these studies was the same type as that used in the propane and benzene exchange studies discussed above, *i.e.*, an American Cyanamid type LAA25 consisting of a mixture of 25 mg/cm^2

platinum black and 25 wt. % teflon on a tantalum screen. The type of
electrochemical cell used in these experiments is shown in Fig. 9.

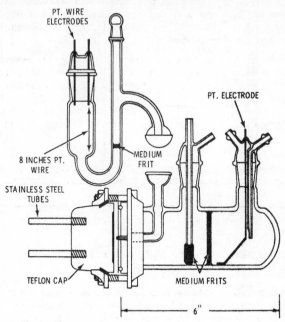

FIG. 9 *Schematic diagram of electrochemical cell.*

A potential step regime, Fig. 10, was used to maintain as reproducible a catalyst surface as possible. With helium passing through
the gas cavity, the working electrode was held at 1.35V *vs.* DHE until
the current observed was less than 10 ma; this was to oxidize the surface and remove any adsorbed hydrocarbons. Next, the potential was
stepped to 0.05V and held for twenty minutes to reduce the oxide film.
The adsorption potential was then imposed on the electrode and the
hydrocarbon reactant entrained in the helium stream was passed over
the working electrode for one hour at which time steady state had been

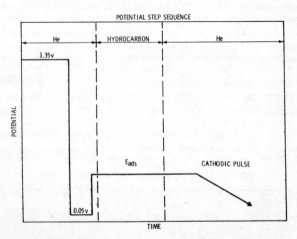

FIG. 10 *Potential step sequence.*

reached. The system was then swept with inert gas for thirty minutes
after which time only trace amounts of reactant could be detected by
gas chromatography. The working electrode was galvanostatically, cathodically pulsed at 1.0 amp and the resulting, evolved products were
either trapped in a solvent (2,2-dimethylbutane or octane) or passed
directly into the gas chromatograph.

The reduced product isomer ratios obtained under various conditions
for the cyclopentenes are shown in Table I. The similarity of results
is probably due to the fact that an equilibrium concentration of the
two is established over the electrode surface no matter which one is
the starting reactant. This equilibrium is reached with or without an
applied potential. After adsorption at 0.30V, cathodic desorption

TABLE I

Product Ratios for the Dimethylcyclopentene Isomers

Adsorption Products	Reactants	1,2 DMCP/2,3 DMCP During Adsorption	CIS/TRANS During Adsorption	CIS/TRANS After Pulse
.30V	1,2 DMCP	7.78		0.53
	2,3 DMCP	7.53		0.51
.50V	1,2 DMCP	6.50		0.77
	2,3 DMCP			
.05V	1,2 DMCP		0.46	
	2,3 DMCP		0.58	

yielded a mean cis/trans ratio of 0.52. The cyclopentane and methyl-
cyclopentane (Table II) obtained along with the dimethylcyclopentane
isomers amount to approximately 15% of the total pulse products. When the
working electrode was potentiostated at 0.05V, passing either of the
dimethylcyclopropanes over it resulted in a mean cis/trans ratio of
0.47. A mean cis/trans ratio of 0.51 obtained at 0.20V substantiated
the potential independence of this ratio.

TABLE II

Observed Products from Dimethylcyclopentene Isomers

Reactants	Products During Adsorption	After Pulse
1,2 Dimethyl-cyclopentene	1,2 Dimethyl-cyclopentene	cis 1,2 Dimethyl-cyclopentane
2,3 Dimethyl-cyclopentene	2,3 Dimethyl-cyclopentene	trans 1,2 Dimethyl-cyclopentane
	cis 1,2 Dimethyl-cyclopentane	Methylcyclopentane
	trans 1,2 Dimethyl-cyclopentane	Cyclopentane

Similar experiments were carried out using dimethylcyclohexenes and
the isomer product ratios obtained are listed in Table III. An equilibrium situation like that of the dimethylcyclopentenes was noted with

TABLE III

Product Ratios for the Dimethylcyclohexene Isomers

Adsorption Potential	Reactants	1,2 DMCH/2,3 DMCH During Adsorption	CIS/TRANS During Adsorption	CIS/TRANS After Pulse
.30V	1,2 DMCH	2.48		2.21
	2,3 DMCH	2.44		2.18
.50V	1,2 DMCH	2.57		2.47
	2,3 DMCH	2.52		2.53
.05V	1,2 DMCH		1.12	
	2,3 DMCH		1.17	

these molecules also, thus the cis/trans ratios found were independent of starting material. There are two features of interest in Table III. First, the cis/trans ratios show a higher percentage of the cis isomer upon cathodic desorption than was shown for the cyclopentenes. Second, the ratio at 0.05V was approximately half that at 0.30V and 0.50V indicating a potential dependence of product formation in this case. The first phenomenon is similar to observations made during studies of dimethylcyclohexene and dimethylcyclopentene under normal heterogeneous catalysis conditions [10,11].

The products obtained during adsorption and after cathodic pulse (Table IV) again include the monosubstituted cycloalkane and the cycloalkane. An aromatic compound, ortho-xylene was also found in substantial quantities (2-10%) at all potentials studied.

TABLE IV

Observed Products from Dimethylcyclohexene Isomers

Reactants	Products During Adsorption	After Pulse
1,2 Dimethyl-cyclohexene	1,2 Dimethyl-cyclohexene	cis 1,2 Dimethyl-cyclohexane
2,3 Dimethyl-cyclohexene	2,3 Dimethyl-cyclohexene	trans 1,2 Dimethyl-cyclohexane
	cis 1,2 Dimethyl-cyclohexane	Methylcyclohexane
	trans 1,2 Dimethyl-cyclohexane	Cyclohexane
	o-xylene	o-xylene

Finding o-xylene made it necessary to determine its influence on the isomer ratios, thus this compound was studied under the same conditions and the results are shown in Table V. Since the ratios obtained were the same as those found for the dimethylcycloalkanes, it does not appear that the presence of o-xylene would have any significant affect on the results.

TABLE V

Comparison of the Product Ratios for
Dimethylcyclohexene Isomers and o-xylene

Adsorption Potential	Reactants	1,2 DMCH/2,3 DMCH During Adsorption	CIS/TRANS During Adsorption	CIS/TRANS After Pulse
.30V	1,2 DMCH	2.48		2.21
	2,3 DMCH	2.44		2.18
	o-xylene			2.20
.50V	1,2 DMCH	2.57		2.47
	2,3 DMCH	2.52		2.53
	o-xylene			2.54
.05V	1,2 DMCH		1.12	
	2,3 DMCH		1.17	
	o-xylene		1.51	

Interpretation of the results obtained is not as simple as was hoped; however some general conclusions can be made.

In the case of the cyclopentenes, the cis/trans ratio is considerably lower than would be predicted from the 1,2/2,3 dimethylcyclopentene ratio. Apparently, the ratio is not determined solely by the equilibrium conditions but may also be a consequence of adsorption geometry of the intermediate and the thermodynamic stabilities of the isomers. Fig. 11 illustrates a possible mechanism to account for increased trans isomer percentage. This involves partial desorption and free rotation about a single bond thereby allowing formation of the trans isomer. The apparent driving force for this reaction could be the greater thermodynamic stability of the trans isomer. It should also be noted that the formation of the monosubstituted cycloalkane and the cycloalkane attaches some importance to the interaction of the methyl group with the catalyst during the adsorption process.

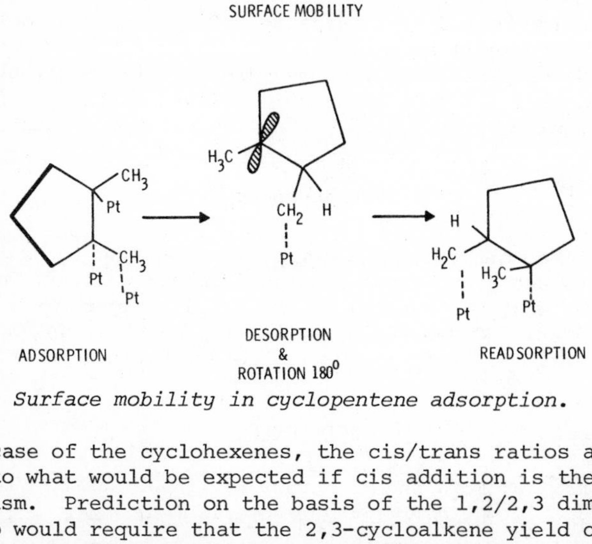

SURFACE MOBILITY

FIG. 11 *Surface mobility in cyclopentene adsorption.*

In the case of the cyclohexenes, the cis/trans ratios again do not correspond to what would be expected if cis addition is the hydrogenation mechanism. Prediction on the basis of the 1,2/2,3 dimethylcyclohexene ratio would require that the 2,3-cycloalkene yield only trans isomer which seems unlikely. The ortho-xylene results support the idea that both types of isomer are formed from each species. Again, the presence of methylcyclohexene and cyclohexane supports the importance of platinum-methyl interactions.

The high cis/trans ratio for the dimethylcyclohexenes indicates that these molecules are more strongly adsorbed than the cyclopentenes thus restricting desorption and rotation. This is substantiated by the potential dependence of the isomer ratio; more strongly interacting intermediate at more anodic potentials.

It is readily apparent in all of the foregoing that the unexplained portions of catalytic reactions and the nature of active sites is mainly due to the fact that until recently no direct technique was available for the investigation and definition of the surface structures of solids. The exact surface structure and composition of the catalyst must be known before adsorption phenomena can be interpreted in terms of the participation of active sites and the establishment of a detailed description of the mechanism of catalytic reactions.

Developments in low energy electron diffraction over the past decade [12-14], have led to instruments and techniques that are applicable to our present problems. Variations in the energies of the electrons (from 5 to 500 electron volts) varies their penetration depths and thus, by successively increasing the electron beam voltage, structural studies of the first, second, third, etc. atomic layers of solid surfaces are permitted.

Equipment has been obtained and placed in operation. The investigation of catalytic reactions by means of LEED will be carried out in three steps:
1) Investigation and characterization of pure surface of catalysts, e.g., the various single crystal surfaces of Pt.
2) Study of adsorption of selected single gases with the aim of characterizing and evaluating structural rearrangement and faceting of surfaces in view of possible creation or stabilization of active sites.
3) Adsorption and desorption studies with two gases. Here, the relative strengths of bonding to the catalyst surface and possible effects on work functions will be investigated.

In conclusion, the direct oxidation of hydrocarbon fuels at a fuel cell electrode is still some time away. However, the present variety of approaches, each offering its own pertinent information, hold much promise of success and indicate that, at least at low power levels, operational fuel cells are possible in the not too distant future.

ACKNOWLEDGMENTS

I would like to thank Dr. S. B. Brummer of Tyco Laboratories for allowing me to use a number of figures from his work. I would also like to acknowledge the efforts of Drs. H. J. Barger, Jr., R. J. York and J. A. Joebstl and Messers G. W. Walker and A. J. Coleman who along with Dr. Brummer performed all of the research cited in this paper.

REFERENCES

[1] S. Brummer, J. Ford and M. Turner, *J. Phys. Chem.*, <u>69</u>, 3424 (1965).

[2] W.T. Grubb, in *International Symposium on Fuel Cells*, Dresden, Germany, 1967, by K. Schwabe (Ed.), Akademie-Verlag, Berlin (1968) p. 237.

[3] L. W. Niedrach and M. Tochner, *J. Electrochem. Soc.*, 114, 17
     (1967).

[4] H. J. Barger, Jr., and M. L. Savitz, *J. Electrochem. Soc.*, 115,
     686 (1968).

[5] H. J. Barger, Jr., and A. J. Coleman, *J. Phys. Chem.*, 72, 2285
     (1968).

[6] H. J. Barger, Jr., and A. J. Coleman, *J. Phys. Chem.*, 74, 880
     (1970).

[7] J. Giner, *J. Electrochem. Soc.*, 111, 376 (1964).

[8] H. J. Barger, Jr., G. W. Walker, and R. J. York, presented at the
     138th National Meeting of the Electrochemical Society, Atlantic
     City, New Jersey, October, 1970.

[9] F. Hartog and P. Zwietering, *J. Catalysis*, 2, 79 (1963).

[10] S. Siegel and B. Dmuchovsky, *J. Am. Chem. Soc.*, 86, 2194 (1964).

[11] S. Siegel, P. Thomas and J. Holt, *J. Catalysis*, 4, 73 (1965).

[12] E. J. Scheibner, L. H. Germer and C. D. Hartman, *Rev. Sci. Instr.*,
     31, 112 (1960).

[13] R. L. Park and H. E. Farnsworth, *J. Chem. Phys.*, 40, 2354 (1964).

[14] A. U. MacRae, *Surface Sci.*, 1, 319 (1964).

RANEY CATALYSTS

H. Binder, A. Köhling, and G. Sandstede

Battelle-Institut e.V., Frankfurt am Main, Germany

ABSTRACT

*In our investigations of Raney catalysts for fuel cells, Raney pal-
ladium turned out to be as satisfactory a catalyst as platinum in the
oxidation of methanol in an alkaline electrolyte. Raney nickel is an
extremely poor catalyst, whereas Raney palladium-silver is one of the
most appropriate catalysts for methanol oxidation in alkaline electrolyte. Raney platinum was found to be an active catalyst in the oxidation of hydrocarbons and their derivatives in acid electrolytes. In
investigations including galvanostatic current-voltage plots, cyclic
voltammetry, and BET measurements we have been able to show that the
current densities attained in relation to the true electrode surface
area during methanol oxidation in dilute sulfuric acid differ only
slightly for the various platinum catalysts investigated, whereas the
current densities at smooth platinum are smaller by a factor of 2 to 3.
In the case of smooth platinum thorough purification of the electrode
and the electrolyte is absolutely necessary. In addition, the consecutive products of methanol oxidation have to be present in the electrolyte; this can be achieved by pre-electrolysis on a Raney platinum electrode for several hours. Tafel slopes for smooth platinum differ only
slightly from those for porous electrodes. The activation energy for
the oxidation of methanol on Raney platinum is smaller by about 1 kcal
per mole than the energy required on smooth platinum. Raney catalysts
are more stable than other catalysts with respect to recrystallization
in sulfuric acid at elevated temperatures.*

1. INTRODUCTION

Electrodes for fuel cells with aqueous electrolytes in general require the use of a material acting as a catalyst in order to reduce the
overvoltage encountered in the electrode process. This catalytically
active material has to be finely divided in an appropriate electrode
matrix in order to achieve a maximum current density per unit of projected electrode surface area facing the electrolyte.

Although Travers and Aubry [1] discovered already in 1938 that at
Raney nickel [2] in alkaline solution the hydrogen potential is reversibly established, it was not until 1954 that Justi [3] and co-workers
applied Raney nickel in a fuel cell electrode. Raney nickel has mainly
been used in hydrogen electrodes [4]. Raney-type metals other than
nickel have been described [5] but seldom been used in fuel cell work.

Since 1959 when we started our investigations on fuel cells we
have mainly used the Raney method for preparing metal catalysts [6].
As described by Raney [2] this method requires the preparation of starting alloys, which usually consist of aluminum and a more noble metal
acting as the catalyst. In our investigations we normally used electrodes with a gold skeleton in the form of porous disks prepared by

15

compressing a mixture of gold powder, the powdered aluminum alloy, and an auxiliary substance to obtain macropores, *e.g.* sodium sulfate. In general, the aluminum of the Raney alloy was dissolved by treating the electrode with potassium hydroxide solution at a temperature of 80°C [7]. It should be noted that most of these experimental electrodes contained much higher amounts of the catalytically active material than can be tolerated in a practical fuel cell. Nevertheless, the current-voltage plots and the other data derived can be used as a basis for comparison of the various catalysts.

2. PALLADIUM, NICKEL, AND SILVER AS RANEY CATALYSTS

In investigations of the oxidation of methanol in alkaline electrolyte it turned out that palladium is as good a catalyst as platinum [7]. Fig. 1 shows the current-voltage plots obtained on Raney palladium for methanol and formate. The most striking feature is that formate is

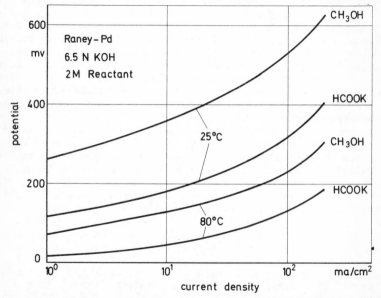

FIG. 1 *Stationary galvanostatic current-voltage curves for methanol and formate oxidation on Raney palladium in alkaline electrolyte.*

oxidized at palladium at higher rate than methanol, even at a temperature of 25°C, whereas at platinum (Fig. 2) a distinct limiting current is observed in the oxidation of formate at 25°C. Fig. 3 shows the current-voltage plots obtained on Raney platinum at 80°C.

Raney nickel proved to be an extremely poor catalyst for the oxidation of methanol although hydrogen is readily oxidized on Raney nickel. The high catalytic activity of Raney nickel in the oxidation of hydrogen may be attributed to the fact that finely divided nickel is capable of storing a large amount of hydrogen [8], so that the material can be regarded as a nickel-hydrogen phase whose high catalytic activity may be due to the decrease in paramagnetism. At the potentials required for the activation of methanol this phase is destroyed by oxidation. According to Sokolsky and Omarova [9] the activity of Raney nickel is determined by the hydrogen content.

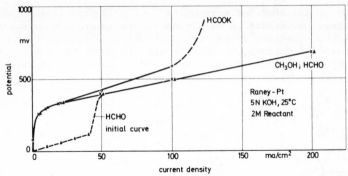

FIG. 2 *Stationary galvanostatic current-voltage curves for metha-nol, formaldehyde and formate oxidation on Raney platinum in alkaline electrolyte.*

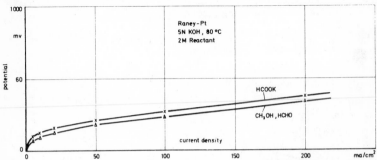

FIG. 3 *Stationary galvanostatic current-voltage curves for metha-nol, formaldehyde and formate oxidation on Raney platinum in alkaline electrolyte.*

It should be noted in this connection that we found Raney palladium-silver to be an appropriate catalyst for the oxidation of methanol in alkaline electrolyte [10]. Alloy catalysts for methanol oxidation are discussed in detail in two other contributions of this symposium [10, 11].

In alkaline electrolyte silver is a very good catalyst for oxygen reduction. The Raney method can also be used for the preparation of finely divided silver [5]. But in the case of silver the stacking fault energy is sufficiently high to cause recrystallization [12], even at room temperature, in the course of which black Raney silver turns bright. Thus, the Raney method is not superior to other methods, *e.g.* decomposition of silver compounds such as silver carbonate [10] or silver oxalate [13]. Raney silver is used in the batteries made by Siemens and Varta.

3.   RANEY PLATINUM

3.1   DEGREE OF CONVERSION

Raney platinum is a catalyst with a high specific surface area, which is resistant to the attack of both strong caustics and strong acids.

In alkaline electrolyte the oxidation of normal alcohols stops at the stage of the respective carboxylic acid which, with the exception

of formic acid, does not undergo further oxidation [14]. Table I shows
the degrees of conversion of various fuels in alkaline electrolyte at
25 and 80°C. The stationary current-voltage curves for the oxidation
of methanol, formaldehyde, and formate are shown in Figs. 2 and 3.
Saturated hydrocarbons are not oxidized in alkaline electrolytes [15],
whereas ethylene can be oxidized [16,17,15].

TABLE I

Coulometric Investigation in 6.5 N KOH

Reactant	Temp. °C	Pot.* mV	Degree of Conv. %	Notes
Methanol	25	300	98.0	
	80	300	97.5	
Formic acid	25	500	99.7	
Ethanol	25	600	33.3	$CH_3COOH \triangleq 33.3\%$
	80	300	33.5	
Glycol	25	600	81.5	$(COOH)_2 \triangleq 80\%$
	80	300	84.0	
n-Propanol	25	600	21.9	$C_2H_5COOH \triangleq 22.2\%$
	80	600	22.9	
Isopropanol	25	600	9.5	$(CH_3)_2CO \triangleq 11.1\%$
	80	600	14.0	
	25-60	700	44.5+	+estimated
Acetone	60	700	34.0	$CH_3COCOOH \triangleq 37.5\%$
Glycerol	25	400	52.5	$CH_3COOH \triangleq 43\%$
	80	300	51.6	$(COOH)_2CHOH \triangleq 57\%$

*vs. H.E. in the same solution

In sulfuric acid at about 100°C even saturated hydrocarbons can
readily be oxidized to give CO_2 and H_2O [18,19]. Fig. 4 shows that

TABLE II

Degree of Conversion of Some Alkanes to CO_2 in 3 N H_2SO_4 at 90°C

Reactant	Potential mv	φCO_2 percent
CH_4	380	100.2
	620	100.1
C_2H_6	450	99.8
	550	100.5
C_3H_8	320	100.5
	450	99.5
C_4H_{10}	460	101.3
	550	100.8

the best performance is achieved with ethane and propane. Table II
summarizes the results obtained on hydrocarbons by coulometric measurements.. In all cases the degree of conversion is 100 percent. At
room temperature only extremely small current densities can be achieved
with hydrocarbons. The oxidation starts at about 50°C. Ethylene and
benzene can also be converted with high current densities, but in the
case of benzene polarization is high [20]. In phosphoric acid at 110°C
differences in the activity of saturated hydrocarbons (Fig. 5) are less
pronounced than in sulfuric acid [21].

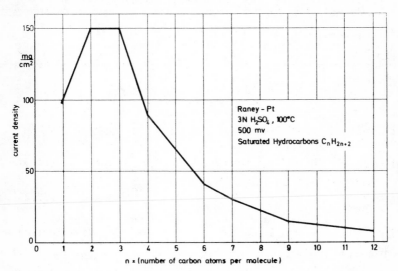

FIG. 4 *Rate of electrooxidation of saturated hydrocarbons at Raney
platinum as a function of the number of carbon atoms per molecule.*

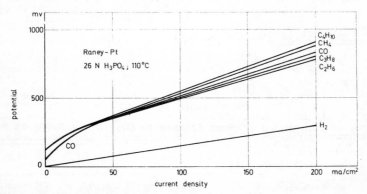

FIG. 5 *Stationary galvanostatic current-voltage curves for the
oxidation of saturated hydrocarbons and carbon monoxide on Raney
platinum in phosphoric acid.*

The results of coulometric measurements on oxidation of derivatives
of methane, ethane, and propane are summarized in Table III. At a temperature of 25°C ethanol is oxidized to the acetic acid stage, but even
acetic acid can be oxidized at 25°C at a low rate if the potential is
about 500 mV. At higher temperatures the oxidation rate of acetic acid
is higher than that of ethanol (Fig. 6), but a distinct limiting current
density is observed.

TABLE III

Coulometric Investigations in 4.5 N H_2SO_4

Reactant	Temp. °C	Pot.* mv	Degree of Conv. %	Notes
Methanol	25	400	98.5	
	70	700	99.2	
	80	300	98.2	
Formic acid	25	500	98.5	
	80	300	99.3	
Ethanol	25	500	39.6	still 0.2 ma/cm^2
	90	700	86.3	
	40-90	700	94.4	
	25-90	700	98.2	
Acetic acid	90	700	99.8	
Glycol	25	500	87.5	still 0.1 ma/cm^2
	25	700	98.0	
	75	500	97.6	
Oxalic acid	80	600	97.5	
n-Propanol	75	500	94.5	
	80	600	89.5	
	25-80		100+	+estimated
Isopropanol	25	500	12.8	Acetone $\hat{=}$ 11.1%
	80	600	58.0	
	25-90		100+	+estimated
Acetone	60	700	40.0	→ 90°C further oxidation
	60-90		100+	+estimated
Diacetone alcohol	80	600	85	
	60	700	100+	+estimated
Propionic acid	80	500	99.8	
Glycerol	25	700	97.0	
	80	500	100.0	

*vs. H.E. in the same solution

TABLE IV

Specific Surface Area of Platinum Electrodes Determined by Anodic H_{ad} Stripping (O_Δ) and by Nitrogen Adsorption (O_{BET})

	before aging			after aging		
	O_Δ m^2/gPt	O_{BET} m^2/gPt	O_Δ m^2/gPt	O_Δ m^2/gPt	O_{BET} m^2/gPt	O_Δ m^2/gPt
Pt black	16.3	8.0	9.8	6.3	6.7	7.4
Pt from PtAl$_2$	23.9	23.0	23.2	19.9	18.3	19.9
Pt from PtAl$_3$	31.0	33.0	29.8	24.7	27.6	27.7
Pt from PtAl$_4$	39.7	43.9	39.5	32.4	34.0	32.3

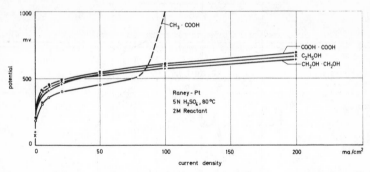

FIG. 6 *Stationary galvanostatic current-voltage curves for the oxi-dation of ethane derivatives on Raney platinum.*

3.2 SPECIFIC SURFACE AREA AND PORE SIZE DISTRIBUTION

In order to investigate whether a Raney catalyst has peculiar qual-ities compared with catalysts prepared by other methods and with smooth platinum, we determined the true current densities in methanol oxida-tion. To this end we measured the specific surface area of the various platinum catalysts. These measurements were performed by the BET method and by oxidative stripping of H_{ad} in dilute sulfuric acid [22] the cat-alysts being contained in gold electrodes. For the preparation of var-ious Raney platinum catalysts we started from the Al-rich phases of the Pt-Al system ($PtAl$, Pt_2Al_3, $PtAl_2$, $PtAl_3$, $PtAl_4$, and "$PtAl_6$", the latter being not a single phase but a mixture of $PtAl_4$ and Al).

A brief summary of the results is given in Table IV. The data in the three columns at left represent the virgin state of the catalysts, those in the three columns at right were measured after a period of several days in which the electrodes were used for methanol oxidation in sulfuric acid at 25°C and 70°C. O_Δ indicates the specific surface area determined by oxidative stripping of H_{ad} before (left-hand column) and after (right-hand column) the BET measurement. The electrochemical data are in good agreement with the BET figures. The decrease of the specific surface area reveals that during methanol oxidation an aging effect occurs.

Fig. 7 shows the specific surface area of the platinum samples as a function of the aluminum content of the starting Raney alloy. The initial surface area varies between 12 m^2/g (Pt from PtAl) and 40 m^2/g (Pt from $PtAl_4$). The smaller figures obtained after the aging process have been used for evaluating the true current density in the electro-chemical oxidation of methanol.

The aging effect occurs only in acid electrolytes at elevated tem-peratures. It is due to the migration of platinum ad-atoms or, perhaps, platinum ad-ions in the catalyst surface in the presence of the electro-lyte, and may be regarded as a recrystallization process [22]. The aging effect is smaller in the case of the Raney catalysts than with platinum black. The recrystallization is coupled with a change in the pore size distribution and an increase in the pore radii (Fig. 8). It should be noted that the catalyst from $PtAl_2$ has the narrowest pore size distribution. Most of the pores of this catalyst have a radius of about 10 Å (the pores of the catalysts from PtAl and Pt_2Al_3 alloys, which have not been measured, should be even narrower), whereas the pore radius of the other catalysts is about twice that value.

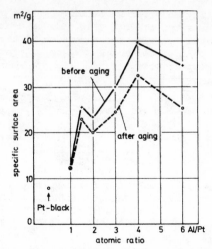

FIG. 7 *Specific surface area of Raney platinum as a function of
the aluminum content of the starting alloy (and of platinum black).*

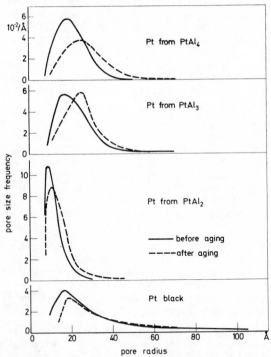

FIG. 8 *Pore size distribution of Raney platinum and platinum black.*

3.3 TRUE CURRENT DENSITY IN METHANOL OXIDATION

Fig. 9 shows, on a semilogarithmic scale, the current-voltage plots
obtained for the oxidation of methanol on the different platinum catalysts. Tafel-like behavior is observed at larger polarization, the
slopes being about 90 mV and 60 mV per decade for 70°C and 25°C, respectively.

The current density observed in relation to the measured surface area (true current density) at 70°C (Fig. 10) is of the same order of magnitude for all catalysts with the exception of those from PtAl and PtAl$_2$: about 10 μA/cm$^2_{real}$ and about 3 μA/cm$^2_{real}$ at a potential of 450 mV and 400 mV, respectively.

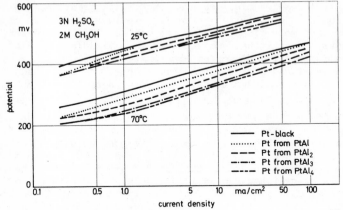

FIG. 9 *Stationary galvanostatic current-voltage curves for methanol oxidation on Raney platinum and platinum black in dilute sulfuric acid.*

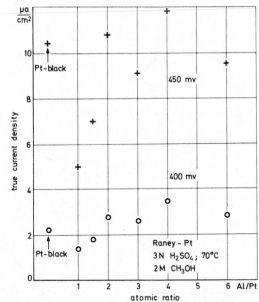

FIG. 10 *True current density for methanol oxidation on Raney platinum and platinum black.*

Thus, with respect to the true current density attained in methanol oxidation, Raney platinum catalysts are not superior to platinum catalysts prepared by other methods. Nevertheless, they have advantages over other platinum catalysts: 1) they have a larger specific surface area and 2) they do not recrystallize in sulfuric acid to such an extent as, for example, platinum black or platinum from PtO$_2$.

4. SMOOTH PLATINUM

4.1 PLATINUM SHEET

4.1.1 Experimental Conditions

For reference, we also measured the true current density on smooth
platinum using a cold-rolled, annealed platinum sheet. In order to
obtain an extremely clean surface, as was absolutely necessary for
this comparison, the sheet was treated in a mixed solution of concentrated sulfuric acid and sodium bichromate and subsequently rinsed in
distilled water and the sulfuric acid electrolyte to be used. Oxide
surface layers formed during this treatment were removed by cathodic
reduction.

The surface area (measured by H_{ad} stripping) of this sheet amounted
to 22 cm^2. The periodic potentiodynamic current-voltage plot of this
sample is shown in Fig. 11, where a corresponding plot of a Raney electrode is superposed. Since the surface area of the platinum in this
electrode was about 100 times that of the smooth sample, the voltage
speed was chosen at a ratio of 1:100.

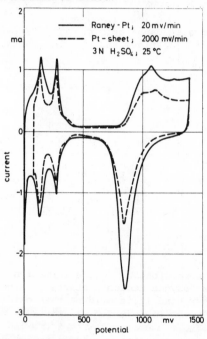

FIG. 11 *Periodic potentiodynamic current-voltage curves using a
platinum sheet and Raney platinum.*

One can see that the shapes of the plots match fairly well. This
fact confirms that the treatment of the platinum sheet and that of the
electrolyte* has been sufficient.

*The electrolyte was prepared with bi-distilled water from a quartz
apparatus, which was then further distilled over an alkaline solution
of $KMnO_4$. The sulfuric acid used was supplied by Merck (Darmstadt),
grade "Suprapur". After preparation of the sulfuric acid solution the
electrolyte was pre-electrolyzed at a current of 200 mA in an atmosphere*

4.1.2 Periodic Potentiodynamic Current-Voltage Plots

Oxidation of methanol on this pretreated sheet during periodic potentiodynamic measurement is shown in the next two Figures. These
curves plotted directly on a semi-logarithmic scale by an electrical
converter reveal much more details of the behavior of platinum in the
presence and absence of methanol than the curves measured on a linear
scale. The curves for increasing voltage are plotted in Fig. 12. It
can be seen from the curve determined in the absence of methanol that

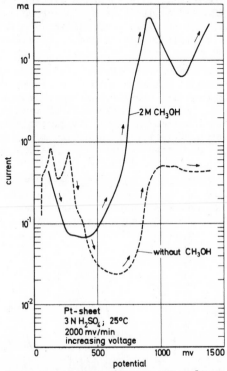

FIG. 12 *Periodic potentiodynamic current-voltage curves (increas-*
ing voltage) for methanol oxidation on a platinum sheet.

oxidation of water corresponding to the formation of an OH_{ad} layer (or
more general of the "oxygen region") starts at a potential of about
700 mV. On the other hand it is evident that oxidation of methanol
already starts at a potential ($\approx$ 450 mV) where normally the electrode
surface is free from H_{ad}-atoms, *i.e.* where the double layer region begins. This observation is in good agreement with Giner's results [23].
No "Pt oxide" is present on the platinum surface at this very moment
[23] irrespective of whether we are dealing with a real new phase which
according to ellipsometric measurements after Bockris *et al.* [24] starts
to form at a potential of about 950 mV or with a layer of adsorbed oxygen

of pure nitrogen for at least 24 h in a vessel containing two platinum
electrodes. The hydrogen and oxygen formed during this pre-electrolysis
were removed by boiling the electrolyte for several hours. The remain-
ing hydrogen peroxide was electrolytically reduced at a platinum elec-
trode cycled in the potential range between 50 and 1400 mV still under
nitrogen.

species which cannot be detected by ellipsometry but merely by coulometric investigations [25]. Hence it must be concluded from our results
that the mechanisms proposed by Wroblowa *et al.* [26] for oxidation of
ethylene on platinum which includes the formation of adsorbed oxygen
cannot be applied to the oxidation of methanol on platinum below a
potential of, say, 700 mV. The postulation of a mechanism for the
methanol oxidation below that potential which includes OH-radicals [27]
is arbitrary or perhaps even erroneous. In the oxidation process under
non-steady-state conditions only adsorbed methanol molecules (possibly
dissociated from the very beginning) and adsorbed and properly polarized
water molecules are necessary and present at the platinum surface.
Under steady-state conditions the presence of chemisorbed species of
methanol have to be taken into consideration.

In our potentiodynamic curve (still Fig. 12) a first maximum is
observed at about 900 mV somewhat below the first maximum of the oxygen
region which means that the formation of the oxide phase (Bockris *et al.*
[24]: "Oxide film") inhibits the oxidation of methanol. This finding
is consistent with observations in the oxidation kinetics of saturated
hydrocarbons [19] or hydrogen [28] on platinum. However, the oxidation
of methanol is not completely inhibited by the formation of the oxide
layer, as can be seen from the subsequent increase in current density.

For the decreasing voltage sweep the curves of Fig. 13 are obtained.
Starting at 1,400 mV as the potential of reversal, a decrease of the
oxidation current is observed until a potential of 1,000 mV is reached,
when the current starts to increase again. Under the conditions chosen
this increase is connected with the electrochemical rather than the

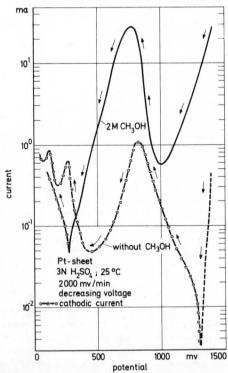

FIG. 13 *Periodic potentiodynamic current-voltage curves (decreasing voltage) for methanol oxidation on a platinum sheet.*

chemical reduction of the oxygen chemisorbate layer. A maximum current is reached when most of this layer has been cathodically removed. Then, the current decreases as the potential is reduced.

The course of the methanol oxidation on smooth platinum is in agreement with our findings with Raney platinum [29].

4.1.3 Galvanostatic Measurements (Steady State)

During galvanostatic measurements a steady state is reached only when an electrolyte-methanol solution is used in which a platinum electrode (in our case a Raney electrode) has been previously operated for several hours. Hence, a real steady state during oxidation of methanol, which is very important for the practical fuel cell in mind, is reached only in the presence of the secondary oxidation products: formaldehyde and formic acid. Such a steady state certainly corresponds to the quasi-steady state which has been defined by Bockris [30] as "that condition when the current changes less than 10 percent per hour at constant potential, or when the potential changes less than 10 mV h^{-1} at constant current. This is not an arbitrary definition; it is the result of experimental observations on the behavior of current-potential relations determined at various times of approach to the steady state."

Even in such an electrolyte the steady state is observed only after one day's operation of the smooth platinum electrode (Fig. 14). At a potential of 450 mV the current density evaluated for the platinum sheet by dividing the current observed by the surface area of 22 cm^2 amounts to 3 $\mu A/cm^2$. This is in satisfactory agreement with the values observed on Raney platinum catalysts.

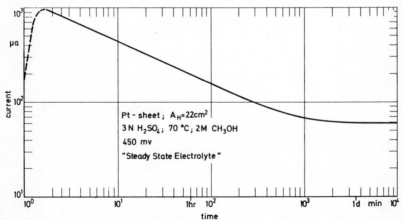

FIG. 14 *Current-time curve for methanol oxidation on a platinum sheet.*

On the other hand, if there are no secondary products present in the electrolyte, the following data are obtained for the oxidation of methanol on smooth platinum (Table V). At a potential of 450 mV and a temperature of 70°C a "steady state" current density of only 0.55 $\mu A/cm^2$ is observed after 16 h instead of 3 $\mu A/cm^2$ in the case of the "steady state electrolyte".

TABLE V

Current Densities in Methanol Oxidation on a Platinum-Sheet in a Fresh Electrolyte Consisting of 3N H_2SO_4 + 2M CH_3OH

Temperature °C	Potential mV	True Current Density Initial μA/cm^2	After 16 h μA/cm^2
70	450	8.0	0.55
	555	160	11.1
	650	1550	108
25	500	4.0	0.18
	600	55	2.5

4.2 PLATINUM GAUZE

Furthermore we investigated a coiled platinum gauze with an electrochemically evaluated surface area of 0.20 m^2 which gave us the following stationary current-voltage plots upon oxidation of methanol (Fig. 15). A coiled platinum gauze like this represents an electrode

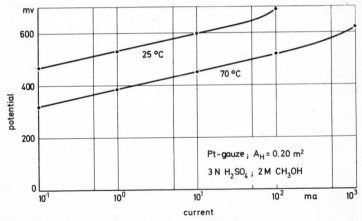

FIG. 15 *Stationary galvanostatic current-voltage curves for methanol oxidation on platinum gauze.*

with a relatively large surface area although we are dealing with the smooth surface of a platinum wire. At a potential of 450 mV and a temperature of 70°C a current of 10 mA is reached, which corresponds to a true current density of 5 μA/cm^2.

5. CONCLUSIONS

Table VI summarizes the results of methanol oxidation obtained with different platinum samples (catalysts and smooth platinum) in a "steady-state electrolyte". It turns out that the extrapolated fictitious and arbitrary (U_H = OmV) "exchange current density" at smooth platinum and a temperature of 70°C is very low and about one order of magnitude smaller than that at a catalyst material. The extremely low value for the extrapolated "exchange current density" indicates that under

steady-state conditions oxidation of methanol on platinum is governed
by the presence of a strongly adsorbed chemisorption layer. The oxidation mechanism involved differs drastically from that observed in transient measurements or from other values derived for the oxidation at
an almost bare platinum surface.

TABLE VI

*True Current Densities and Tafel Slopes for Various Platinum
Electrodes Upon Oxidation of Methanol in Dilute Sulfuric
Acid: "Steady-State Electrolyte"*

	"i_0" at U_H = 0 mV (extrapol.) [A/cm^2]		i_{400} at U_H = 400 mV [A/cm^2]		i_{450} at U_H = 450 mV [A/cm^2]		Tafel slope [mV/Dec.]	
	25°C	70°C	25°C	70°C	25°C	70°C	25°C	70°C
Pt-sheet	--	$5 \cdot 10^{-12}$	--	$7 \cdot 10^{-7}$	--	$3 \cdot 10^{-6}$	--	80
Pt-gauze	$5 \cdot 10^{-15}$	$5 \cdot 10^{-12}$	--	$1 \cdot 10^{-6}$	$5 \cdot 10^{-8}$	$5 \cdot 10^{-6}$	65	70
Raney-Pt*	$2 \cdot 10^{-15}$	$5 \cdot 10^{-11}$	--	$3 \cdot 10^{-6}$	$1.5 \cdot 10^{-7}$	$1 \cdot 10^{-5}$	60	90
Pt-black	$3 \cdot 10^{-15}$	$2 \cdot 10^{-11}$	--	$2 \cdot 10^{-6}$	$1.5 \cdot 10^{-7}$	$1 \cdot 10^{-5}$	60	80

*from PtAl$_3$

The current densities achieved with smooth platinum at, *e.g.*, 450
mV and 70°C, amount to about one third to one half of the value observed
at Raney platinum or platinum black.

The most striking feature of this Table is that the Tafel slopes at
a given temperature differ only slightly, although for porous electrodes
twice the normal slope has been theoretically predicted for flooded
electrodes [31]. They are in good agreement with the Tafel slopes
observed by Petry *et al.* [32] for the oxidation of methanol on platinized platinum in acid solution and also under steady state conditions.
For a more detailed discussion of the significance of Tafel slopes in
such a complex oxidation reaction compare Breiter [33] or Petry [32].

Nevertheless, it may be concluded from these results that the
activity of Raney platinum and other platinum catalysts upon oxidation
of methanol is somewhat superior to that of smooth platinum samples.
From the data of Table VI the apparent heat of activation for the methanol oxidation at a given potential can be estimated to be about 1 kcal
per gram atom smaller for Raney platinum than for smooth platinum.
This difference may be attributed to the influence of lattice distortions in the catalyst samples. According to the analysis of X-ray
line profiles of Raney platinum (from PtAl$_4$) microstresses of about
240 kg/mm^2 are derived which correspond to about 5.5 kcal per gram
atom of platinum.

Summarizing, it may be stated that Raney catalysts are substances
with a high specific surface area so that in electrochemical oxidation
or reduction the current densities generally achieved are higher than
with other catalysts of the same material. The more important feature,
for instance in the case of Raney platinum in acid, is the greater
stability to recrystallization.

ACKNOWLEDGMENT

    This work was conducted under a contract of the Robert Bosch GmbH,
Stuttgart, Germany. Its permission to publish these results is gratefully acknowledged.

REFERENCES

[1] A. Travers and J. Aubry, *Atti del X° Congresso Internationali di
     Chimica*, Rome (1938).

[2] M. Raney, *Ind. Eng. Chem.*, $\underline{32}$, 1190 (1940).

[3] E. Justi, *Jahrbuch*, Akademie Wissenschaften Literatur, Mainz,
     (1955) p. 200.

[4] E. Justi, M. Pilkuhn, W. Scheibe and A. Winsel, "Hochbelastbare
     Wasserstoff-Diffusionselektroden für Betrieb bei Umgebungstemper-
     atur und Niederdruck", *Abh. Mainzer Akad.*, No. 8, Franz Steiner
     Verlag, Wiesbaden (1959).

[5] E. Justi and A. Winsel, *Kalte Verbrennung - Fuel Cells*, Franz
     Steiner Verlag, Wiesbaden (1962).

[6] H. Krupp, H. Rabenhorst, G. Sandstede, G. Walter and R. McJones,
     J. Electrochem. Soc., $\underline{109}$, 553 (1962).

[7] H. Binder, A. Köhling, H. Krupp, K. Richter and G. Sandstede,
     "Fuel Cell Systems", *Advances in Chemistry Series*, $\underline{47}$, 269, Am.
     Chem. Soc., Washington D. C. (1965).

[8] M. Pilkuhn and A. Winsel, *Z. Elektrochem.*, $\underline{63}$, 1056 (1959).

[9] D. V. Sokolsky and S. R. Omarova, *Dokl. Akad. Nauk, SSSR*, $\underline{102}$,
     977 (1955).

[10] H. Binder, A. Köhling, W. H. Kuhn, W. Lindner and G. Sandstede,
     these proceedings, p. 131, see III/1.

[11] H. Binder, A. Köhling and G. Sandstede, "Effects of Alloying Com-
     ponents on the Catalytic Activity of Platinum in the Case of
     Carbonaceous Fuels", these proceedings, p. 43, see I/4.

[12] F. R. L. Schoening and J. N. van Niekerk, *Acta Cryst.*, $\underline{3}$, 10 (1955).

[13] J. E. Schroeder, D. Pouli and H. J. Seim, "Fuel Cell Systems - II",
     Advances in Chemistry Series, $\underline{90}$, 93, Am. Chem. Soc., Washington,
     D. C. (1969).

[14] H. Binder, A. Köhling and G. Sandstede, "Fuel Cell Systems",
     Advances in Chemistry Series, $\underline{47}$, 283, Am. Chem. Soc., Washington,
     D. C. (1965).

[15] H. Binder, A. Köhling and G. Sandstede, *Adv. Energy Conv.*, $\underline{6}$, 135
     (1966); *Comptes Rendus, Deuxièmes Journées Internat. d'Etude de
     Piles a Combustibles*, Brussels (1967) p. 173.

[16] M. Green, J. Weber and V. Drazic, *J. Electrochem. Soc.*, $\underline{111}$, 721
     (1964).

[17] J. O'M. Bockris, H. Wroblowa, E. Gileadi and B. J. Piersma, *Trans.
     Faraday Soc.*, $\underline{61}$, 2531 (1965).

[18] H. Binder, A. Köhling, H. Krupp, K. Richter and G. Sandstede, *J.
     Electrochem. Soc.*, $\underline{112}$, 355 (1965).

[19] H. Binder, A. Köhling and G. Sandstede, *Comptes Rendus, Journées Internat. d'Etude de Piles a Combustibles, Revue Energie Primaire,* I, 74 Brussels (1965).

[20] G. Sandstede, *Chemie-Ing.-Technik,* 37, 632, 782 (1965).

[21] H. Binder, A. Köhling and G. Sandstede, paper presented at the Meeting of the Electrochemical Society, Washington, D. C., Oct., 1964, Extended Abstracts, Battery Division, p. 25.

[22] H. Binder, A. Köhling, K. Metzelthin, G. Sandstede and M.-L. Schrecker, *Chemie-Ing.-Technik,* 40, 586 (1967).

[23] J. Giner, *Electrochim. Acta,* 9, 63 (1964).

[24] A. K. N. Reddy, M. A. Genshaw and J. O'M. Bockris, *J. Chem. Phys.,* 48, 671 (1967).

[25] See *e.g.* H. Wroblowa, M. L. B. Rao, A. Damjanovic and J. O'M. Bockris, *J. Electroanal. Chem.,* 15, 135 (1967).

[26] H. Wroblowa, B. J. Piersma and J. O'M. Bockris, *J. Electroanal. Chem.,* 6, 401 (1963).

[27] V. S. Bagotzky and Yu. B. Vasilyev, *Electrochim. Acta,* 9, 869 (1964).

[28] M. W. Breiter, *Electrochim. Acta,* 7, 601 (1962).

[29] H. Binder, A. Köhling and G. Sandstede, *Adv. Energy Conv.,* 6, 135 (1966).

[30] J. O'M. Bockris and S. Srinivasan, *Fuel Cells: Their Electrochemistry,* McGraw-Hill Book Co., New York (1969) p. 391.

[31] L. G. Austin, *Trans. Faraday Soc.,* 60, 1319 (1964); *Handbook of Fuel Cell Technology,* by C. Berger (Ed.) Prentice Hall Inc., Englewood Cliffs, New Jersey (1968) Chapter 1.

[32] O. H. Petry, B. I. Podlovchenko, A. N. Frumkin and Hira Lal, *J. Electroanal. Chem.,* 10, 253 (1965).

[33] M. W. Breiter, *Electrochemical Processes in Fuel Cells,* Springer-Verlag, Berlin, Heidelberg, New York (1969) p. 36.

FUEL CELLS AND THE THEORY OF METALS[*]

C. V. Bocciarelli

Pennsylvania Research Associates, Inc.

ABSTRACT

The first and last events in a fuel cell are the electronic ex-changes between metal and ions. These do not appear to have been con-sidered from the metal's viewpoint, although the metals impose their own limitations as well as open new possibilities. Metal theory is applied to obtain information about the role of metal catalysts in electro-catalysis, particularly in reference to alkaline hydrogen-oxy-gen fuel cells. Recourse is made to a simple model, analogous to that used to interpret field emission in vacuum. The following results have been obtained:

Theoretical values for all the quantities in Tafel equation in terms of bulk properties of the metal catalysts, such as free electron densities and Fermi level, and a clear understanding of the reasons why some processes are reversible (H-electrodes) and some irreversible (O-electrodes).

Selection rules for desirable properties of catalytic materials. Accordingly alloys of an intermetallic have been produced with varying Fermi levels, with and without added d levels, which performed as predicted in actual cells.

From the preceding results the ability to improve the current volt-age characteristics of alkaline fuel cells. Examples of initial per-formance of small fuel cells utilizing these catalysts are given.

1. BACKGROUND

In the following a metal is described using Fermi-Dirac statistics according to Sommerfeld. This treatment is known to be inaccurate; however for similar metals (*i.e.* noble metals) it ought to be relatively accurate; in any case it turns out to be a very useful criterium.

Fig. 1 shows schematically a metal at a temperature greater than 0° Kelvin. The ordinate is in electron volts, the abscissa, the den-sity of free electrons per unit volume, n. The Fermi level E_F is the top of the distribution at 0°K and the last energy level for which Fermi-Dirac statistics hold. It is proportional to $n^{2/3}$.

The practical importance of the Fermi level is this: If two metals are brought into electrical contact, the voltage difference between the two as measured by a voltmeter is the difference in their Fermi levels. The d levels are present only in transition elements. The density of d levels may be comparable to the density of free electrons. Above the Fermi level is the so-called Boltzmann tail, which is formed by

[*]*This work is supported in part by National Heart Institute Con-tract PH43-68-1417 and NASA Contract NAS7-574.*

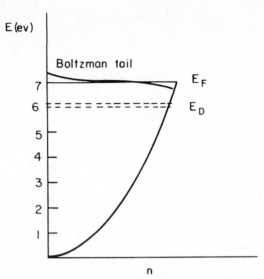

FIG. 1 *Schematic of electron density vs. energy for a metal at ~ 300°K.*

electrons in thermal equilibrium with the metal lattice — a function only of temperature.

2. AVAILABILITY OF ELECTRONS INDEPENDENTLY OF MECHANISMS OF TRANSFER

Consider the Tafel equation — this can be inverted in a more useful form. The right hand side of the expression $\Delta E = a + b \ln i$, *i.e.* a Tafel line may be put into a more compact form:

$$\Delta E = \ln[mi^b] = \ln m + b \ln i = a + b \ln i$$

(for $a = \ln m$; $e^a = m$). In this form it becomes easier to interpret physically

$$i = [\frac{1}{m}]^{1/b} e^{\Delta E/b}$$

and as $m = e^a$

$$i = e^{\frac{-a + \Delta E}{b}}. \tag{1}$$

The following expression is easily derived from Eq. 236 of Ref. [1]

$$n_\xi = c_1 n \xi^{1/2} e^{-\xi/kT} \tag{2}$$

where c_1 is a constant, n the total population density, k the Boltzmann constant, T is the temperature in °K, and n_ξ is the population of electrons of energy ξ. It is the limiting case of the Fermi distribution for sufficiently high temperatures and/or sufficiently low densities; the distribution is Maxwellian.

As $1/2\ mv^2 = \xi$ (where m is the mass, v the velocity of the electron), $v \sim \xi^{1/2}$ and as $n \sim E_F^{3/2}$, the current may be expressed as

$$i_\xi \simeq n_\xi v_\xi = c_2 E_F^{5/2} e^{\dfrac{-\xi}{kT}} \tag{3}$$

assuming $E_F \simeq \xi$ as reasonable for the Tafel region (for instance $E_F = 6V$, $\Delta E_F \leq 200$ mV). Then for the interval ΔE

$$i = c_2 E_F^{5/2} \int_{-\Delta E}^{0} e^{-\xi/kT} d\xi = c_2 kT E_F^{5/2} (e^{\Delta E/kT} - 1)$$

$$= I_0 (e^{\Delta E/kT} - 1) \quad . \tag{4}$$

Comparing Eqs. (1) and (4) we obtain

$$b = kT \quad , \tag{5}$$

$$I_0 = c_2 kT E_F^{5/2} \quad , \tag{6}$$

and

$$a = -kT \ln I_0 \quad . \tag{7}$$

Eq. (5) shows that b has a theoretical value equal to kT or about 0.025V at room temperature, a value approached by Pt and Pd (b = 0.03V) [2].

Eq. (6) gives the value of I_0 in terms of a fundamental catalyst parameter E_F. I_0 here is a current density for a real surface as in a perfect crystal face. The apparent value of I_0 will increase with comminution inversely as the particle size — and diminish if some of the surface is blocked by inactive layers.

The height of the Boltzmann tail theoretically extends to infinity. Hence it must be evaluated to some energy level whose population density is capable of emitting a predetermined small current that may be chosen equal to I_0.

Substituting in Eq. (3) the value of I_0 from Eq. (6) one obtains, for $I_0 = i_\xi$,

$$kT = e^{\dfrac{-\xi}{kT}}$$

where ξ is now a specific energy E_0 for $i = I_0$. Hence

$$E_0 = -kT \ln kT \tag{8}$$

where E_0 is positive because kT < 1; this shows that the height of the Boltzmann distribution to I_0 is dependent only on the absolute temperature. The values of E_0 vary from 85 mV at room temperature to 120 mV at 100°C. These values are smaller by about half than those previously assumed.

3. CONDITIONS IN PRESENCE OF THE ELECTROLYTE

It is a fundamental quantum mechanical rule that transitions are only possible between two states of equal energy, one of which is not occupied. Thus for a transfer of charge to occur between the metal and the ion the acceptor or donor level in the ion must be at the same energy level as the one in the metal where an available electron (at the cathode) or a hole (at the anode) exists.

In a vaccum the donor level of H — or the acceptor level of O — is 13.5 eV below vacuum level. Fig. 2 shows diagrammatically the change in these levels between the bulk and the metal face as the dielectric constant changes continuously between 80 in the bulk and 1 in the metal. To satisfy the above rule in the area of the Boltzmann tail, a dielec-tric constant between 1.645 and 1.675 will be required.

It is nearly meaningless to talk about a dielectric constant (an extensive property) when it varies from 1 to 80 in about 10 Å. If a meaning is assigned to it, then it must be that the distance from the surface of the catalyst must be virtually nil and consistent only with an adsorbed ion. As a consequence the tunneling distance is so short (of the order of interatomic distances in the metal) that the electron barrier, independently of its shape, must be considered to be fully transparent; in effect the ions are in "metallic" contact with the metal. Therefore at high currents, when the system is locked at the

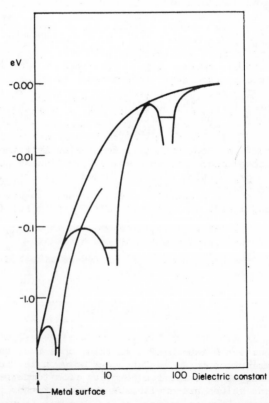

FIG. 2 *Change in the donor level of H — or acceptor level of O —*
between the bulk electrolyte and the metal surface.

Fermi level, the only resistance to the flow of current is derived from the relatively low conductance of the electrolytic paths.

On the other hand at the cathode at low currents, in the logarithmic region, the resistance of the source can be quite high; in this region the overall current-voltage relations of a fuel cell are nearly fully characterized by the properties of the cathode (the ensemble of catalyst, gas, and ions) as the resistance of the electrolyte is vanishingly small compared with that of the source.

In all the above it has been tacitly assumed that the supply of gas is adequate, which can generally be the case. Of course the external circuit resistances must be added to the electrolyte resistance. At the anode the conditions are different. The holes are emitted (or identically the electrons adsorbed) from a single D band. The voltage loss implied by the Boltzmann tail (*i.e.* irreversibility) is not present here; hence the anode is reversible.

4. APPLICATION TO A HIGH POWER FUEL CELL

From the preceding the conditions at high currents are these: the cathode is "locked" at the Fermi level, the anode at the d level, E_H, that can be considered a Fermi level for holes. The current capabilities at both ends are enormous. The corresponding voltages do not change as long as the electrons can be supplied by the rest of the system. The current voltage relationship is of the form $E = IR$ where E is fixed, hence I depends only on R. E on the other hand depends on the difference between E_F at the cathode and E_H at the anode. The situation is analogous to that of a P - N junction in transistor technology where, to get high voltages, one shifts from lower to higher gap materials, as from Ge to Si.

Clearly if this difference can be increased there will result an increase in power, because the same current is generated at higher voltages. Increases to about 150 mV have been obtained; these correspond to similar increases in open circuit voltages as a consequence of the fact that the height of the Boltzmann tail is invariant. There are two concluding observations:

First: The current capabilities at the Fermi level are so large for any metal that all metals will, at the Fermi level, produce essentially the same current — but of course at very different voltages.

Second: In a given fuel cell the open voltage is uniquely a property of the catalysts used — the higher the better.

Conversely, for equal open voltages the relative currents are a measure only of electrode fabrication.

5. MATERIAL STUDIES

The choice of material is limited; Au and Ag were considered and all that follows could have been done using them as a basis. In the case of noble metals (including Cu) it is known that the addition of polyvalent substitutional metal impurities will increase the free electron density n in proportion as each polyvalent metal furnishes say 3 electrons to the lattice instead of 1. Further it was thought that the addition of metals such as Co and Fe could introduce the desired d levels and produce a substitute for Pt at the anode.

Fig. 3 shows that this is indeed the case. Both curves were obtained using Pt black (AB-40 of American Cyanamid) as a cathode; the
lower curve is for AuCu + 15 at% Al, the upper is for AuCu + 15 at% Co.
Note the enormous overall difference for basically very similar materials and the difference in open circuit voltage.

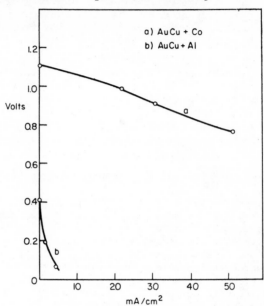

FIG. 3 *Anode performance of Co- and Al-doped AuCu vs. AB-40 (Pt
black).*

To return to the choice of material, Cu would be a very attractive
choice — it has a higher E_F than either Au or Ag which have about the
same. Unfortunately Cu is of low chemical stability. However, the
intermetallic AuCu has great resistance to chemical attack — much
greater than Au — and an E_F intermediate between Cu and Au.

6. METALLURGY

Briefly, small single crystal (200 μ) spheres of AuCu suitably
doped are produced in a shot tower. These are then ground in a gas-
tight ball mill under argon. Grinding in air results in partial oxidation. The black oxide of copper, CuO, is soluble in KOH and must be
avoided. Before adopting the gas-tight form of ball mill, this caused
much trouble and some confusion. The AuCu was doped with Al to produce
a specific oxygen catalyst, with Co and Fe to produce an hydrogen catalyst, always at 15 at%*. As catalysts, these were evaluated *vs.* AB-40
on the following basis.

As previously mentioned, the open circuit voltage is a figure of
merit for the catalyst. It has the great advantage of being relatively
independent from electrode fabrication. One measures the height of the

---

*This percentage is somewhat arbitrary. However for a trivalent
dopant at 22 at%, a change in crystal symmetry is obtained from f.c.c.
to b.c.c.*

Boltzmann tail, a constant at a given temperature, plus the difference
between the Fermi level of the cathode material and the effective d
level in the anode.   These were as follows:

	Cathode	Anode	Open Voltage	ΔE
1.	Pt	Pt	1.060	
2.	AuCu + Al	Pt	1.150	+ .090
3.	AuCu + Co	Pt	1.110	+ .050
4.	Pt	AuCu + Co	1.118	+ .060
5.	AuCu + Fe	Pt	1.105	+ .045
6.	Pt	AuCu + Fe	1.125	+ .065

These results show an improvement on the cathode side (2) as well
as on the anode side (6); used together these will result in an open
circuit voltage of about 1.210 V — very nearly the theoretical value.
Unfortunately this experiment could not be done in time for this meeting.

Finally, Fig. 4 shows the performance of an Al-doped AuCu as a
cathode *vs.* AB-40.

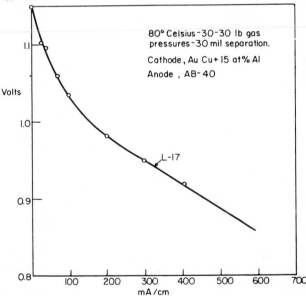

FIG. 4 *Cathode performance of Al doped AuCu vs. AB-40 Pt black.*

7. EXPERIMENTAL EVIDENCE

Figs. 5 to 7 were obtained from identical electrodes, save as noted.
The electrodes were reproducible to within ± 10 mVolt.

Fig. 5 shows currents for constant voltage for materials of different electron densities, n.   The currents vary proportionately to n.
This result, while instinctively correct, is difficult to prove as the
voltages vary in significance for materials of different E_F.

Fig. 6 shows voltages for constant currents. These vary as expected, with $n^{2/3}$, $i.e.$ with the Fermi level E_F.

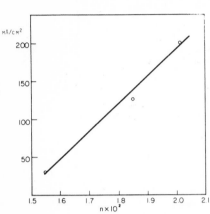
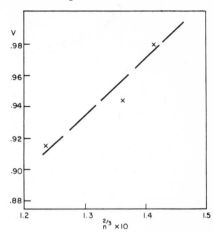

FIG. 5 *Current density vs.*   FIG. 6 *Voltages at constant*
n at constant voltage (0.900 V)   *current (60 mA/cm^2) vs. $n^{2/3}$ × 10.*

Fig. 7 shows variations of currents with grain size. These vary inversely with grain diameter as expected. It is worth noting that the grinding technique offers straightforward opportunities for increasing current densities by decreasing grain size.

Fig. 8 is reproduced from an American Cyanamid report [3]. Note that the characteristics are linear over most of the graphs; over this section an $E = IR$ relationship obtains; R is about 0.2 Ω/cm^2. This verifies the prediction made for the current voltage relationship at high currents. The prolongation of these straight portions to the voltage axis gives the value of E, and the difference between this value and E at open voltage is the quantity referred to as E_0 in this paper. E_0 is very close to the theoretically deduced value.

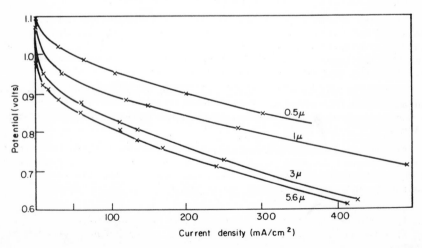

FIG. 7 *Performance vs. powder size AuCu 15 at% Al; 30 mil. asbestos,*
80°C, 8 n KOH, anode: AB-40 Pt black.

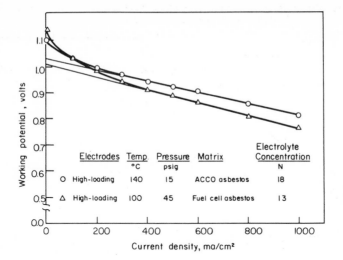

FIG. 8 *Maximum performance with high-loading electrodes.*

8.   CONCLUSIONS

Improvements in E_F, in R, in current density, and, because of life requirements, lower operating temperatures, are needed.

It would seem reasonable at present to entertain the possibility of currents of the order of 1 Amp/cm^2 at voltages between 0.9 and 0.95 V at temperatures of the order of 60°C.

To recapitulate, the preceding argument is based on the following considerations.

1) At the anode only catalysts having d character work. Hence it is necessary to ascribe the anodic action to the transfer of electrons via the d level. These are below the Fermi level.

2) There exists a great deal of symmetry between the situation at the anode and cathode save that at the cathode metals without any d character whatever are satisfactory catalysts. It is logical to assume that the electrons originate from the levels where available electrons are, from the Boltzmann tail and the Fermi level.

3) At high levels of current, *i.e.*, when the catalysts at cathode and anode are in good electrical contact, the voltage driving the cell is the difference in the voltage between the Fermi level at the cathode and the d level at the anode.

4) A mechanism for charge transfer has been suggested, but note that the above considerations are independent of the precise mechanism of charge transfer. However mechanisms implying that the charges acquire or lose energy are not acceptable.

Based on the above:
1) A figure of merit for catalysts has been suggested;
2) A method for increasing the E_F of catalysts has been demonstrated;
3) A method for developing materials having d character from material not having such a character has been shown.

REFERENCES

[1]   Richtmeyer, Kennard, and Lauritzen, *Introduction to Modern Physics*, Fifth Edition, McGraw-Hill, New York (1955) p. 417.

[2]   E. C. Potter, *Electrochemistry*, Cleaver-Hume Press Ltd., London (1961) p. 134.

[3]   "High Performance Light Weight Fuel Cell Electrodes", American Cyanamid Co., Stanford Research Lab., Final Report on Contract No. NASA 3-2786, Figs. 5-21, p. 90.

EFFECT OF ALLOYING COMPONENTS ON THE CATALYTIC ACTIVITY OF PLATINUM IN THE CASE OF CARBONACEOUS FUELS

H. Binder, A. Köhling, and G. Sandstede

Battelle-Institut e.V., Frankfurt am Main, Germany

ABSTRACT

The anodic oxidation of methanol at platinum metals in sulfuric acid at 25°C involves substantial polarization, even if platinum is used as catalyst. At 80°C polarization is still very high, osmium now being the most active catalyst. Some of the Raney platinum metal alloy catalysts were superior to the pure metals. The most active alloy proved to be platinum-ruthenium, where at a current density of 50 mA/cm^2 and a temperature of 80°C a decrease in polarization by about 140 mV with respect to pure platinum has been observed. Such a Raney platinum-ruthenium catalyst is stable over a long period of time. After a period of 3,500 h the drop in the current density amounted to about 50 percent. In investigations using sheets of cold-rolled platinum-ruthenium alloys the alloy $Pt_{0.5}Ru_{0.5}$ showed maximum activity. The lattice spacings of the corresponding Raney alloy catalysts are in good agreement with those of the compact samples. This indicates that the Raney method is suitable for alloy preparation at room temperature. Periodic voltammetric curves reveal that the composition at the surface is different from that of the bulk alloy. Current density with respect to the true surface area of the Raney platinum-ruthenium catalyst is about thirty times that of a Raney platinum electrode. This enhancement cannot be explained by extreme lattice distortions as was demonstrated by the line width of the X-ray patterns. The rate of formic acid oxidation is not affected by alloying platinum with ruthenium. Conclusions concerning the oxidation mechanism of methanol have been drawn from results of chemisorption and desorption experiments. A method for the determination of the surface area of platinum-ruthenium alloys by cyclic voltammetry is described. It is assumed that the higher activity of the alloy catalysts is due to a change in the galvanomagnetic properties of the bulk alloy which in turn leads to weaker adsorption of methanol and the secondary products. The apparent heat of activation for the oxidation of methanol on platinum-ruthenium is smaller by about 4 kcal/mole than the value for platinum.

1. INTRODUCTION

In heterogeneous catalysis metal alloys have often played an important part, and sometimes surprising synergistic effects have been observed. The preparation of finely divided powders is one of the most important problems. In order to obtain maximum specific activity they should be prepared at as low a temperature as possible. Platinum metal alloys are often prepared according to the Adams procedure [1] which yields extremely fine catalyst powders that sometimes contain varying amounts of oxides [2]. Fine catalyst powders of metals can also be obtained on the basis of the procedure by Brown and Brown [3]

who used sodium borohydride as reducing agent for the precipitation of
alloys from a solution of metal salts. This method was used, *e.g.*, by
McKee and Norton [4] for the preparation of platinum-ruthenium alloy
catalysts.

In the course of our investigations on the activity of platinum
metal catalysts for the oxidation of carbonaceous fuels and for methanol in particular we came to the conclusion that the Raney method is
suitable for the preparation of alloy catalysts [5].

2.   RANEY PLATINUM METALS

First a short survey is given of some of our earlier results.
Fig. 1 shows the current-voltage plots obtained upon oxidation of methanol on Raney platinum metals in 4.5 N H_2SO_4 at 25°C, the best catalysts being Rh, Pt and Ir. However, polarization proved to be considerable. Even at 80°C (Fig. 2) polarization was still very high, but
the order of the metals with respect to their activity was changed,
osmium now being the most active catalyst.

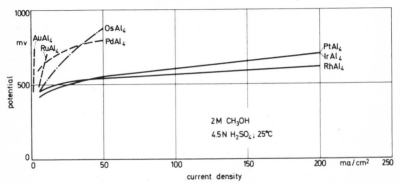

FIG. 1 *Stationary galvanostatic current-voltage curves for methanol oxidation on Raney metals (made from the aluminum alloys indicated)
in sulfuric acid at 25°C.*

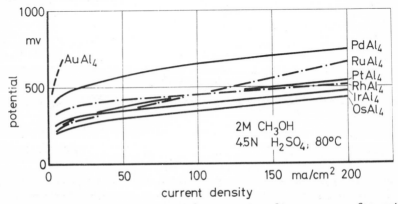

FIG. 2 *Stationary galvanostatic current-voltage curves for methanol oxidation on Raney metals (made from the aluminum alloys indicated)
in sulfuric acid at 80°C.*

3.   RANEY ALLOY CATALYSTS

In our comparative appraisal of alloy catalysts the proportion of catalyst was kept constant in all electrodes.   The alloy catalysts were prepared *in situ* by treating Raney alloys of type $A_{0.5}B_{0.5}Al_4$ (A and B being metals of the platinum group) with potassium hydroxide solution in the preformed electrode disk with a gold skeleton.

Fig. 3 shows the current-voltage plots taken at 25°C with the platinum alloy catalysts.   It is obvious that the most active alloy is

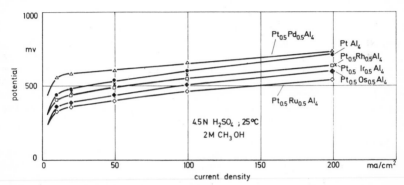

FIG. 3 *Stationary galvanostatic current-voltage curves for methanol oxidation on Raney platinum alloys (made from the aluminum alloys indicated) in sulfuric acid at 25°C.*

that containing ruthenium.   Fig. 4 shows the corresponding figures for 80°C.   At a current density of 50 mA/cm^2 ruthenium caused the polarization to decrease by 140 mV (170 mV at 100 mA/cm^2) compared with pure Raney platinum under the same conditions.   The finding that Raney platinum-ruthenium alloys are excellent catalysts for the oxidation of methanol [5] was in good agreement with alloys prepared by other methods (see above) described, for instance, by Heath [6] and by Frumkin, *et al.* [7], [8].

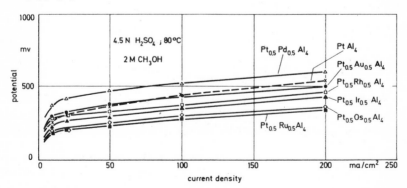

FIG. 4 *Stationary galvanostatic current-voltage curves for methanol oxidation on Raney platinum alloys (made from the aluminum alloys indicated) in sulfuric acid at 80°C.*

A platinum-ruthenium alloy like this appears to be extremely stable (Fig. 5).   At a current density of 2 A/cm^2 and a temperature of 80°C, after an initial deterioration during the first two days, no further

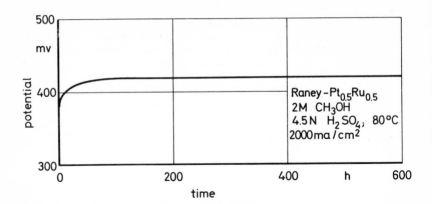

FIG. 5 *Potential-time curve for methanol oxidation on a Raney platinum-ruthenium electrode in sulfuric acid.*

increase in potential was observed over a period of more than 600 hours. However, these catalysts too are subject to aging with a consequent drop in the current density of about 50 percent at a given potential during a period of 3,500 h. Current densities up to 10 A/cm^2 can be reached with methanol on such a platinum-ruthenium electrode with a gold skeleton. Fig. 6 shows the Tafel plots where at high current densities the deviation from a straight line is mostly due to the ohmic drop which has not been accounted for in the plot. In the region of low overvoltage a Tafel slope of 35 mV is observed. Above a potential of, say, 200 mV, a value of about 65 mV can be found if the ohmic drop

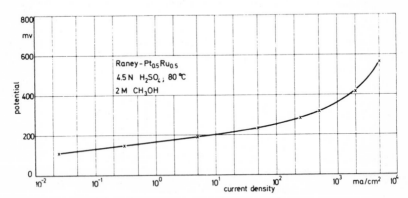

FIG. 6 *Stationary galvanostatic current-voltage curve for methanol oxidation on a Raney platinum-ruthenium electrode in sulfuric acid.*

has been allowed for. It is worth mentioning that in the case of platinum-ruthenium alloys the Tafel slope at 25°C tends to be larger than that at 70°C.

The results for all of our 50:50 alloys investigated in 4.5 N sulfuric acid are compiled in Table I. It shows the potentials for a current density of 50 mA/cm^2. Further investigations have been carried out with alloys of various compositions. The effect of the composition on the potential at a current density of 50 mA/cm^2 is shown in Fig. 7.

TABLE I

Potentials (at a Current Density of 50 mA/cm^2) for Methanol Oxidation on Raney Platinum Metal Alloys (and Gold Alloys) $A_{0.5}B_{0.5}$ at 80°C (Upper Figures) and 25°C (Lower Figures)

2M CH$_3$OH
4.5N H$_2$SO$_4$
50 ma/cm^2

	Ru	Rh	Pd	Os	Jr	Pt	Au
Ru	350 / >900	**Rh**					
Rh	300 / 500	420 / 540	**Pd**				
Pd	290 / 490	520 / 580	570 / 800	**Os**			
Os	290 / >900	320 / 540	340 / 720	320 / 880	**Jr**		
Jr	300 / 440	380 / 520	400 / 580	350 / 500	350 / 560	**Pt**	
Pt	230 / 400	330 / 510	480 / 610	250 / 440	290 / 490	370 / 540	**Au**
Au	310 / >900	460 / 580	650 / 780	400 / >900	300 / 540	380 / 580	>900 / >900

When considering the magnetic susceptibility of binary platinum metal systems (Fig. 8) it would appear [5] that this parameter, which to a certain extent is related to the number of unpaired electrons in the metal, has a major effect, on methanol oxidation. The information derived here is of a strictly qualitative nature. In various series of experimental investigations we therefore studied in more detail the influence of ruthenium on the activity of platinum in the oxidation of methanol.

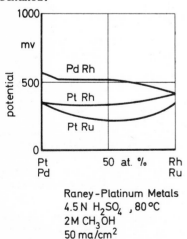

FIG. 7 *Potential for methanol oxidation on Raney platinum metal alloys as a function of composition.*

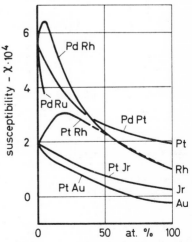

FIG. 8 *Magnetic susceptibility of platinum metal alloys (and gold alloys) as a function of composition (literature data).*

4. PLATINUM RUTHENIUM ALLOYS

4.1 COLD ROLLED SHEETS

In order to find out about the existence of surface layers in dilute sulfuric acid, we investigated a number of cold-rolled and polished platinum-ruthenium alloys (approximately 1 cm^2) with different
ruthenium contents. Fig. 9 shows the periodic potentiodynamic current-
voltage plots of six of these samples in the potential range between
50 and 600 mV. The lower curves reflect the behavior of ruthenium,
the upper curves that of platinum. It is worth noting that according
to X-ray measurements of our samples, the alloy with a platinum content
of 40 percent consists of a single phase (the cubic phase of platinum)
although the potentiodynamic curve rather corresponds to that of ruthenium sheet.

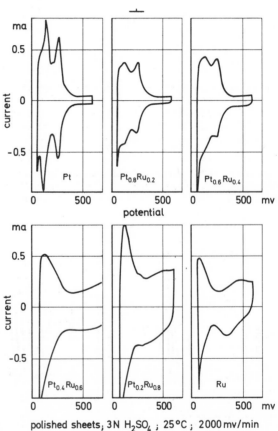

FIG. 9 *Periodic potentiodynamic current-voltage curves using polished platinum-ruthenium sheets in sulfuric acid.*

Fig. 10 shows the lattice spacings of all the platinum-ruthenium
alloys investigated. For the sheets in the range of the substitutional
alloys the constant a_o of the f.c.c. lattice falls as expected, devi-
ating only slightly from Vegard's law. The spacings for the Raney
platinum-ruthenium catalysts investigated by us are added for comparison.

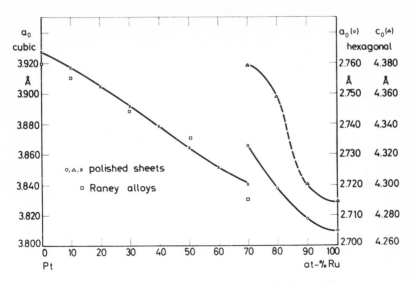

FIG. 10 *Lattice constants of platinum-ruthenium alloys.*

As the deviation is sufficiently small, the Raney method may be considered adequate for the formation of alloys at room temperature.

Fig. 11 depicts parts of the periodic potentiodynamic current-voltage plots, *i.e.* the branches for increasing potential, for the oxidation of methanol on the smooth platinum-ruthenium sheets. On the left-
hand side are shown the plots for a voltage speed of 2000 mV/min, on
the right-hand those for a speed of 100 mV/min. The best activity is

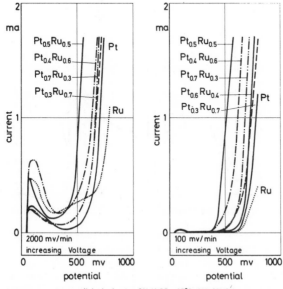

polished sheets, 3N H₂SO₄, 25°C, 2M CH₃OH

FIG. 11 *Periodic potentiodynamic current-voltage curves (increasing voltage) for methanol oxidation on polished platinum-ruthenium
sheets in sulfuric acid.*

observed for the alloy with a ruthenium content of 50 percent, followed
by that containing 60 percent ruthenium. It should be noted, however,
that the roughness factor for the latter alloy appears to be larger
than that for the alloy with a ruthenium content of only 30 percent.
The order might thus be changed if conditions were completely commensu-
rable.

4.2   RANEY CATALYSTS

Fig. 12 shows periodic potentiodynamic current-voltage curves obtained with Raney platinum-ruthenium alloys. Only the Raney alloy with
10 percent ruthenium showed a platinum-like behavior, whereas with the
polished sheets (see Fig. 9), even at a ruthenium content of more than
50 percent (not shown in the figure), the curve resembles that for

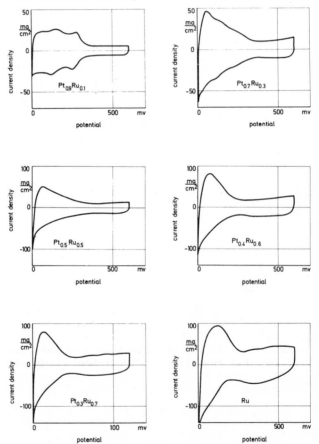

Raney alloys , 3N H_2SO_4 , 25°C , 40 mv/min

FIG. 12 *Periodic potentiodynamic current-voltage curves using Raney
platinum-ruthenium alloys in sulfuric acid.*

platinum. This finding should be taken as an indication that the surface properties of an alloy as revealed by electrochemical measurements
do not by any means permit conclusions to be drawn as to its bulk
properties (see [9]).

With regard to the true current densities (Fig. 13), which have been derived from stationary galvanostatic current-voltage curves for methanol oxidation, it is observed that at higher potentials the maximum is reached at a ruthenium content of about 30 percent (both at 25°C and 70°C). With decreasing potential the maximum tends to be shifted towards the 50:50 alloy. At 70°C and a potential of 350 mV, for instance, the maximum current is about 30 times that achieved with the Raney platinum catalyst. Heath [6] observed a maximum for alloys with a ruthenium content of 40 atom percent, while the maxima found by Entina and Petry [10,18] are at about 25 percent and 40 percent for 25°C and 75°C, respectively.

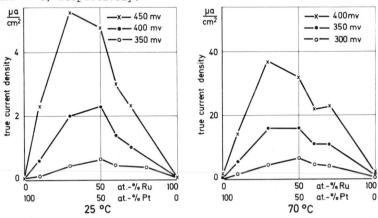

Raney alloys; 3N H₂SO₄ ; 2M CH₃OH

FIG. 13 *True current density for methanol oxidation on Raney platinum-ruthenium as a function of composition.*

5.   RESULTS CONCERNING THE OXIDATION MECHANISM

5.1   GENERAL PROBLEMS

Since we are dealing with current densities related to the true surface area, the enhancement of the activity must be due to a factor other than roughness. One might think that lattice distortions have a bearing on the activity, but the line widths of the X-ray patterns indicate (Fig. 14) that the alloy with a ruthenium content of 60 percent should have a higher activity than the alloys containing 50 percent or 40 percent ruthenium.

Some authors have speculated on a redox couple taking part in the oxidation mechanism, see *e.g.* [6]. Such a mechanism has the disadvantage of requiring the assumption that a layer of chemi-sorbed oxygen or even an oxide film, *e.g.*, on platinum and its alloys, participates in the oxidation mechanism of methanol. In the case of platinum it has been shown [11] that the oxidation of methanol tends to be inhibited rather than enhanced by the oxide layer formed on the surface.

We would also be interested in knowing why pure ruthenium is nearly as poor a catalyst as pure platinum or even a poorer one at a temperature of 25°C, although the formation of an oxide layer takes place at a more favorable potential than on platinum. And why does the addition of ruthenium to platinum not influence its activity for the oxidation of formaldehyde or formic acid, whereas the addition of tin does [13]?

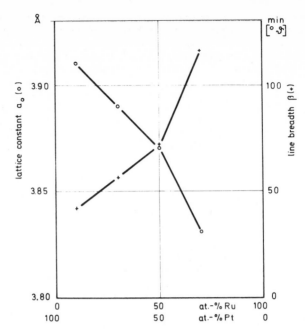

FIG. 14 *Lattice constants and line widths of Raney platinum-ruthenium alloys.*

5.2   METHANOL ADSORPTION ON PLATINUM

Let us consider the case of pure platinum (Fig. 15). At a Raney platinum electrode freed from adsorbed species — except water and electrolyte — a potential drop is observed immediately after addition of methanol to the electrolyte; see also [8]. A minimum potential of about 100 mV is reached after a few minutes, and after several hours a steady-state value of about 400 mV is observed. In this state the platinum catalyst is almost completely covered by a chemisorbed product [8,16], the amount of the H_{ad} species being reduced to less than

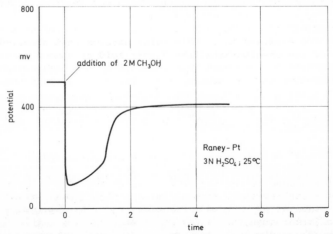

FIG. 15 *Potential-time curve of a Raney platinum electrode in sulfuric acid after addition of methanol.*

10 percent of the initial value (Fig. 16, dotted cyclic curve). During
oxidation of this chemisorbate a maximum is observed at a potential of
670 mV.

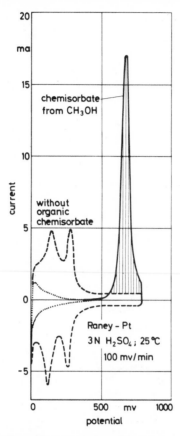

FIG. 16 *Potentiodynamic current-voltage curve (anodic sweep) for
the oxidation of the chemisorbate from methanol on Raney platinum.*

The charge required corresponds to about 1.8 e⁻ per platinum site
covered by the chemisorbate. This value exceeds that observed in the
oxidation of the chemisorbate from formic acid or carbon monoxide
which, according to our measurements, amounts to not more than 1.6 e⁻
per platinum atom in the surface [12]. The difference in the charge
and the fact that the value of the maximum potential in the oxidation
of the formic acid chemisorbate or that from carbon monoxide is 630 mV
indicate that the chemisorbates from methanol and formic acid need not
be completely identical.

5.3 METHANOL ADSORPTION ON PLATINUM-RUTHENIUM

From the foregoing section it is obvious that the oxidation of methanol below a potential of, say, 600 mV and under steady-state conditions
must take place at a platinum surface covered with the strongly bonded
methanol chemisorbate; only a few platinum sites are available for the
adsorption of water which is required during methanol oxidation. This
fact has mostly been disregarded in the discussions on the mechanism

of methanol oxidation on platinum. We therefore assume that, in the
case of platinum-ruthenium alloys, the change in the "electronic
factor" results in a weaker bond strength between the adsorbed methanol and the catalyst; the adsorbed methanol may be regarded as already
being a secondary product formed by dehydrogenation [8]. The hydrogen
atoms appear to be attached to the platinum atoms, whereas the organic
group is attached to the ruthenium atoms.

Evidence supporting this assumption is produced by adsorption measurements. Fig. 17 shows the potentiodynamic curve for the oxidation
of adsorbed species from methanol on a platinum-ruthenium electrode
(50:50) at 25°C, the curve for pure sulfuric acid being included for
comparison. We note the two characteristic peaks for the oxidation of
H_{ad}-atoms on platinum and the maximum for the oxidation of the carbonaceous group which is shifted to a potential of 500 mV instead of
670 mV on pure platinum.

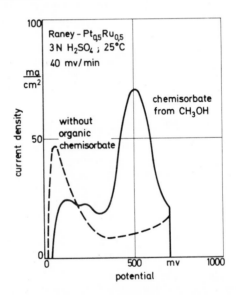

FIG. 17 *Potentiodynamic current-voltage curve (anodic sweep) for
the oxidation of the chemisorbate from methanol on Raney platinum-
ruthenium.*

5.4 DESORPTION OF INTERMEDIATE PRODUCTS

The first stable oxidation product, *i.e.* formaldehyde, is released
from the platinum-ruthenium surface at a higher rate than from a platinum surface.

According to our measurements, formaldehyde is released from the
electrode to such an extent that at 70°C we never obtained a yield of
100 percent carbon dioxide during the oxidation of methanol. This loss
is due to volatization of formaldehyde.

At room temperature the yield of carbon dioxide is also less than
100 percent when the electrolysis is started, even in an electrolyte
saturated with CO_2. After one hour of operation CO_2 is found in an
amount of only 87 percent of the theoretical value. This means that
the intermediate product released from the surface must be enriched in

the bulk electrolyte. Only at steady-state conditions does the current yield amount to 100 percent with respect to CO_2 evolution.

Since the rate of formic acid and formaldehyde oxidation is hardly affected by the addition of ruthenium to the platinum catalyst, the important function of ruthenium must be seen in the oxidation of methanol itself.

5.5 HEAT OF ACTIVATION

It should be noted that, according to our measurements, the apparent heat of activation for the overall reaction in the oxidation of methanol on Raney platinum-ruthenium (50:50) calculated from the slope of the Arrhenius plot for the current density at a given potential is smaller by about 4 kcal per mole than the activation energy for the oxidation on platinum at the same potential. At the other compositions of the platinum-ruthenium system, the difference is smaller.

For a potential of 350 mV the apparent heat of activation for platinum amounts to about 2 kcal/mole and the value for $Pt_{0.5}Ru_{0.5}$ is 20 kcal/mole. These values are in good agreement with that obtained for platinum-ruthenium (20 atom percent Ru) by Entina and Petry [10], who found 20.3 kcal/mole and with that obtained for platinum by Stenin and Podlovchenko [14], who found 21.8 to 23.4 kcal/mole at the same potential.

However, things are complicated by the observation reported earlier in this paper that, *e.g.*, in the case of the alloy $Pt_{0.5}Ru_{0.5}$, Tafel slopes at 25°C tend to be rather larger than those at 70°C, *cf*. [10]. This means that there could be a change in the oxidation mechanism which may be caused by a change in oxygen coverage. Perhaps, oxidation involves several parallel reaction paths, as has been emphasized, *e.g.*, by Breiter [15].

APPENDIX:

ELECTROCHEMICAL DETERMINATION OF THE SURFACE AREA OF PLATINUM-RUTHENIUM ALLOYS

On a ruthenium electrode the formation of the O_{ad}-layer is highly irreversible. Fig. 18 shows the behavior of pure ruthenium sheet at different potentials of reversal. If the lower potential of reversal is changed and the upper one is kept constant at a value of 1,000 mV (Fig. 19), reduction of the oxide layer decreases until finally in the potential range between 700 and 1,000 mV only a very small charge is exchanged. The same behavior is encountered with Raney platinum-ruthenium electrodes.

Fig. 20 shows the potential range between 810 and 900 mV. Under the assumption that this charge is required for the double layer capacity and that this is independent of the potential, we can derive from this charge the value for the double layer capacity. In Fig. 21 this value is used for subtracting the double layer capacity from the complete amount of charge exchanged in the potential range between zero and 700 mV at the same electrode.

It can be seen that in the minimum the current observed is larger than that required for charging the double layer capacity. It may be

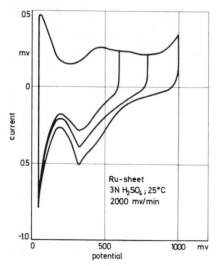

FIG. 18 *Periodic potentiody-*
namic current-voltage curves using
a ruthenium sheet at different
upper potentials of reversal.

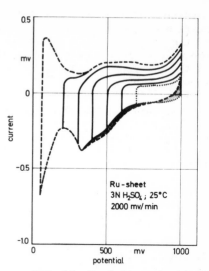

FIG. 19 *Periodic potentiody-*
namic current-voltage curves us-
ing a ruthenium sheet at differ-
ent lower potentials of reversal.

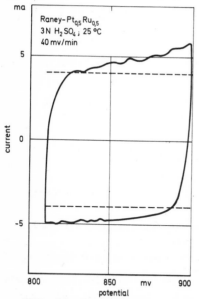

FIG. 20 *Periodic potentiody-*
namic current-voltage curve using
Raney platinum-ruthenium in the
double layer region.

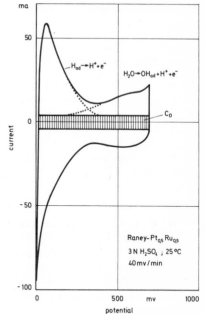

FIG. 21 *Periodic potentiody-*
namic current-voltage curve using
Raney platinum-ruthenium.

assumed that the areas representing the oxidation of H_{ad}-atoms and the
formation of O_{ad}-layers overlap. The presumable curves are shown.

Under the assumption that in the potential range between zero and
300 mV the area not hatched can be exclusively attributed to the

oxidation of H_{ad}-atoms and that each platinum or ruthenium atom in the surface of the catalyst is covered with one H_{ad}-atom, the surface area of the catalyst can be evaluated. This method is very similar to the one we used in the case of pure platinum. Since the lattice constant of the alloy is smaller than that of pure platinum by about 2 percent, we used a value of 210 $\mu C/cm^2$ for the evaluation of the surface area of the alloy instead of 200 $\mu C/cm^2$ for pure platinum [17].

The values derived by this electrochemical method are in fair agreement with those derived by BET measurements. A typical value of the specific surface area of Raney platinum-ruthenium measured after aging of the electrode is about 35 m^2/g. From this value the double layer capacity for the platinum-ruthenium electrode under the conditions chosen can be calculated to be 105 $\mu F/cm^2$.

ACKNOWLEDGMENT

This work was conducted under a contract of the Robert Bosch GmbH, Stuttgart, Germany. Its permission to publish these results is gratefully acknowledged.

REFERENCES

[1]   R. Adams and R. L. Schriner, *J. Am. Chem. Soc.*, 45, 2171 (1923).

[2]   See *e.g.* L. W. Niedrach, D. W. McKee, J. Paynter, and I. F. Danzig, *Electrochem. Techn.*, 5, 318 (1967).

[3]   H. C. Brown and C. A. Brown, *J. Am. Chem. Soc.*, 84, 1493 (1962).

[4]   D. W. McKee and F. J. Norton, *J. Phys. Chem.*, 68, 481 (1964).

[5]   H. Binder, A. Köhling, and G. Sandstede, in *Hydrocarbon Fuel Cell Technology* by B. S. Baker (Ed.) Academic Press Inc., New York (1965) p. 91.

[6]   C. E. Heath, in *Comptes Rendus, Journées International d'Etude des Piles à Combustible*, *Revue Energie Primaire*, Vol. I, p. 99 Brussels (1965).

[7]   A. N. Frumkin in *Batteries 2* by D. H. Collins (Ed.), Pergamon Press, London (1965).

[8]   O. A. Petry, B. I. Podlovchenko, A. N. Frumkin, and Hira Lal, *J. Electroanal. Chem.*, 10, 253 (1965).

[9]   V. S. Entina and O. A. Petry, *Elektrokhimiya*, 4, 111 (1968).

[10] V. S. Entina and O. A. Petry, *ibid.*, 4, 678 (1968).

[11] H. Binder, A. Köhling and G. Sandstede, "Raney Catalysts", these proceedings, p. 15.

[12] H. Binder, A. Köhling and G. Sandstede, "Platinum Catalysts ...", these proceedings, p. 59.

[13] K. J. Cathro, *J. Electrochem. Soc.: Electrochem. Techn.*, 116, 1608 (1969).

[14] V. F. Stenin and B. I. Podlovchenko, *Elektrokhimiya*, 3, 481 (1967).

[15] M. W. Breiter, *Discuss. Faraday Soc.*, 45, 79 (1968).

[16] J. Giner, *Electrochim. Acta*, 9, 63 (1964).

[17] H. Binder, A. Köhling, K. Metzelthin, G. Sandstede, and M.-L. Schrecker, *Chemie Ing. Technik*, 40, 586 (1968).

[18] O. A. Petry, *Dokl. Akad. Nauk*, *SSSR*, 160, 871 (1965).

PLATINUM CATALYSTS MODIFIED BY ADSORPTION
OR MIXING WITH INORGANIC SUBSTANCES

H. Binder, A. Köhling, and G. Sandstede

Battelle-Institut e.V., Frankfurt am Main, Germany

ABSTRACT

The activity of platinum catalysts for the conversion of organic
fuels could be changed by addition of inorganic substances. The adsorption of organic substances and elements on smooth platinum (platinum
gauze) and porous platinum (Raney platinum) has been investigated by
means of the linear voltage sweep method. The charge per platinum atom
required for desorption of the chemisorbates was found to involve 1.5
to 1.6 electrons for carbon monoxide, formic acid and reduced carbon
dioxide, and 1.8 electrons for methanol. From the position of the
anodic stripping curve of the chemisorbates, it may be concluded that
anodic oxidation of the fuels at low potentials takes place in the
presence of an almost complete monolayer of the chemisorbate. More or
less complete monolayers of sulfur, selenium, tellurium, lead and other
elements greatly accelerate the anodic oxidation of carbon monoxide,
formic acid and formaldehyde, but inhibit the conversion of methanol
and hydrocarbons. The conversion of methanol is, however, accelerated
in the presence of lead, molybdenum sulfide or tungsten sulfide. The
rate-increasing or rate-decreasing effects are also produced if platinum powder is mixed with the elements or with specific sulfides or
other heavy metal compounds. The results suggest that a sorption layer
is formed on the platinum *via* the solution phase. With each fuel mentioned the reaction is inhibited in potassium hydroxide solution.
Tafel slopes and Arrhenius slopes of the anodic oxidation curves are
changed by the adsorbed layers. The stability of the absorbed layers
and the dependence of the rate increase on the degree of coverage are
described. The state of the platinum surface and the mechanism of
anodic oxidation in the presence of an adsorbed layer are discussed.

INTRODUCTION

In research on catalysts for fuel cells, attempts have been made
for several years to modify or replace the platinum in order to obtain
better catalysts for the conversion of organic fuels, especially in
acid electrolytes. As a result of these attempts, various other anode
catalysts have been found; the activity of platinum has been improved
by alloying with ignoble or other noble metals (references in [1,2])
and even inorganic catalysts have been discovered (references in [3]).
Up to now inorganic catalysts, however, have had only a minor activity
for the conversion of organic fuel. The composition of the surface of
the inorganic catalysts in electrolyte solutions is different from the
composition of the bulk. The surface layer may result from adsorption
of substances or chemical modification. The activity of, for example,
tungsten carbide (WC) is determined by a surface layer, which can be
formed by controlled oxidation.

59

Adsorbed covering layers play an important role in the conversion on platinum electrodes. Owing to the adsorbed layers, catalysis-inhibiting effects are observed, but depending on the nature of the adsorbed species, even rate-increasing effects have recently become known. In the absence of any fuel or other foreign substances — such as impurities — platinum electrodes in acid or alkaline electrolytes may be covered with the following adsorbates: hydroxide radicals, oxygen atoms, hydroxide ions or other anions, hydrogen atoms and water molecules. The last can be distinguished by the flop-down and the flip-up state [1]. In the presence of organic substances the reactants, the intermediates, and the reaction or dissociation products may be adsorbed as well. In addition to these adsorbates, ions or elements may be added deliberately in order to change the properties of the platinum surface.

On the other hand, modification by alloying changes not only the bulk properties of the platinum, but also its surface properties. Perhaps it would be sufficient in many cases to change the surface properties in order to increase the catalytic activity. Despite the fact that a great deal is known about the electrocatalytic properties of platinum, the reaction mechanism of the anodic oxidation of organic substances has still not been adequately investigated. However, it would be worthwhile to know the behavior of organic substances on platinum in greater detail. The results may then be applied by analogy to other catalysts, or used as a basis for their investigation. This would be valuable, since platinum is the only catalyst that is perfectly stable throughout the potential range under consideration.

Before describing the change in the electrocatalytic properties of platinum caused by deliberately adsorbed surface layers, we will consider the adsorbed layers, which result from the chemisorption of some organic fuels.

ORGANIC CHEMISORBATES ON THE PLATINUM SURFACE

The chemisorption of carbon-containing substances on the platinum surface has been investigated by many researchers; for example, Bagotzky, Bockris, Breiter, Brummer, Frumkin, Gileadi, Gilman, Giner, Grubb, Niedrach, Pavela, Piersma, Shlygin, Vielstich [1,4-6] and ourselves [7,8]. The results are still controversial. It is not possible to summarize them here. Some of our older and more recent results are reported followed by a discussion of the effect of elemental adsorbates.

The chemisorbed amounts were measured potentiodynamically by applying an anodic potential sweep and recording the anodic stripping curve. In the investigations described later on, cyclic voltammetry was also used, and stationary current-voltage plots were recorded as well. The potentials were measured against an autogenous hydrogen electrode in the same solution at the same temperature. Smooth platinum and Raney platinum were used as test electrodes.

Fig. 1 shows the potentiodynamic current-voltage curve (anodic stripping curve) of Raney platinum covered by a chemisorbate formed by adsorption of carbon monoxide. Adsorption was effected at open circuit, the potentials being about 250 mV at 30°C, and about 150 mV at 70°C. The absence of the hydrogen oxidation peaks indicates that the surface is completely covered with the chemisorbate which is not removed by rinsing with oxygen-free electrolyte. The charge required for oxidation of the chemisorbate is about 1.6 electrons per platinum surface atom, independent of temperature. It is assumed for this calculation, that the $Pt:H_{ad}$ ratio equals 1.

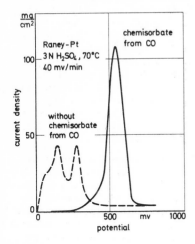

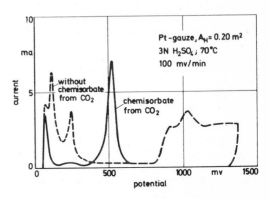

FIG. 1 *Potentiodynamic current-voltage curve (anodic sweep) for the oxidation of chemisorbate from carbon monoxide.*

FIG. 2 *Potentiodynamic current-voltage curve (anodic sweep) for the oxidation of chemisorbate from reduced carbon dioxide.*

A similar potentiodynamic current-voltage curve is obtained when the chemisorbate consists of reduced CO_2 (Fig. 2). The fact that on platinum CO_2 is reduced already at a potential more positive by about 200 mV than the reversible hydrogen potential was reported for the first time by Giner [9]. The oxidation peak of the reduced CO_2 chemisorbate is observed at about the same potential as the oxidation peak of the chemisorbate from CO. About 1.5 electrons per platinum surface atom flow during oxidation of the reduced CO_2 chemisorbate. This result was obtained on smooth platinum using a platinum gauze with a surface area A_H of about 0.20 m^2, as determined by anodic desorption of the H_{ad} atoms.

A charge of 1.8 electrons per platinum atom was found [2] for the oxidation of the chemisorbate formed by the adsorption of methanol at open circuit (*ca.* 400 mV) and room temperature.

As can be seen from Fig. 3, similar potentiodynamic current voltage curves are obtained for a chemisorbate resulting from formic acid adsorption. The adsorption again took place at open circuit, the potential being about 200 mV at room temperature and about 100 mV at 70°C. When the temperature is decreased, the oxidation peak is shifted toward higher potentials. The charge required for the oxidation of the chemisorbate is 1.6 electrons per platinum atom. In addition to the anodic stripping curve, Fig. 3 shows a cyclic potentiodynamic current-voltage curve (dotted line), which may be of importance to the interpretation of the anodic oxidation mechanism. The area included by the cyclic curve indicates that a small part of the platinum surface can be charged and discharged without affecting the anodic stripping curve. This may be due not only to H_{ad} formation (a few percent of the total adsorbable amount), but also to reduction and oxidation of the chemisorbate.

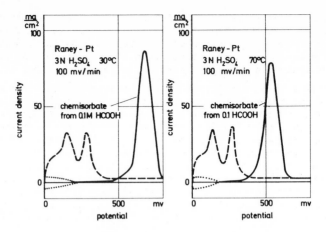

FIG. 3 *Potentiodynamic current-voltage curve (anodic sweep) for the oxidation of chemisorbate from formic acid.*

Another result which may elucidate the reaction mechanism can be derived from Fig. 4. It shows a steady state current-voltage plot of the oxidation of formic acid together with the potentiodynamic oxidation curve of the chemisorbate. The potentiodynamic curve clearly indicates that the platinum surface is covered completely or almost completely with the chemisorbate and that anodic oxidation occurs despite this fact. The same applies to the anodic oxidation of carbon monoxide,

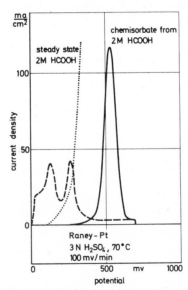

FIG. 4 *Potentiodynamic current-voltage curve (anodic sweep) for the oxidation of chemisorbate from formic acid and a stationary current voltage curve for formic acid oxidation.*

formaldehyde and methanol. This raises the question as to whether anodic oxidation at low potentials takes place at the surface of the chemisorbate or at certain sites of the platinum surface that are not covered by the chemisorbate. If the latter assumption is true, only a minute fraction of the platinum surface is involved in the electrochemical reaction. (In the case of methanol, however, we found that about 10 percent of the H_{ad} coverage can be charged and discharged.) The answer to this question cannot be given at this stage. At any rate, the chemisorbate does not take part in the anodic oxidation of the fuel.

There is a great similarity in the behavior of the four organic substances and the reduced CO_2. The oxidation peaks of their chemisorbates are observed at about the same potential and the charges per platinum surface atom required for the oxidation of the chemisorbates are not very different. This suggests that they consist of the same species in slightly different ratios to one another. From an inspection of the simplest possible radicals, one arrives at the carboxyl radical which releases one electrom per molecule upon oxidation and a COH radical which releases three electrons per molecule upon oxidation. The latter may be a formaldehyde radical or a carbinol radical. A $C(OH)_2$ radical may also exist, at least as intermediate.

In the case of CO_2 the radicals may be produced by reaction with H_{ad} atoms at a potentiostatically fixed potential of O mV as follows:

$$CO_2 + H_{ad} \longrightarrow COOH_{ad} \tag{1}$$

$$COOH_{ad} + H_{ad} \longrightarrow COHOH_{ad} \tag{2}$$

$$COHOH_{ad} + H_{ad} \longrightarrow COH_{ad} + H_2O \quad . \tag{3}$$

In the case of the fuels, the H_{ad} atoms are displaced so that the formation of the chemisorbate may take place at any potential up to its oxidation potential. The steps for the formation of the carbon monoxide chemisorbate might be:

$$CO + H_2O_{ad} \longrightarrow COOH_{ad} + H_{ad} \tag{4}$$

$$CO + H_{ad} \longrightarrow COH_{ad} \quad . \tag{5}$$

Depending on the ratio of $COOH_{ad}:COH_{ad}$, part of the H_{ad} atoms may give hydrogen:

$$2\ H_{ad} \longrightarrow H_2 \quad . \tag{6}$$

At any rate, hydrogen is evolved in the case of formic acid chemisorption:

$$HCOOH \longrightarrow COOH_{ad} + H_{ad} \tag{7}$$

$$HCOOH + H_{ad} \longrightarrow COH_{ad} + H_2O \tag{8}$$

$$2\ H_{ad} \longrightarrow H_2 \quad . \tag{9} = (6)$$

For methanol more steps would be necessary in order to produce the radicals in question:

$$CH_3OH \longrightarrow CH_2OH_{ad} + H_{ad} \tag{10}$$

$$CH_2OH_{ad} \longrightarrow CHOH_{ad} + H_{ad} \tag{11}$$

$$CHOH_{ad} \longrightarrow COH_{ad} + H_{ad} \tag{12}$$

$$COH_{ad} + H_2O_{ad} \longrightarrow COHOH_{ad} + H_{ad} \qquad\qquad (13)$$

$$COHOH_{ad} \longrightarrow COOH_{ad} + H_{ad} \qquad\qquad (14)$$

$$2\ H_{ad} \longrightarrow H_2 \qquad . \qquad (15) = (9) = (6)$$

The amount of H_{ad} on the platinum surface depends on the potential, no matter if potentiostatically adjusted or open circuit. The amount of the individual types of radicals may also depend on the potential.

With respect to the distribution of the radicals on the surface, it is of interest to consider the result achieved by Breiter [4], who reported that 2.1 electrons per C atom are released upon oxidation to CO_2 of the chemisorbate resulting from methanol. Together with the figure of 1.8 electrons per platinum atom, this would mean that there are some bare platinum atoms on the surface, provided that the Pt/CO_xH_y ratio is equal to one. Further results are necessary in order to know whether there is space left on the platinum surface for the intermediate adsorption of fuel molecules during the anodic oxidation.

CHEMISORBATE OF SULFUR AND OTHER ELEMENTS ON THE PLATINUM SURFACE

As we found an influence of elemental adsorbates on the activity of the platinum catalyst, we investigated the stability of such adsorbates by cyclic voltammetry.

In Fig. 5, we note from the low current density in the hydrogen oxidation range that sulfur is adsorbed on the platinum surface, which is anodically oxidized at potentials above 650 mV to form SO_2. Thermodynamically, the oxidation potential of sulfur is 540 mV, and the sulfur/hydrogen sulfide equilibrium is established at 140 mV. Even at 0 mV, however, the sulfur adsorbate is not removed by reduction.

Fig. 6 shows the periodic current-voltage curves after adsorption of sulfur on the surface of a platinum gauze. From these curves the degree of coverage can be determined. At best, a monolayer can be formed. The curves show that the first hydrogen peak, i.e., the less firmly bound H_{ad}, reappears first as sulfur coverage is decreased by oxidation. The potential, at which oxidation of the sulfur adsorbate starts, depends on the degree of coverage. At a temperature of 70°C and a high degree of coverage, it is about 600 mV; at a low degree of coverage, about 700 mV. At room temperature the potential is about 50 to 100 mV higher.

In phosphoric acid at 155°C, the sulfur adsorbate is also stable up to a potential of 650 mV [10]. In potassium hydroxide solution the sulfur adsorbate is likewise stable, even at 70°C. However, anodic oxidation is already initiated at potentials above 500 mV. Thermodynamically, elemental sulfur cannot exist in potassium hydroxide solution. Sulfide ions should be oxidized directly to sulfate ions. That means that the sulfur adsorbate is different from elemental sulfur.

Besides sulfur other adsorbates can be deposited on the platinum surface, e.g., selenium, tellurium, lead, etc. [11-13]. A selenium adsorbate is stable at potentials up to 900 to 950 mV; whereas the oxidation potential of elemental selenium amounts to 740 mV.

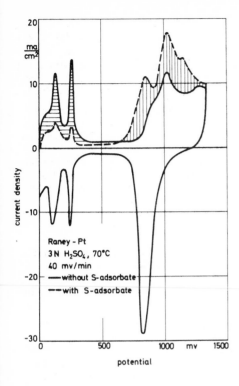

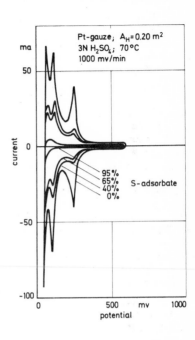

FIG. 5 *Periodic potentio-*
dynamic current-voltage curve
with platinum, and potentio-
dynamic current-voltage curve
(anodic sweep) with platinum
partly covered by sulfur.

FIG. 6 *Periodic potentiodynamic*
current-voltage curves with platinum
partly covered by sulfur.

RATE-INCREASING EFFECT BY ADSORBED LAYERS

Starting from a clean platinum surface, the initial current density for the oxidation of various organic fuels is relatively high. Yet because of the increasing coverage of the platinum surface with organic adsorbate, the current density decreases considerably with time. Only at potentials between about 700 and 900 mV is the initial current practically maintained, because at these potentials an organic chemisorbate is no longer formed. It has become apparent that it is possible to prevent the inhibition by sulfur and other elements adsorbed on the platinum surface so that the high initial current densities are maintained. This means that an element adsorbed on the platinum surface can significantly speed up the anodic oxidation of various fuels [8,10, 13-16].

Fig. 7 illustrates the acceleration of the oxidation of carbon monoxide on a platinum electrode covered with a monolayer of sulfur.

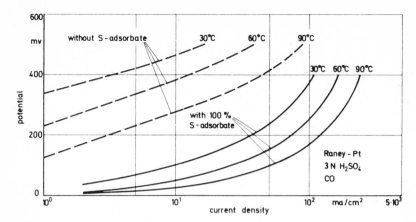

FIG. 7 *Stationary galvanostatic current-voltage curves for carbon monoxide oxidation on platinum with and without sulfur adsorbate.*

The dashed curves were measured in the absence of sulfur, the solid curves in the presence of sulfur, both in 3N H_2SO_4. Fig 8 shows the dependence of the rate increase on the degree of coverage of the platinum surface. The highest acceleration is attained with a complete monolayer of sulfur.

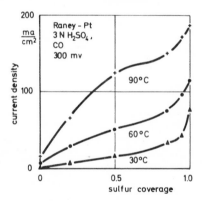

FIG. 8 *Rate of electrooxidation of carbon monoxide on platinum as a function of sulfur coverage.*

The rate of the electrooxidation of carbon monoxide is also increased in phosphoric acid. The current density in 44N H_3PO_4 at 155°C is increased about threefold in the presence of a sulfur adsorbate [10].

In the case of formic acid the acceleration is even higher than with carbon monoxide. Fig. 9 shows stationary current-voltage curves measured in the absence and presence of sulfur. The current density at a potential of 200 mV is increased by a factor of approximately 100. As can be seen from Fig. 9, the Tafel slope increases as the degree of sulfur coverage increases, indicating a change in the reaction mechanism. In addition, the diagram shows that the reaction rate decreases at a high degree of coverage.

By comparing Fig. 9 with Fig. 10, it is noted that the Tafel slopes for porous electrodes, in this case Raney platinum containing 75 mg

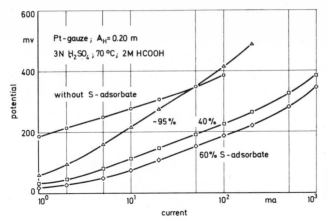

FIG. 9 *Stationary galvanostatic current-voltage curves for formic acid oxidation on platinum with and without sulfur adsorbate.*

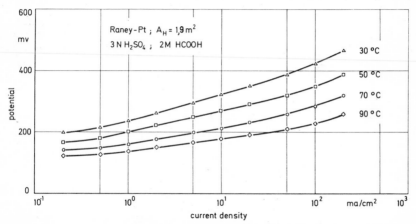

FIG. 10 *Stationary galvanostatic current-voltage curves for formic acid oxidation on platinum.*

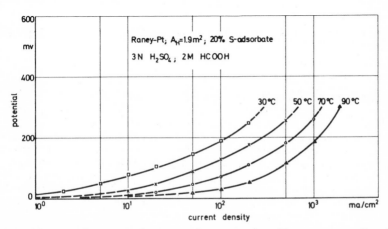

FIG. 11 *Stationary galvanostatic current-voltage curves for formic acid oxidation on platinum with sulfur adsorbate.*

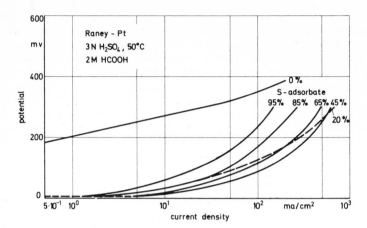

FIG. 12 *Stationary galvanostatic current-voltage curves for formic acid oxidation on platinum with and without sulfur adsorbate in different coverages.*

platinum per cm^2, are slightly less steep than those for platinum gauze electrodes. This might be explained in terms of a temperature effect, since relatively high currents flow through a large internal surface area electrode, which has a very low volume. Fig. 11 shows the steady-state current-voltage curves measured with the same Raney platinum electrode in the presence of sulfur. Here the curves start at an open circuit potential of nearly 0 mV. The variation in the appearance of the curves with increasing sulfur coverage is shown in Fig. 12.

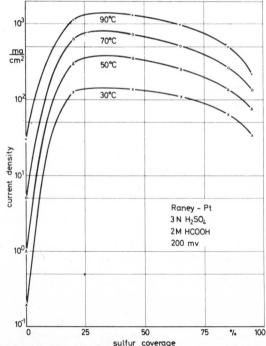

FIG. 13 *Rate of electrooxidation of formic acid on platinum as a function of sulfur coverage.*

   The accelerating effect of sulfur on Raney platinum corresponds to
that on smooth platinum. The maximum is merely shifted to lower degrees
of coverage. It can be seen more clearly from Fig. 13 that the increase
in the conversion rate of formic acid has a pronounced maximum at a
medium sulfur coverage of about 40 percent.

   The change in the anodic oxidation mechanism in the presence of
sulfur not only is indicated by the change of the Tafel slopes, but
also is obvious from the decrease of the activation energy, as may
be seen from Fig. 14. As it is not possible to determine the exchange-
current density for formic acid oxidation at the porous electrodes, we

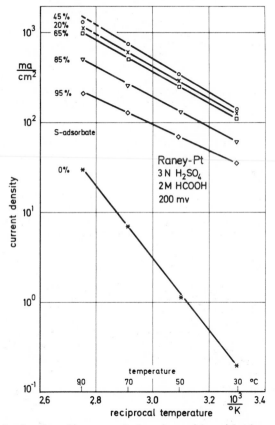

FIG. 14 *Arrhenius diagram of formic acid oxidation on platinum
with and without sulfur adsorbate in different coverages.*

plotted the current density at a potential of 200 mV (overpotential
of 396 mV). Of course, the "apparent activation energy" derived from
the curves includes — in addition to the equilibrium potential term —
also the overpotential term with possibly different values of the
transfer coefficient. Therefore the differences of the values of the
apparent activation energy not only refer to a change of the chemical
activation energy, but also tell us something about what has happened
in integral energetic terms. As can be derived from the Fig. 14, the
difference between the values for the absence of sulfur and for 45
percent coverage amounts to approximately 8 kcal/mole. This coverage
gives maximum current density, although the activation energy decreases
further with increasing coverage.

Fig. 15 shows the Arrhenius curves obtained for carbon monoxide. In this case too the slope of the curves is smallest at 100 percent coverage, but the reaction rate is highest. The reduction of the activation energy amounts to about 7 kcal/mole.

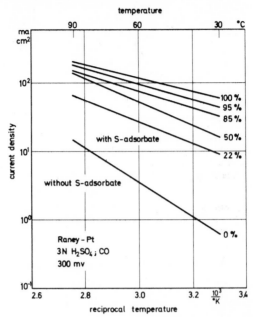

FIG. 15 *Arrhenius diagram of carbon monoxide oxidation on platinum with and without sulfur adsorbate in different coverages.*

RATE-INCREASING EFFECT BY MIXING PLATINUM BLACK WITH INORGANIC SUBSTANCES

The rate of fuel conversion can also be increased by using mixtures of platinum black either with the elements mentioned or with certain inorganic substances [13]. Fig. 16 shows the results of formic

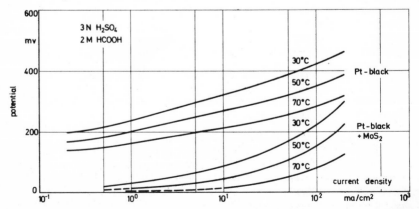

FIG. 16 *Stationary galvanostatic current-voltage curves for formic acid oxidation on platinum black and on platinum black mixed with molybdenum sulfide.*

acid oxidation obtained with platinum black and with a mixture of platinum black and molybdenum sulfide. The rate increase is more than 100-fold at 200 mV.

Fig. 17 shows the corresponding Arrhenius diagram. The difference in the apparent activation energy amounts to about 8 kcal/mole, and hence is similar to the results obtained in the presence of a sulfur adsorbate.

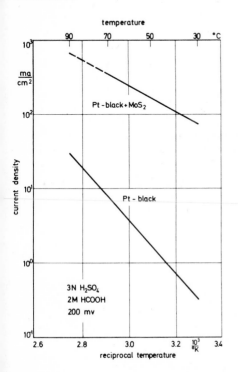

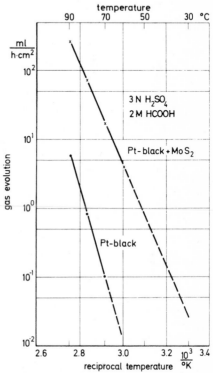

FIG. 17 *Arrhenius diagram of formic acid oxidation on platinum black and on platinum black mixed with molybdenum sulfide.*

FIG. 18 *Arrhenius diagram of formic acid decomposition on platinum black and on platinum black mixed with molybdenum sulfide.*

Decomposition of formic acid at open circuit is also highly enhanced. The results are shown in Fig. 18, where they are presented in an Arrhenius diagram. The electrode, containing approximately 100 mg of platinum, was the same as that used for the measurement of the current-voltage plots. The evolution of 1 ml of gas ($H_2 + CO_2$) corresponds approximately to a current of 1 mA. As can be seen from Fig. 18, the decomposition rate is increased by a factor of more than 100. The activation energies are much larger than those for the anodic oxidation, and the difference in activation energy amounts to about 20 kcal/mole — a value more than twice that derived for the anodic oxidation. From these results, conclusions as to the decomposition mechanism may be drawn. If we assume the following reactions taking place:

$$HCOOH \longrightarrow (COOH)_{ad} + (H)_{ad} \qquad\qquad (16) \triangleq (7)$$

$$(COOH)_{ad} \longrightarrow CO_2 + (H)_{ad} \tag{17}$$

$$2 (H)_{ad} \longrightarrow H_2 \tag{18} \; \hat{=} \; (9)$$

then the last one, the combination of two $(H)_{ad}$ atoms to give hydrogen, might be inhibited by the presence of the organic radicals on the platinum surface. It should be noted that $(H)_{ad}$ is not firmly adsorbed and not chemisorbed — as is H_{ad} — because it cannot be detected on a platinum surface covered with organic chemisorbate at open circuit. $(COOH)_{ad}$ must also be different from the tenaciously chemisorbed $COOH_{ad}$ radical (see also reactions 19 to 21).

Not only molybdenum sulfide, but also other sulfides exhibit a rate-increasing effect, as can be seen from Fig. 19. Part of the effect is certainly due to sulfur which has been transferred from the sulfides to the platinum surface; however, the kind of metal ion of the sulfides also seems to have an influence on the decrease in polarization.

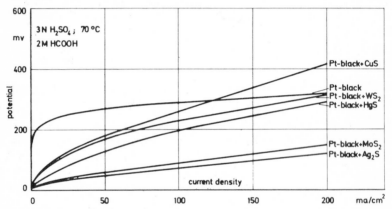

FIG. 19 *Stationary galvanostatic current-voltage curves for formic acid oxidation on platinum black and on platinum black mixed with various sulfides.*

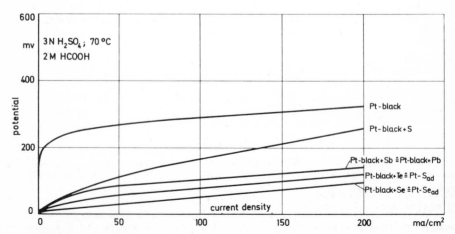

FIG. 20 *Stationary galvanostatic current-voltage curves for formic acid oxidation on platinum black and on platinum black mixed with various elements as well as on platinum with a 50 percent sulfur or selenium adsorbate.*

The current-voltage curves shown in Fig. 20 were measured for mixtures with different elements. It can be seen from the lower curve that it makes no difference whether the selenium used is adsorbed on, or mixed with platinum black. This suggests that the platinum is covered with a sorption layer *via* the solution phase. This can also be inferred from Fig. 21, which shows that the potential drops if the electrodes are allowed to be immersed in the electrolyte for an extended period of time.

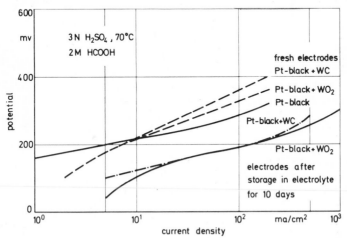

FIG. 21 *Stationary galvanostatic current-voltage curves for formic acid oxidation on platinum black and on platinum black mixed with tungsten carbide or tungsten oxide.*

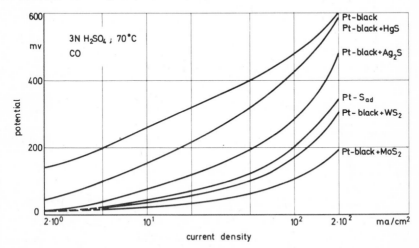

FIG. 22 *Stationary galvanostatic current-voltage curves for carbon monoxide oxidation on platinum black and on platinum black mixed with various sulfides.*

The accelerating effect of inorganic substances for the anodic oxidation is also found with other fuels. Fig. 22 shows the effect of sulfides for carbon monoxide conversion. Comparing this with Fig. 19 shows that although the sequence obtained for the various sulfides is not the same as with formic acid, molybdenum sulfide is also best.

The rate is also increased in the case of formaldehyde. Fig. 23
shows the results obtained with mixtures of platinum black with selenium and tellurium. Admixed elemental sulfur likewise accelerates
the anodic oxidation.

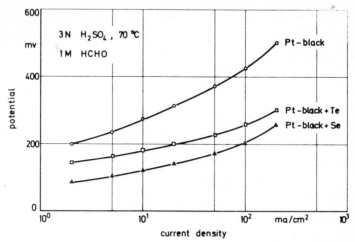

FIG. 23 *Stationary galvanostatic current-voltage curves for formaldehyde oxidation on platinum black and on platinum black mixed with
tellurium or selenium.*

The conversion of methanol can be accelerated by using lead, tungsten sulfide and molybdenum sulfide (Fig. 24). In the case of tungsten sulfide and molybdenum sulfide a combined effect may be assumed:
sulfur increases the oxidation rate of formic acid, and tungsten and
molybdenum accelerate the oxidation of methanol. The redox effect of
molybdate is known from the investigations by Shropshire [17] and
Heath [18].

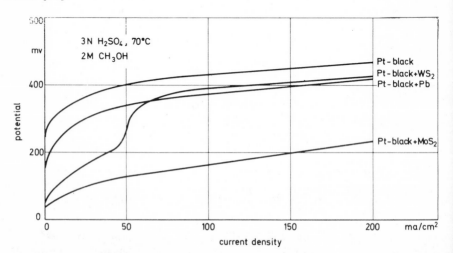

FIG. 24 *Stationary galvanostatic current-voltage curves for methanol oxidation on platinum black and on platinum black mixed with lead,
molybdenum sulfide or tungsten sulfide.*

INHIBITING EFFECT OF ADSORBED LAYERS

In the presence of a sulfur adsorbate the oxidation rate of methanol is not increased, not even at a platinum-ruthenium alloy, as may be seen from Fig. 25. Rather, we observe a decrease of the current density approximately proportional to the coverage of the platinum surface. The oxidation rate of formic acid is not increased in the presence of ruthenium, as methanol oxidation is; whereas the presence of sulfur on the platinum-ruthenium surface does increase the oxidation rate of formic acid (Fig. 25).

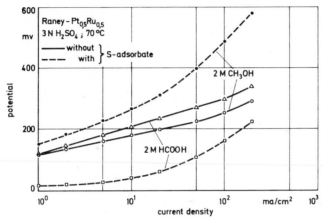

FIG. 25 *Stationary galvanostatic current-voltage curves for methanol and formic acid oxidation on platinum-ruthenium with and without sulfur adsorbate.*

Sulfur has also an inhibiting effect on the anodic oxidation of saturated hydrocarbons. Fig. 26 shows periodic current-voltage curves obtained with ethane and ethylene. The oxidation rate of ethane is reduced drastically, whereas that of ethylene is only slightly affected. There might even exist a favorable influence at lower sulfur coverage.

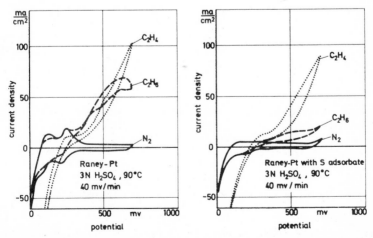

FIG. 26 *Periodic potentiodynamic current-voltage curves for hydrocarbon oxidation on platinum with and without sulfur adsorbate.*

In the case of potassium hydroxide solution as an electrolyte, however, the reaction of ethylene is inhibited by a sulfur adsorbate (Fig. 27). This means that ethylene cannot be adsorbed on the sulfur layer in the presence of OH⁻ ions.

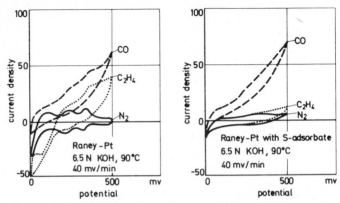

FIG. 27 *Periodic potentiodynamic current–voltage curves for carbon monoxide and ethylene oxidation on platinum with and without sulfur adsorbate in alkaline electrolyte.*

Also included in Fig. 27 are the periodic current–voltage curves for carbon monoxide. They show practically no effect of sulfur in alkaline electrolyte, which indicates that adsorption of carbon monoxide strong enough for the reaction occurs also on the sulfur layer.

The presence of sulfur, on the other hand, inhibits the conversion of formic acid in potassium hydroxide solution, which suggests an entirely different oxidation mechanism in alkaline solutions from that in acids. The reaction rate is decreased proportionally to the sulfur coverage.

CONSIDERATIONS PERTAINING TO THE MECHANISM OF THE ANODIC OXIDATION OF FORMIC ACID AND CARBON MONOXIDE

As was shown before, the electrooxidation of organic substances under steady state conditions at low potentials takes place in the presence of a nearly complete organic chemisorbate monolayer on the platinum surface. This is also true if the platinum surface is partly covered with a sulfur adsorbate. In this case too, the platinum surface, which is still free from sulfur, is covered by an organic chemisorbate. Fig. 28 shows a potentiodynamic current–voltage curve (for the oxidation of the organic chemisorbate) which results from a linear potential sweep applied to a platinum electrode with sulfur adsorbate. The degree of coverage with sulfur can be determined from the potentiodynamic curve in the pure electrolyte. The oxidation curve of the organic chemisorbate corresponds exactly to that in absence of sulfur (Fig. 4). The specific charge required for the oxidation of the organic chemisorbate merely leads to a minor increase with increasing degree of sulfur coverage: 1.8 electrons per platinum atom of the sulfur-free surface are obtained at 40 percent sulfur coverage.

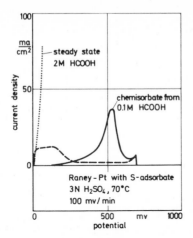

FIG. 28 *Potentiodynamic current-voltage curve (anodic sweep) for the oxidation of chemisorbate from formic acid and a stationary current voltage curve for formic acid oxidation on platinum partly covered with sulfur.*

Carbon monoxide shows an analogous behavior. There is also an organic chemisorbate on those sites of the platinum surface which are not covered with sulfur. However, no chemisorption takes place if the monolayer of sulfur is complete. On the other hand, physisorption on the sulfur layer must occur, because otherwise, anodic oxidation would not be possible.

Turning back to Fig. 28, which also shows a steady state current-voltage curve for formic acid oxidation, we have to conclude that the electrooxidation takes place at the adsorbed monolayer consisting of a mixture of sulfur, COOH (carboxyl) radicals and COH radicals. In the following, an attempt is made to sketch a highly simplified picture of the state of the platinum surface in the presence of sulfur and formic acid chemisorbate and of the way of the electrooxidation of arriving molecules. In the case of formic acid, it is favorable if an adsorbed carboxyl radical (or COH radical) is present so that its OH group can form a hydrogen bond with the CO group of formic acid.

By interaction of the formic acid molecule thus adsorbed with an adsorbed sulfur atom, the C-H bond is weakened. Subsequently, the hydrogen will lose an electron *via* a sulfur atom so that a proton is emitted into the electrolyte. The OH group of the remaining carboxyl radical, which is weakly bound to the OH group of the chemisorbed carboxyl radical on the platinum, also interacts with a neighboring sulfur atom so that another electron transfer is caused and the second hydrogen atom is emitted into the electrolyte in the form of a proton. At this very moment desorption of the carbon dioxide occurs. The OH group of the carboxyl radical already chemisorbed on the platinum does not interact with a sulfur atom; thus it cannot be split, because the high potential of the oxidation peak of the chemisorbate is not changed by sulfur, as may be inferred from the anodic stripping curve. The reaction sequence would then read (brackets mean weak adsorption):

$$HCOOH \longrightarrow (HCOOH)_{ad} \tag{19}$$

$$(HCOOH)_{ad} \longrightarrow (COOH)_{ad} + H^+ + e^- \tag{20}$$

$$(COOH)_{ad} \longrightarrow CO_2 + H^+ + e^- \tag{21}$$

The same reaction path might be persued in the absence of sulfur: the formic acid molecule is weakly adsorbed in the same way on the chemisorbed carboxyl radical (or COH radical). The effect of the sulfur would then be the acceleration of step (20) and step (21). Instead of the simultaneous de-electronation according to reaction (20) and (21), an intermediate $(H)_{ad}$ may be formed so that the charge transfer step may be the de-electronation of this weakly adsorbed H atom:

$$(H)_{ad} \longrightarrow H^+ + e^- \tag{22}$$

as this must have a certain lifetime, at least at open circuit (see reactions 16 to 19).

Carbon monoxide oxidation, too, involves weakly adsorbed intermediates. This holds both in the presence and in the absence of a sulfur layer. It is therefore likely that carbon monoxide reacts with adsorbed water molecules:

$$CO + H_2O_{ad} \longrightarrow (COOH)_{ad} + H^+ + e^- \tag{23}$$

$$(COOH)_{ad} \longrightarrow CO_2 + H^+ + e^- \tag{24}=(21)$$

The reactions may also be formulated with $(H)_{ad}$ as an intermediate, which then reacts according to equation (22).

At open circuit the shift reaction occurs by which hydrogen is formed according to equation (18). The rate of the shift reaction is increased roughly tenfold if the platinum surface is covered with a sulfur monolayer. This conversion rate is however small compared with the rate of anodic oxidation of the carbon monoxide at reasonable potentials.

The discussion of carbon monoxide and formic acid oxidation mechanism is mainly based on the measured stability of the organic chemisorbates, but it must be supplemented by further results. The electrode kinetics, especially, for these and other organic fuels is largely unexplained. Further diagnostic parameters have to be measured, in particular at smooth surfaces.

Apart from conclusions on the oxidation mechanism, it is obvious that the effect of intentionally deposited sorption layers is of great importance for practical fuel cells. Even cations, which are added to the electrolyte, may produce an effect, as was recently found by Brummer [19] in investigations on hydrocarbons. It is not yet known whether they are accumulated in the double layer or are active in another way.

Much remains to be done in order to clarify the effect of chemisorbed layers on the activity of catalysts, even including platinum. However, the information available on platinum is sufficient to form the basis for intensive investigations into the effect of adsorbates also on other catalyst surfaces.

ACKNOWLEDGMENT

The assistance of Frau S. Ammende in the experimental work is greatly appreciated.

REFERENCES

[1]   J. O'M. Bockris and S. Srinivasan, *Fuel Cells: Their Electro-chemistry*, McGraw-Hill, New York, (1969).

[2]   H. Binder, A. Köhling, and G. Sandstede, "Effect of Alloying...", these proceedings p. 43.

[3]   K. v. Benda, H. Binder, A. Köhling, and G. Sandstede, these proceedings, p. 87.

[4]   M. W. Breiter, *Electrochemical Processes in Fuel Cells*, Springer Verlag, Berlin (1961).

[5]   E. Gileadi (Ed.) *Electrosorption*, Plenum Press, New York (1969).

[6]   W. Vielstich, *Fuel Cells*, Wiley Interscience, New York (1970).

[7]   H. Binder, A. Köhling, and G. Sandstede, *J. Electroanal. Chem. and Interfacial Electrochem.*, 17, 111 (1968).

[8]   H. Binder, A. Köhling, and G. Sandstede, in: "Fuel Cell Systems II," *Advances in Chemistry Series*, 90, 128, American Chemical Society, Washington D. C. (1969).

[9]   J. Giner, *Electrochimica Acta*, 9, 63 (1964).

[10] H. Biner, A. Köhling, and G. Sandstede, *Comptes Rendus, Deuxièmes Journées Internat. d'Etudes des Piles à Combustibles*, Societé Commerciale d'Applications Scientifiques (COMASCI), Bruxelles (1967) p. 173.

[11] H. Binder, A. Köhling, and G. Sandstede, *Angew. Chem.*, 79, 903 (1967); *Angew. Chem., Int. Ed.* 6, 884 (1967).

[12] E. Schwarzer, and W. Vielstich, *Comptes Rendus, Troisièmes Journées Internat. d'Etude des Piles à Combustibles*, Presses Academiques Européennes, Bruxelles (1969) p. 220.

[13] H. Binder, A. Köhling, and G. Sandstede, *Energy Conversion*, 11, 17 (1971).

[14] H. Binder, A. Köhling, and G. Sandstede, *Adv. Energy Conv.*, 7, 77 (1967).

[15] H. Binder, A. Köhling, and G. Sandstede, *Adv. Energy Conv.*, 7, 121 (1967).

[16] H. Binder, A. Köhling, and G. Sandstede, *Nature*, 214, 268 (1967).

[17] J. A. Shropshire, *J. Electrochem. Soc.*, 112, 465 (1965).

[18] C. E. Heath, *Comptes Rendus, Journées Internat. d'Etude des Piles à Combustibles, Revue Energie Primaire*, Vol I, p. 99, Bruxelles (1965).

[19] J. Huff, these proceedings, p. 3.

SUMMARY OF DISCUSSION AND COMMENTS: SESSION CHAIRMAN I

J. O'M. Bockris

University of Pennsylvania, Philadelphia, Pennsylvania

The following comments refer to the papers of Session I in the order they were given:

1. Paper of Dr. J. R. Huff.

The mechanism which Dr. Huff is quoting here differs from the one which has been derived by Bockris, Gileadi and Stoner [1], who, in the examination of propane in concentrated solutions of phosphoric acid, came to the conclusion that the combination of an organic radical with adsorbed oxygen, (early in the consecutive sequence) was the rate determining step. The hydrocarbon gets adsorbed on the electrode with ionization giving off a proton water; water discharges to give OH_{ads}; and there is a rate determining reaction between the adsorbed organic radical and adsorbed OH from the water. Why there is a difference between the mechanism derived from Brummer's work, and that of our own work, is not at first clear. The Brummer mechanism would not seem to be consistent with the observed values of $\partial \log i/\partial V$.

A tentative explanation is as follows:

The potential sweep method used in the work of Brummer which Dr. Huff quoted seems to me to need careful evaluation with respect to the significance of its data. The trouble is that the observed currents are not always steady state ones. This does not imply a mere academic objection concerning a few percent. It is possible that the rate determining step is not yet that for the steady state at the time of measurement. Then, of course, the mechanism derived would not be relevant, *e.g.*, *vs.* the performance of fuel cells.

These controversies about the rate determining step in the mechanism of the oxidation of the hydrocarbons seem exceedingly important in respect to what should be done to a catalyst for hydrocarbon oxidation. One of the missing pieces of evidence is the nature of the intermediates which predominate on the surface. At the moment, our approach to these intermediates involves much guess work. We need some new methods. One might be the interpretation of the adsorption coefficient of light on the surface of the electrode.

Dr. Huff's remarks about LEED are very encouraging. It is interesting to report briefly that at the University of Pennsylvania we have tried some Auger studies. One difficulty is the transference of the sample to the spectroscope without oxidation, etc., and a possible technique is to freeze a layer of ice on to the sample *in situ* and then unfreeze it in the apparatus under vacuum with appropriate liquid air trap protection.

2. Paper of Dr. H. Binder.

It is important to distinguish the "platinum oxide" to which Dr. Binder refers from adsorbed oxygen. Such distinctions have not been easily made in the past but the matter is becoming disentangled by

81

the use of optical methods, particularly ellipsometry [2]. The oxide
is adsorbed oxygen up to about 0.9V N.H.S. and afterwards an oxide
forms. It is the adsorbed oxygen which reacts with adsorbed organic
species in organic reaction mechanisms.

There seems to be a discrepancy in the value of the exchange current density which Dr. Binder quotes for the methanol oxidation: on
plainer electrodes, the value given at room temperature is $ca.$ 10^{-6}
amps. cm^{-2}. It is possible that an extrapolation of the current-
potential curve has been made from a region of potential at which the
methanol is reacting on the oxide, $i.e.$, when the rate constant is
likely to be smaller than for the reaction on the oxide.

These comments illustrate the necessity of attempting to work increasingly in the future with optical monitoring of the surface, along
with electrochemical kinetic measurements.

3. Paper of Dr. C. V. Bocciarelli

Dr. Bocciarelli's remarks that electrochemists can't get along
without physicists and solid state scientists is well taken. However,
they must be physicists who understand the situation. Dr. Bocciarelli
does not himself fully comprehend the potential which acts across the
interface and controls the rate of an electrocatalytic reaction. This
is not a simple thing but, in a sense, involves also the metal-metal
junction in the cell; that which is made up of the test electrodes
and the metal of the reference electrode. This is a problem which has
been written about earlier by Frumkin and by Parsons. See also [3]
and [4].

Thus, the rate of an electrocatalytic reaction is measured at the
reversible potential. Unfortunately — and confusedly for many — it
does not mean that the potential at the interface is the same one for
each catalyst used, although the apparent potential measured in an
electrochemical cell may well be the same for all metals. In fact,
the potential at the interface will be completely different for every
catalyst used; it is only the potential of the overall electrochemical
cell which is the same, and this has not been taken into account in
Dr. Bocciarelli's formulation, which remains, consequently, without a
sound physical basis.

The other central factor which Dr. Bocciarelli has neglected in his
analysis, is the chemical bonding of the adsorbed O radicals, perhaps
of oxygen atoms, to the surface of metals. Chemisorption of a radical
is the principle physical cause of electrocatalysis, although other
factors, for example, the number of kink sites on the surface of the
electrode, may have some part to play.

It is important, in view of Dr. Bocciarelli's contribution, to make
clear that there is no mystery whatsoever in principle about electrocatalysis and its origin; about "reversible" and "irreversible" electrode reactions, etc. Thus, there is no direct influence of dielectric
constant in the way that Dr. Bocciarelli surmises. There is a rate determining step in any reaction which is controlled in rate by the
strength of the chemical bonding before and after the transition state.
The detailed equation for these bond strengths and how they effect
catalysis depends on what the rate determining step is. Other factors
influence electrocatalysis relatively little.

This is not to say that we are near "a good understanding" of electrocatalysis. We are in a state of too great simplicity, even naiveté,

and lack of detailed knowledge. We know so few details; not even, for example, the strength of the M-O bonds in the presence of adsorbed water. But the principles of electrocatalysis have been understood for at least eight years.

Now one must turn to Dr. Bocciarelli's gold-copper catalyst and ask ourselves what the explanation for its good behavior is, if it cannot have the sort of explanation which Dr. Bocciarelli wants to give it. The major thing to recall here is that gold's reputation for being such a poor catalyst for oxygen reduction in acid solutions totally changes in alkaline solutions (cf. Dr. Bocciarelli's use of it) where it is an excellent catalyst [5]. Adding copper to it would be expected to im-prove catalysis if the copper Raneyates, and it may do this because, in view of the work of Pickering and Wagner [6], there is dissolution of Cu-Au alloys in some potential ranges.

4. Paper of Dr. H. Binder (Effect of Alloying).

The interesting information which the authors give us must, perhaps, be looked at with a slightly skeptical eye when one realizes that it has all been done by potential sweep work. The doubtful point about the use of potential sweeps in electrode kinetic and electrocatalytic studies (as opposed to studies of adsorption and approximate electro-analytical uses of the potentiodynamic method) is, has it been estab-lished that, for the conditions used, the steady state is being ob-served? There is a certain danger in using the potential sweep method for more than an orientive investigation. It is that the state of the surface under reaction sequence has not settled down (i.e., the rela-tive concentration of the radicals has not yet reached that state which will correspond to the steady state mechanism, so that the rate deter-mining step may be other than the rate determining step in the steady state). Then, any catalytic information obtained will, of course, be about a different situation altogether and its direct relevance to studies of the catalysis involved will be tenuous. Conversely, the use of the galvanostatic steady states at very long times (e.g., 30 to 60 minutes), also seems difficult to interpret be-cause of the many adventitious factors which effect the "final value" of the overpotential, which is in any case arbitrary.

It seems to me that the best way to carry out these electrocatalytic, — indeed many electrode kinetic, — investigations is by potentiostatic transients. Then one has a simple algebra for the analysis of the re-sults because the potential dependent part of the reaction rate is in-dependent of time. One also obtains information about the low time transient behavior, and thus, about the build-up of the radicals on the surface, with an unambiguous indication of what the steady state is when it is a "rational steady state". That is, when it has come not to change more at low times (often in the region of 0.01-1 sec. for planar electrodes), so that one does not mix up the problem with that of the "aging of the catalyst".

5. Paper of Dr. G. Sandstede.

The observations concerning the Tafel slopes and their change with adsorbed layers can be interpreted well in terms of a hypothesis first stated in the thesis of S. Srinivasan [7]. The area available for re-action depends upon coverage by an adsorbed (impurity) radical, this depends upon potential, and consequently an extra component above that given by normal activation kinetics, occurs in determining the gradient

of the current-potential line. The observed value of the Tafel slope
is hence not that expected from the simple theory.

Evidence in favor of the hypothesis arises from Srinivasan's work,
(and others have observed the same earlier) in which he found that if
the solution is suitably cleaned up, the Tafel slope reverts to the expected "normal" one. Further, a crude calculation of the coverage dependence with Langmuirian isotherm conditions shows that there is a
reasonable numerical consistency between observation and hypothesis.

REFERENCES

[1] J. O'M. Bockris, E. Gileadi, G. E. Stoner, *J. Phys. Chem.*, 73,
     427 (1969).

[2] J. O'M. Bockris, A.K.N. Reddy and M. A. Genshaw, *J. Chem. Phys.*,
     48, 671 (1968).

[3] A. T. Kuhn, H. Wroblowa and J. O'M. Bockris, *Trans. Faraday Soc.*,
     63, 1458 (1967).

[4] J. O'M. Bockris, J. McHardy and R. Sen, these proceedings, p. 385.

[5] M. A. Genshaw, *Thesis*, University of Pennsylvania (1967).

[6] H. W. Pickering and C. Wagner, *J. Electrochem. Soc.*, 114, 698
     (1967).

[7] S. Srinivasan, *Thesis*, University of Pennsylvania (1963).

SESSION II

INORGANIC AND ORGANIC SUBSTANCES AS CATALYSTS FOR FUEL CELLS

Chairman

C. E. Heath

ELECTROCHEMICAL BEHAVIOR OF TUNGSTEN CARBIDE ELECTRODES

K. von Benda, H. Binder, A. Köhling and G. Sandstede

Battelle-Institut e.V., Frankfurt am Main, Germany

ABSTRACT

The electrochemical oxidation of various fuels at porous sintered tungsten carbide (WC) electrodes was investigated in the following electrolytes: sulfuric acid, potassium hydroxide solution, and phosphoric acid. In sulfuric acid, WC is a catalyst for the conversion of hydrogen, hydrazine, formaldehyde, acetaldehyde, formic acid, and carbon monoxide. Hydrogen conversion involves current densities of technical importance and is greatly enhanced by previous anodization in the presence of a reducing agent. This process has been designated as "activation" of the electrode. Hydrazine is not decomposed at open circuit potential and polarization is high. The oxidation of formaldehyde nearly stops at the stage of formic acid, which is slowly oxidized. Acetaldehyde conversion is slow and stops at the stage of acetic acid. At 90°C, formic acid is dehydrogenated under open circuit conditions. In phosphoric acid, the current densities for hydrogen conversion are smaller than in sulfuric acid. In 85 percent acid at 150°C, carbon monoxide is oxidized only if the electrode is supplied additionally with water vapor. In potassium hydroxide, hydrogen conversion has a maximum current density at about 300 mV vs. HE in the same solution. The conversion of carbon monoxide and formaldehyde is inhibited by reaction products. Hydrazine is rapidly oxidized at potentials of about 100 mV; at lower potentials, cathodic currents are measured. Activation of the electrodes in an alkaline electrolyte is possible and yields even better performances for hydrogen conversion in sulfuric acid than activation in an acid electrolyte. The behavior of WC in potassium hydroxide shows the existence of different inactive and active surface species. As methanol is not directly converted at WC electrodes, catalytic cracking at 220°C was applied. An outlook on other inorganic catalysts for the anode is given.

INTRODUCTION

In the last few years, tungsten carbide (WC) has gained widespread interest as a highly effective catalyst for hydrogen conversion in acid electrolytes. Böhm and Pohl [1] were the first to publish results on WC electrodes for hydrogen conversion. In our investigations we also arrived at WC as an active electrocatalyst for hydrogen. We found that the performance of WC can be improved by an activation process [2,3,4] and that not only hydrogen but also other fuels [2,3,5] can be oxidized at WC anodes. In addition, we investigated the behavior of WC electrodes in potassium hydroxide electrolyte [6] and phosphoric acid in order to obtain further information about the activation process.

This paper gives a short summary of WC behavior in acid and alkaline media. New results are reported, and the use of cracked methanol is discussed as well.

EXPERIMENTAL CONDITIONS

All measurements were made in a half-cell arrangement. The reference electrode was an autogenous hydrogen electrode in the same electrolyte and at the same temperature as the counter electrode (graphite) and the test electrode. Oxygen from the air was excluded by saturating the electrolyte with pure nitrogen. In most cases the electrolyte was kept at 70°C.

The test electrodes were porous sintered WC tablets obtained by compacting an intimate mixture of tungsten powder, carbon black and ammonium carbonate, decomposing the ammonium carbonate under reduced pressure at 200°C, and firing the tablet in a hydrogen atmosphere at 1300°C. This method of preparation was preferred to carburization with methane or carbon monoxide because in this way it is possible to produce porous WC electrodes directly from the elements, and because in addition the metal-to-carbon ratios could be varied easily.

In addition to the sintered electrodes, plastic-bonded hydrophobic electrodes were fabricated. The best performance was achieved with WC whose carbon content was somewhat smaller than that corresponding to the stoichiometric composition.

RESULTS IN SULFURIC ACID

The investigations were performed in 2N sulfuric acid. In this acid electrolyte, corrosion is not observed up to 300 mV, but if the potential is raised to 500 mV some corrosion occurs. This has been demonstrated by cyclic voltammetry. Fig. 1 shows an example of an electrode. The corrosion current depends on the preparation conditions of the WC electrode. After some sweeps the corrosion current decreases and becomes negligible after some hours. While corrosion is taking place, some gas is evolved at the anode. This is due to the anodic oxidation of carbidic carbon to carbon dioxide.

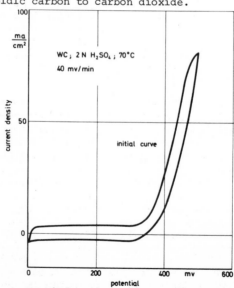

FIG. 1 *Cyclic potentiodynamic current-voltage curve of a WC electrode in dilute sulfuric acid showing initial corrosion.*

After corrosion has ceased, the electrocatalytic activity for hydrogen conversion is smaller than before. Fig. 2 shows the decrease in activity of an electrode which has been exposed to corrosion for 17 hours at a potential of 700 mV. In this case the current density has decreased by about 20 percent of the initial value. The initial activity cannot be restored by flushing the electrode with hydrogen. However, the original activity can be obtained again by a treatment in potassium hydroxide solution.

A very different behavior is observed if the anodic treatment takes place in the presence of a reducing agent; *e.g.*, hydrogen bubbling through the electrode. This treatment increases the activity for hydrogen conversion up to twice the initial value (Fig. 3). Particularly effective is a controlled process in the presence of a dissolved reducing agent such as hydrazine or formaldehyde. We have termed this process "activation" of an electrode. Activated WC is not affected by dilute sulfuric acid up to a potential approximately 700 mV, even in the absence of a reducing agent.

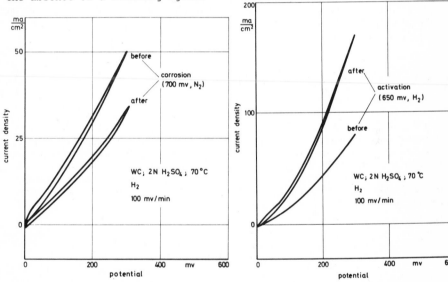

FIG. 2 *Cyclic potentiody-*
namic current-voltage curves of
a WC electrode with hydrogen in
dilute sulfuric acid before and
after corrosion.

FIG. 3 *Cyclic potentiody-*
namic current-voltage curves of
a WC electrode with hydrogen in
dilute sulfuric acid before and
after activation.

The activated state of WC can be illustrated by a cyclic voltammetric curve (Fig. 4). The cyclic charge curve prior to the activation shows no characteristic features. In addition to double layer charging and discharging, it is necessary to consider an oxidation or reduction reaction which is due to a change of the oxidation state of the WC surface. This change may be explained at least in part by deposition and dissolution of adsorbed hydrogen. Potential reversal has to take place at 300 mV because untreated WC is not stable at higher potentials; but after activation no corrosion is observed up to potentials of about 700 mV. The double layer capacity is increased after the activation and turning points are observed. This suggests a new

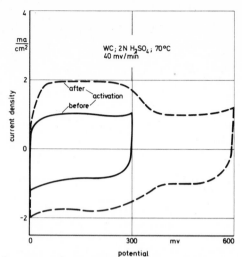

FIG. 4 *Cyclic potentiodynamic current-voltage curves of a WC elec-trode in dilute sulfuric acid before and after activation.*

surface species. The activation of electrodes does not necessarily entail turning points, see results in KOH.

CONVERSION OF VARIOUS FUELS IN SULFURIC ACID

Fig. 5 shows galvanostatic current-voltage curves for hydrogen con-version at different temperatures. From the dependence of the current density on temperature a mean value of 5.8 kcal/mole has been derived for the activation energy at a potential of 300 mV. Our research work aimed at the development of electrodes for industrial fuel cells has resulted in hydrophobic PTFE-bonded WC electrodes. These electrodes are operated with hydrogen at atmospheric pressure. Fig. 6 shows the steady state current-voltage curves of such an electrode [2]. Recently Böhm and Pohl developed electrodes whose performance has been more than doubled [7].

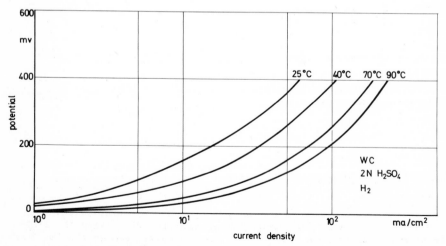

FIG. 5 *Galvanostatic current voltage-curves of a WC electrode with hydrogen in dilute sulfuric acid under steady state conditions.*

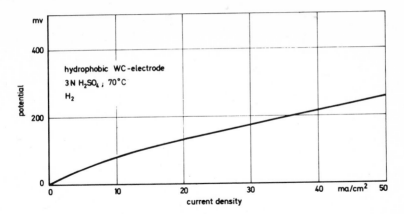

FIG. 6 *Galvanostatic current-voltage curve of a hydrophobic PTFE-bonded WC electrode with hydrogen in dilute sulfuric acid under steady state conditions.*

Fig. 7 shows galvanostatic current-voltage plots for <u>hydrazine con-version</u> at 70°C before and after activation of the electrode. Polariza-tion is rather high because oxidation of hydrazine does not start at potentials below, say, 200 mV. On the other hand, there is no decom-position of the fuel at open-circuit potential. This may be due to the

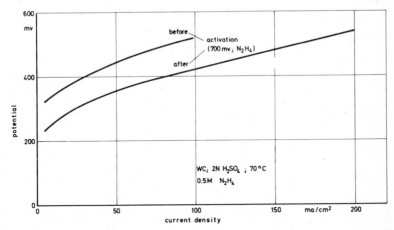

FIG. 7 *Galvanostatic current-voltage curves (steady state) of a WC electrode with hydrazine in dilute sulfuric acid before and after activation.*

formation of hydrazine sulfate in sulfuric acid electrolyte so that the activation energy is increased. As is the case with all dissolved fuels, electrode performance depends on fuel concentration; this, in turn, is determined by the solubility product of hydrazine sulfate. Thus, the best performance is achieved in a solution of hydrazine sul-fate in water as the electrolyte.

The oxidation of dissolved <u>formaldehyde</u> already starts at the po-tential of the hydrogen electrode. Fig. 8 shows current-voltage plots for two different aldehydes at 70°C. The oxidation of formaldehyde nearly stops at the stage of formic acid, which oxidizes, but slowly.

WC electrodes have recently been used in acid formaldehyde cells [8].

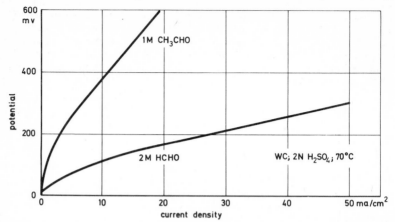

FIG. 8 *Galvanostatic current-voltage curves of a WC electrode with aldehydes in dilute sulfuric acid under steady state conditions.*

Acetaldehyde, too, is oxidized at WC (Fig. 8), but the oxidation completely stops at the stage of acetic acid. This means that at WC electrodes, acetic acid can be formed from acetaldehyde in an electrochemical reaction — a process which might be of interest to the chemical industry. Current densities decrease with the chain length of the aldehydes. This is why only a few milliamperes per cm^2 are obtained with propionaldehyde (Fig. 9).

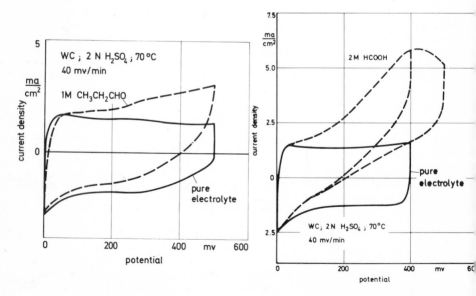

FIG. 9 *Cyclic potentiodynamic current-voltage curve of a WC electrode with propionaldehyde in dilute sulfuric acid.*

FIG. 10 *Cyclic potentiodynamic current-voltage curves of a WC electrode with formic acid in dilute sulfuric acid.*

Fig. 10 shows cyclic current-voltage curves for <u>formic acid</u> conversion. At 70°C and 200 mV we obtained only 1 mA/cm^2. If the potential
is raised above 400 mV, an inhibition is noted (see sweep up to 500 mV
in Fig. 10). This may be due to an intermediate or to a stronger bonding of water molecules to the surface of WC. An interesting feature in
the oxidation of formic acid at 90°C is the observation that the fuel is
dehydrogenated at the WC electrode under open-circuit conditions. Fig.
11 shows galvanostatic current-voltage plots for formic acid conversion

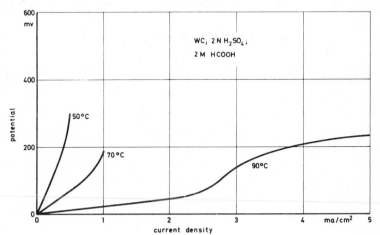

FIG. 11   *Galvanostatic current-voltage curves of a WC electrode
with formic acid in dilute sulfuric acid under steady state conditions.*

at various temperatures. At 90°C, a step is observed in the current-
voltage curve. This can be explained if we assume that at low current
densities the oxidation of hydrogen evolved in the decomposition of
formic acid is the rate-determining step, and that at high current densities this dehydrogenation rate is insufficient, so that direct oxidation of formic acid takes place. This new process requires a higher
potential.

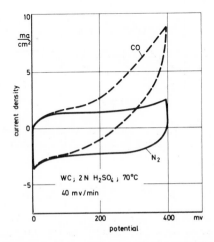

FIG. 12   *Cyclic potentiodynamic current-voltage curve of a WC
electrode with carbon monoxide in dilute sulfuric acid.*

Alcohols, including even methanol, or carboxylic acids other than formic acid cannot be electrochemically oxidized at WC electrodes in sulfuric acid.

Of considerable technical importance is the activity of WC for carbon monoxide conversion, because traces of carbon monoxide contained in crude hydrogen do not poison the catalyst. Fig. 12 shows the cyclic current voltage for carbon monoxide oxidation. Current densities are quite small, but there is no inhibition of the oxidation reaction.

RESULTS OBTAINED IN PHOSPHORIC ACID

If phosphoric acid is used as electrolyte, WC electrodes generally exhibit the same behavior as in sulfuric acid, but current densities for fuel conversion are smaller. The rate of hydrogen conversion in 45 percent (18N) phosphoric acid at 90°C is only one third of that measured in dilute sulfuric acid.

In 85 percent (44N) phosphoric acid ($\hat{=}$ monohydrate) at 150°C, hydrogen is converted at a very low rate, but if the electrode is supplied additionally with water vapor, the conversion rate increases approximately tenfold. Fig. 13 shows the results without H_2O vapor and with a $H_2:H_2O \approx 1:1$ mixture. Carbon monoxide is only oxidized if water vapor is present.

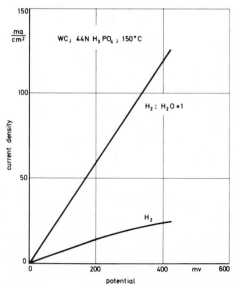

FIG. 13   *Potentiodynamic current-voltage curves (nearly steady state) of a WC electrode with hydrogen in phosphoric acid with and without addition of water vapor.*

WC BEHAVIOR IN ALKALINE ELECTROLYTE

All measurements were made in 6.5N potassium hydroxide solution. Corrosion begins at potentials higher than, say, 450 mV and does not cease after some hours at 500 mV. This is in contrast to the behavior in sulfuric acid. However, current densities for corrosion are very small up to 600 mV (Fig. 14, reference curve with N_2).

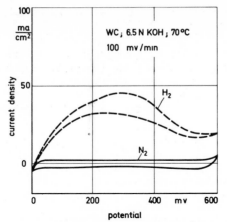

FIG. 14   *Cyclic potentiodynamic current-voltage curve of a WC*
electrode with hydrogen in potassium hydroxide solution.

Hydrogen conversion starts at zero potential, but cyclic voltammetry shows maximum current density in the vicinity of about 300 mV
(Fig. 14).   The current-voltage curves of individual electrodes show
some variations with respect to the position of that maximum; moreover,
the maximum depends on the pretreatment.   Current densities for hydrogen conversion are much smaller in potassium hydroxide than in sulfuric
acid.

On the other hand, carbon monoxide is converted in KOH at a much
higher rate than in sulfuric acid, but this holds only for a fresh WC
surface as the electrode is soon blocked by a reaction product.   Fig.
15 shows this in terms of cyclic voltammetry.   The cyclic current-
voltage curve for carbon monoxide conversion at the blocked electrode

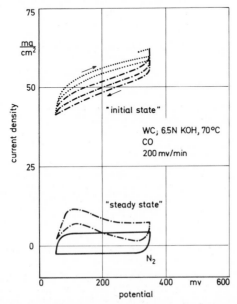

FIG. 15   *Cyclic potentiodynamic current-voltage curves of a WC*
electrode with carbon monoxide in potassium hydroxide solution.

remains constant (denoted as "steady state" in Fig. 15) and exhibits a
maximum, whose position and profile vary strongly as a function of the
preparation conditions of the WC and the electrolyte temperature. The
blocking species proves to be strongly bound to the surface of the electrode because the rapid loss of activity indicated in Fig. 15 cannot be
stopped by rinsing the electrode with boiling water.

For most of our test electrodes the open circuit potential under CO
was below zero, down to a minimum of -60 mV. The thermodynamic equilibrium potential is -500 mV.

Methanol is not converted at WC in potassium hydroxide solution.
If formaldehyde is dissolved in the alkaline electrolyte, an initial
open circuit potential of about -40 mV is measured at WC electrodes,
while hydrogen is evolved even at room temperature. After some hours
the rest potential has risen to zero and hydrogen evolution has ceased.
Under anodic load, very high initial current densities are measured
which decrease very rapidly due to inhibition by a reaction product.
Fig. 16 shows the decrease during cyclic voltammetry; after several

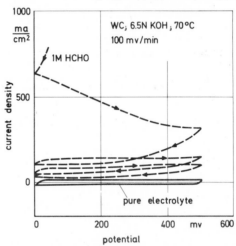

FIG. 16   *Cyclic potentiodynamic current-voltage curves of a WC
electrode with formaldehyde in potassium hydroxide solution.*

cycles formaldehyde conversion has stopped, while hydrogen evolution
continues and diminishes at a much lower rate than fuel conversion.
The inhibiting species is not the formate ion (which is not oxidized
at WC in potassium hydroxide), nor oxalate or glyoxal, because the
initial current density for formaldehyde conversion is not affected by
addition of these species to the electrolyte.

Hydrazine as fuel gives rise to an open circuit potential of about
100 mV. Catalytic decomposition of hydrazine to hydrogen and nitrogen
at the electrode surface should result in the rest potential of the
hydrogen electrode. As this is not the case, other reactions must be
involved. Fig. 17 shows a cyclic current voltage curve for hydrazine
conversion; we note strong cathodic currents at potentials below the
rest potential. It seems likely that hydrogen oxidation is not the
rate-determining reaction for hydrazine conversion in potassium hydroxide. At a potential of 200 mV, a current density as high as 150 mA/cm^2
is measured; up to 500 mV no maximum appears in the current-voltage
plot, in contrast to the hydrogen conversion in KOH solution.

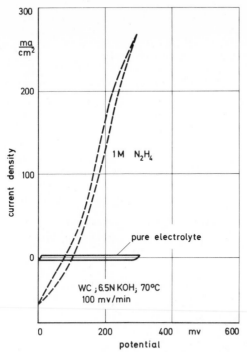

FIG. 17   *Cyclic potentiodynamic current-voltage curve of a WC*
electrode with hydrazine in potassium hydroxide solution.

DISCUSSION OF WC SURFACE PROPERTIES

The behavior of WC in alkaline electrolyte is more complex than in
acid electrolyte. As WC is by no means as inert a material as platinum,
this behavior provokes the question, "What kind of surface species cover
the surface of a working WC electrode?".

From the results obtained so far we may conclude that we have to
discern electrocatalytically active and inactive surface species on WC
electrodes. No doubt anodic treatment in acid produces tungsten oxide,
which is inactive for hydrogen conversion; in the presence of a fuel,
however, a surface species with a high specific activity for hydrogen
oxidation is formed. This might be an oxide with interstitial hydrogen:
a hydrogen tungsten bronze. If an activated electrode is treated for a
few minutes with hot potassium hydroxide solution (70°C), subsequent
cyclic voltammetry in sulfuric acid shows a sharp reduction of the area
enclosed by the current-voltage curve, while the activity for hydrogen
conversion remains unaffected. The obvious explanation is that inactive
tungsten oxide has been dissolved to give tungstate and that the active
species is not attacked by this treatment. The capacity of the electrodes is different for different samples. Fig. 18 shows an example.
Although systematic investigations have not yet been carried out, the
results obtained so far suggest that both active and inactive oxide
species are formed during activation. However, the tungsten bronzes
and especially hydrogen tungsten bronzes are reported to be attacked
by hot alkali.

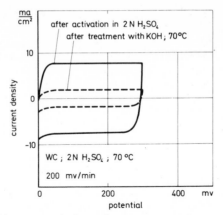

FIG. 18   *Cyclic potentiodynamic current-voltage curves of a WC electrode in dilute sulfuric acid before and after treatment with hot potassium hydroxide solution.*

Surprisingly, WC electrodes are also activated in potassium hydroxide; the activation is particularly effective if formaldehyde is used as the reducing agent. As indicated before, formaldehyde conversion is quickly blocked, but this is overcome if the electrode potential is maintained in the corrosion range above 500 mV.

Activation in alkaline electrolyte significantly improves the conversion of hydrogen in acid electrolyte. In this way we obtained a current density of 290 mA/cm^2 at 300 mV, a performance which has not been reached by activation in acid electrolyte. Probably, the WC surface will be more densely covered with active species in alkaline solution than in acid because the inactive species are disolved.

If the potential during activation in potassium hydroxide is very high, *e.g.*, 800 mV, the formation of active surface species is accompanied by the formation of inactive oxide, whose rate of formation at this potential is higher than its rate of dissolution. Treatment in boiling potassium hydroxide solution removes the inactive oxide species and leaves the active layer intact.

Activation in acid or alkaline media increases the rate of hydrogen oxidation not only in sulfuric acid but also in potassium hydroxide solution.

In contrast to the behavior in acid electrolyte, the initial activity in alkaline electrolyte is not maintained, and the current densities slowly decrease until they become constant. Nevertheless, electrode performance remains higher than before activation.

Curiously, activation in KOH solution reduces current densities for carbon monoxide conversion to about one tenth of the steady state values measured at untreated electrodes; and, vice versa, the rate of hydrogen conversion is reduced to about one half by previous oxidation of carbon monoxide at the electrode. Therefore, we must acknowledge the existence of different surface species with specific electrocatalytic properties at WC electrodes. Our investigations in potassium hydroxide electrolyte have helped to make this clear, but the chemical nature of the active layer on WC is still open to conjecture.

FUEL CELL WITH HYDROGEN AND CARBON MONOXIDE

Tungsten carbide has the advantage of not being poisoned by carbon monoxide in acid electrolyte. Thus, hydrogen conversion is not affected, as already pointed out by Böhm and Pohl [1]. These authors also reported that tungsten sulfide and molybdenum sulfide produce minor activity in the conversion of formic acid and carbon monoxide in acid electrolyte [9]. Thus, it may be possible to develop electrodes which not only are not affected by carbon monoxide, but also convert hydrogen and carbon monoxide simultaneously. Cells of this type would operate on reformed hydrocarbon or methanol.

As methanol cannot yet be converted anodically at non-platinum metal catalysts, it has to be transformed into a hydrogen-containing gas, for example, by reforming with water vapor. Another possibility is cracking into hydrogen and carbon monoxide, as demonstrated by us at the Achema exhibition in Frankfurt (Main), June 1970. A tungsten carbide fuel cell was operated together with the cracker. Methanol vapor is fed into the cracker at a temperature of 220°C and is catalytically cracked. The reaction

$$CH_3OH \longrightarrow 2\ H_2 + CO$$

is complete. No by-products are formed. A supported nickel-copper catalyst is used for the reaction.

The WC electrode did not show any decrease in catalytic activity over several thousand hours, which is in accordance with the findings by Böhm and Pohl [1].

OUTLOOK ON FURTHER DEVELOPMENT WORK

The molybdenum sulfide catalyst referred to in the foregoing also has a minor activity for hydrogen. Recently tungsten carbide has been modified by alloying with silver [10]. Furthermore, cobalt carbide has been found to be highly active for hydrogen conversion [10]. The activity of WC for aldehyde conversion has already been mentioned. Moreover, cobalt sulfides show a certain activity in formic acid conversion [11]; they are, however, not entirely stable. These results show that further inorganic catalysts are in sight so that cheap anodes for acid electrolyte may be developed. The first results with organic fuels are also encouraging and should stimulate future research.

It may be emphasized in this context that besides the electronic properties of the catalysts, the properties of their surfaces play an important role in the electrocatalytic reaction. Therefore, more detailed knowledge is required of the composition of the surface of inorganic catalysts immersed in the electrolyte and with potential applied.

ACKNOWLEDGMENT

We thank W. Faul and H. Schomann for their assistance in the experimental work.

REFERENCES

[1]   H. Böhm and F. A. Pohl, *Wiss. Ber. AEG-Telefunken*, 41, 46 (1968).

[2]   H. Binder, A. Köhling, W. H. Kuhn, W. Lindner and G. Sandstede, *Nature*, 224, 1299 (1969).

[3]   H. Binder, A. Köhling and G. Sandstede, *Preprints of Papers, Amer. Chem. Soc., Meeting*, New York (1969), 13/3, 99; *Advan. Chem. Ser.*, in preparation.

[4]   H. Binder, A. Köhling, W. H. Kuhn, W. Lindner. and G. Sandstede, *Energy Convers.*, 10, 25 (1970).

[5]   H. Binder, A. Köhling, W. H. Kuhn and G. Sandstede, *Angew. Chem.*, 81, 748 (1969).

[6]   K. v. Benda, H. Binder, W. Faul and G. Sandstede, *Chem.-Ing.-Tech.*, 43, No. 16 (1971).

[7]   H. Böhm and F. A. Pohl, *5th Symposium, Energie-Direkt-Umwandlung*, Essen (October 1970).

[8]   H. Jahnke, W. Haecker and M. Schönborn, *5th Symposium, Energie-Direkt-Umwandlung*, Essen (October 1970).

[9]   H. Böhm, W. Diemer, J. Heffler, F. A. Pohl and W. Sigmund, *Energy Convers.*, 10, 119 (1970).

[10] K. Mund, G. Richter and F. v. Sturm, *Comité International de Thermodynamique et de Cinétique Electrochimiques*, Prague, (September 1970).

[11] M. Prigent, A. Sugier and O. Bloch, *Comptes Rendus, Troisièmes Journées Internat. d'Etude des Piles à Combustibles*, Presses Académiques Européenes, Bruxelles (1969) p. 263.

REDUCTION OF OXYGEN AND HYDROGEN PEROXIDE ON SOME TUNGSTEN AND NIOBIUM OXYGENATED BRONZES

M. Voïnov and H. Tannenberger

Institut Battelle, Genève, Switzerland

ABSTRACT

In the search for an electrocatalyst for an acid electrolyte fuel cell, some oxygenated tungsten and niobium bronzes were investigated. Little or no activity was detected for oxygen reduction and hydrogen oxydation. Hydrogen peroxide reduction is catalyzed by these bronzes in acid solution, but the rate of reduction is of the same order of magnitude in alkaline solution. Tungsten and niobium bronzes behave somewhat differently. Ce and Yb tungsten bronzes behave in the same manner.

INTRODUCTION

Given the present state of theory on electrocatalysis, any chemically resistant material possessing sufficient electronic conductivity is worth investigating for catalytic activity. So-called oxygenated bronzes are expected to be stable in acid solution because they are formed by doping acid-stable oxides such as WO_3 or Nb_2O_5 with different elements. Some of these bronzes have been shown to be stable [1]. The doping, besides changing the crystallographic structure of the oxides, imparts a rather good electronic conductivity which, in the case of K or Li tungsten bronzes, can be as high as $2.5 \ 10^4$ mho cm^{-1}. The reason for this good electronic conductivity is still a matter of controversy. The fact is that the alkali-metal tungsten bronzes, at least, show very little or zero Knight shift for both the alkali and the tungsten atoms, and consequently the s orbitals of the alkali and tungsten atoms are not involved in the conduction process. Different models have been proposed for the electronic structure of the bronzes [2,3,4].

Apart from the tungsten bronzes there exist niobium, molybdenum, titanium, tantalum and vanadium bronzes. They are obtained by doping Ta_2O_5, MoO_3, TiO_2, V_2O_5, Nb_2O_5 not only with alkali-metals but with elements such as Al, Pb, Cd and rare earths.

SYNTHESIS OF BRONZES

Different methods are available to synthetize the bronzes under different forms. The solid-state reaction methods, which consist of mixing, compressing and heating a mixture in the right proportion of WO_3 and Na_2O, for instance, plus a reducing agent which can be W, WO_2 or Sn, give powdered bronze. Variants of these methods use the doping element chloride and take advantage of the low vapor pressure of tungsten oxychloride [5]. In another variant the doping element is furnished by the decomposition of its azide [6].

To obtain bronzes suitable for making electrodes for electrocatalytic studies, one can, at least for some tungsten bronzes, use
some sort of chemical vapor deposition in which WO_3 vapor is reacted
and condensed with vapor of the doping element. This method has been
used to make single crystals of Tl_xWO_3 [7]. Another method for making
massive electrodes is electrocrystallization upon electrolysis of a
fused salt mixture [8].

ELECTROCRYSTALLIZATION OF TUNGSTEN BRONZES

When a molten mixture of tungsten trioxide and a doping element
oxide is electrolyzed, bronzes crystallize on the cathode. The
mechanism for the formation of these bronzes is probably very involved,
but for practical purposes it is sufficient to know what influence the
composition of the melt and the electrolyzing temperature have on the
composition of the bronzes.

From a ternary diagram of the system W-O-doping element (Fig. 1),
one might imagine that the bronzes form as a result of a loss of oxygen.
Upon electrolysis, however, oxygen develops on the anode and the bronze
is deposited on the cathode. It is therefore supposed that, upon
electrolysis, the melt around the cathode becomes enriched with the
doping element until a solidus is crossed which corresponds to crystallization of the bronze.

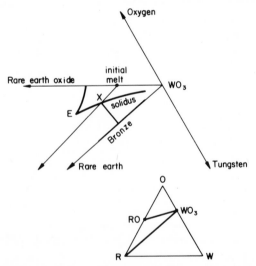

FIG. 1 *Schematic ternary diagram of the system W-O-rare earth.*

Thus, if the concentration of the melt in the doping element increases, the bronze will contain more doping element. If, on the other
hand, tungstate with a high melting point exists (this is the case for
Ce and Yb), we have observed that it can crystallize upon electrolysis.
For a given temperature, therefore, there is an upper composition for
the melt if one wants to crystallize the bronze. The temperature
affects the position of the solidus for the precipitation of bronze
and tungstate.

We have synthetized and investigated three bronzes, namely $Ce_{0.1}WO_3$,
$Yb_{0.1}WO_3$ and $Sr_{0.8}NbO_3$.

Ce and Yb tungsten bronzes have been synthetized by electrolysis under the following conditions:

2M CeO_2 - 8M WO_3 1200°C

1M Yb_2O_3 - 8M WO_3 1250°C

The crucible was glazed alumina and the electrodes were platinum. Tension on the cell was programmed to increase logarithmically, and the cathode rotated at 240 rpm. We obtained crystals of about 1 mm^3.

We could not make Sr_xNbO_3 bronzes by electrolysis, so we first made a porous piece by sintering, then melted it under vacuum.

EXPERIMENTAL PROCEDURE

Small pieces of bronze were heat-pressed in Kel-F (180°C - 10 kg/cm^2). Then the Kel-F was machined to fit into a rotating disk assembly (Fig. 2). The surface of the electrodes was polished mechanically. Electrolytes were made from double-quartz-distilled water.

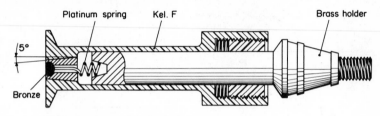

FIG. 2 *Rotating electrode holder.*

The standard measuring procedure was to run a potential sweep (1 mV/s) between 40 and 1200 mV/H$_2$ under nitrogen and then in the presence of different reactants. The rate of the electrode rotation could be up to 25,000 rpm if necessary.

Background current under nitrogen could be as high as 500μA/cm^2 (geometric), and the limit for detection of a reaction current was 10μA/cm^2 (geometric).

ELECTROCHEMICAL BEHAVIOR OF $Ce_{0.1}WO_3$, $Yb_{0.1}WO_3$ AND $Sr_{0.8}NbO_3$

1. ACID SOLUTION: HClO$_4$ 1M

We tested these bronzes for O_2 reduction, H_2, CH_3OH and CO oxydation in HClO$_4$ 1M. With regard to oxydation of H_2, CH_3OH and CO, we could not detect any current higher than 10μA/cm^2 between 0 and 1200 mV/H$_2$ in the same solution. As regards oxygen reduction, we did not detect more than 10μA/cm^2 on Ce and $Yb_{0.1}WO_3$. On the other hand, the extrapolated exchange current for oxygen reduction on $Sr_{0.8}NbO_3$ is quite high (10^{-7} A/cm^2), but the Tafel slope appears to be 600 mV/decade.

The rate of H_2O_2 reduction was measured on these bronzes. The results are very similar for Ce and $Yb_{0.1}WO_3$ (Figs. 3 and 4). The reduction is characterized by two Tafel slopes with a change of slope of around + 250 mV/H$_2$. In the two Tafel regions the current increases linearly with H_2O_2 concentration (Figs. 5 and 6).

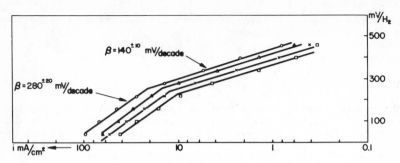

FIG. 3   *Reduction of H_2O_2 (0.2M) in $HClO_4$ (1 M) on $Ce_{0.1}WO_3$.*
□ 31°C   x 41°C   Δ 53°C   0 64°C.

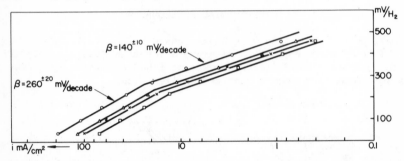

FIG. 4   *Reduction of H_2O_2 (0.2M) in $HClO_4$ (1 M) on $Yb_{0.1}WO_3$.*
□ 31°C   x 42°C   Δ 58°C   0 72°C.

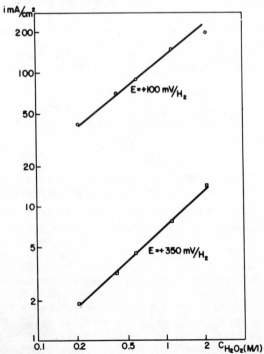

FIG 5.   *Concentration dependence of H_2O_2 reduction current on $Ce_{0.1}WO_3-HClO_4$ 1 M — 35°C — Argon.*

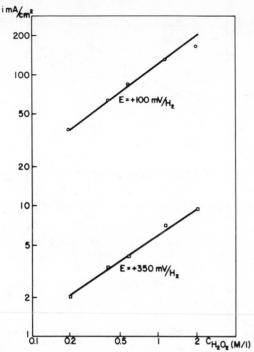

FIG. 6 *Concentration dependence of H_2O_2 reduction current on Yb$_{0.1}WO_3$-HClO$_4$ 1 M - 44°C - Argon.*

On Sr$_{0.8}$NbO$_3$ the rate of H_2O_2 reduction is lower than on tungsten bronzes (Fig. 7); and at constant potential, the current increases with the square root of the H_2O_2 concentration (Fig. 8).

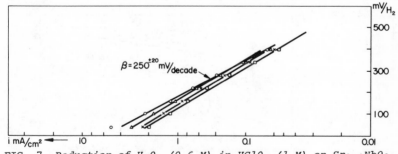

FIG. 7 *Reduction of H_2O_2 (0.6 M) in HClO$_4$ (1 M) on Sr$_{0.8}$NbO$_3$.*
□ 35°C   x 46°C   Δ 55°C   0 65°C.

If we compare this with electrodes of Au and Ag, we find that Sr$_{0.8}$NbO$_3$ is not such a good electrocatalyst for the reduction of H_2O_2. On the other hand, Ce and Yb$_{0.1}WO_3$ appear to be one order of magnitude better.

Thus, according to current thinking, if we had a cheap catalyst for the reduction of O_2 to H_2O_2, we could consider making an air electrode for acid electrolytes. Such a catalyst is as yet unknown. However, it is known that rapid reduction of O_2 to H_2O_2 takes place in an alkaline solution on carbon, but then the reduction of H_2O_2 to water has to be catalysed.

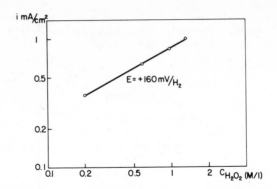

FIG. 8 *Concentration dependence of H_2O_2 reduction current on $Sr_{0.8}NbO_3$-$HClO_4$ 1 M - 35°C - Argon.*

If one of these bronzes were stable in alkaline solution, and if
its catalytic activity were still one order of magnitude higher than
silver activity (as in acid solution), it would certainly be very
attractive.

2. ALKALINE SOLUTION 0.1M NaOH

Potential sweeps were run under N_2 at 6 mV/s between 0 and 1200
mV/H_2, and the dissolution current for all the bronzes was smaller
than 50μA/cm², with no well-defined dissolution potential between 0
and 1200 mV/versus H_2 in the same solution. $Sr_{0.8}NbO_3$ appears to be
stable in NaOH 0.1M, and the results are fairly reproducible.

On $Sr_{0.8}NbO_3$ the rate of reduction of H_2O_2 (Fig. 9) is of the same
order of magnitude as in $HClO_4$ solution, and certainly very low by
comparison with published results on silver.

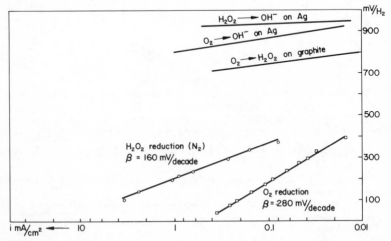

FIG. 9 *Reduction of O_2 (≃ 1 atm) and H_2O_2 (0.04 M/l) on $Sr_{0.8}NbO_3$
in NaOH 0.1 M — 1400 rpm, 23°C, potential sweep: 6.3 mV/s.*

On $Sr_{0.8}NbO_3$ the reduction of O_2 was measurable, but does not
amount to much.

On Ce and Yb in 0.1M NaOH solution the rate of reduction of H_2O_2 is smaller than on Ag. There is a strong hysteresis, which is also found for O_2 reduction (Figs. 10 and 11).

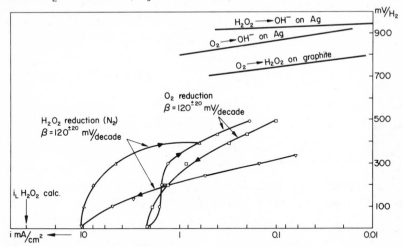

FIG. 10 *Oxygen and H_2O_2 (0.04 M/l) reduction on $Yb_{0.1}WO_3$ in NaOH 0.1 M — 1400 rpm, 23°C, potential sweep: 6.3 mV/s.*

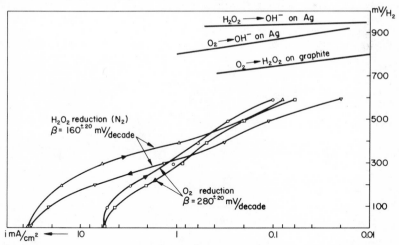

FIG. 11 *Oxygen and H_2O_2 (0.04 M/l) reduction on $Ce_{0.1}WO_3$ in NaOH 0.1 M — 1400 rpm, 23°C, potential sweep: 6.3 mV/s.*

The preceding results represent the difference between the potential sweep taken under nitrogen and under oxygen. For $Yb_{0.1}WO_3$ the hysteresis is even more marked than on $Ce_{0.1}WO_3$, and the activity for H_2O_2 reduction as well as O_2 reduction is much lower than on Ag.

CONCLUSION

Comparing the properties of the three bronzes we have studied, it appears that doping WO_3 with Yb and Ce, which are respectively at the beginning and the end of the rare-earth series, does not influence the electrocatalytic behavior of the bronzes.

On the other hand, $Sr_{0.8}NbO_3$ bronze behaves differently, and a tentative conclusion would be that the electrocatalytic properties of the bronzes are, as a first approximation, mainly dictated by the nature of the oxides.

REFERENCES

[1] A. Damjanovic, D. Sepa and J. O'M. Bockris, *J. Res. Inst. Catal. Hokkaido Univ.*, 16, 1, 1 (1968).

[2] A. R. Mackintosh, *J. Chem. Phys.*, 38, 1991 (1963).

[3] J. B. Goodenough, *Bull. Soc. Chim. Fr.*, p. 1200 (1965).

[4] R. Fuchs, *J. Chem. Phys.*, 42, 3781 (1965).

[5] E. Conroy and G. Podolsky, *Inorg. Chem.*, 7, 614 (1968).

[6] B. L. Chamberland, *Inorg. Chem.*, 8, 1183 (1969).

[7] M. J. Sienko, *J. Amer. Chem. Soc.*, 81, 5556 (1959).

[8] J. H. Ingold and R. C. DeVries, *Acta Met.*, 6, 736 (1958).

ELECTROCATALYSIS OF OXYGEN REDUCTION
BY CUBIC SODIUM TUNGSTEN BRONZE

J. McHardy and J. O'M. Bockris

University of Pennsylvania, Philadelphia, Pennsylvania

ABSTRACT

The promotion of electrocatalytic activity by traces of platinum in the non-stoichiometric compound "sodium tungsten bronze" has been examined. In steady state experiments, 400 ppm of platinum increased the rate of oxygen reduction at a given potential by over three orders of magnitude. A method for preparing platinum-doped crystals is described and an explanation of their activity in terms of semiconductor properties of the surface is indicated.

INTRODUCTION

A search for acid-stable electrocatalysts initiated some years ago led to the finding that sodium tungsten bronze could display remarkable catalytic activity for oxygen reduction [1,2]. Subsequent investigation revealed that the high activity was associated with traces of platinum in the bronze [3,4], and this note summarizes work done to elucidate the promotor action of platinum.

Sodium tungsten bronze is a non-stoichiometric compound with a range of composition Na_xWO_3 where $0 < x < 1$ and which exhibits metallic conductivity when $x > 0.25$. Cubic symmetry is associated with bronze crystals with $x > 0.4$, and such crystals are readily prepared by cathodic deposition from fused $Na_2WO_4 + WO_3$. Platinum "doping" is accomplished by codeposition from the melt, the platinum concentration in the melt being maintained by anodic dissolution of a 1 cm^2 platinum electrode. The balance of the current needed to grow a crystal is provided by an inert anode — usually gold.

RESULTS

Typical conditions for the preparation of a crystal appear in Table 1.

TABLE I

Melt composition	73g Na_2WO_4 (anhydrous); 40g WO_3
Temperature	800°C
Current	15mA
Time	24 hours

The relation between platinum anode current (most of which went for oxygen evolution) and platinum content of one series of crystals is shown in Fig. 1.

109

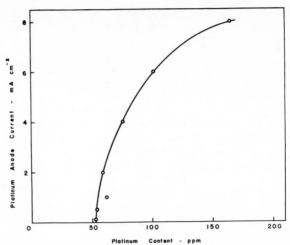

FIG. 1 *Relation between platinum anode current during electrolysis, and platinum content of the crystals deposited.*

Steady state curves of potential *vs.* log (current density) for oxygen reduction on typical crystals are compared with that on metallic platinum in Fig. 2. Included in the figure is the curve obtained on a "pure" (*i.e.*, platinum-free) crystal in argon saturated solution. Clearly, oxygen reduction proceeds at appreciable rates even on a pure crystal. It may be noted that oxygen reduction currents are completely masked on all crystals under non-steady state conditions such as transients or voltage sweeps — apparently by faradaic processes associated with reversible reactions of the bronze surface.

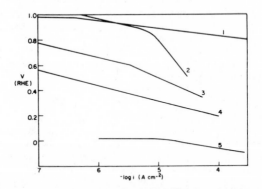

FIG. 2 *Potential vs. log (current density) curves in 0.1N H_2SO_4 solution [6]: (1) Platinum, (2) Sodium Tungsten Bronze (390 ppm Pt), (3) Sodium Tungsten Bronze (55 ppm Pt), (4) Pure Sodium Tungsten Bronze $Na_{0.7}WO_3$, (5) Pure Sodium Tungsten Bronze in Argon Saturated Solution.*

The variation of catalytic activity of Na_xWO_3 with platinum content* is shown in Fig. 3. The activity approaches that of bare platinum at a content of ~ 400 parts per million (ppm) and appears to

Determined by emission spectroscopy.

be leveling off.   Higher platinum contents have proved unobtainable
by the codeposition technique.

      In spite of the metallic nature of the bulk material, the surface
of a bronze electrode exhibits all the properties of a semiconductor.
The most likely explanation is the dissolution of sodium out of the
surface during electrochemical tests, and ion probe microanalysis has
confirmed this to be so [5].   On an undoped crystal, the depleted
layer was found to extend up to 2000 Å into the surface and to have
a composition corresponding to a semiconducting bronze, $i.e.$, $x \leqslant 0.25$.
On a Pt-doped crystal the depleted layer was shallower, but the sodium
content was much lower.   Two reasons exist for believing that this
effect of platinum may account for the enhanced catalytic activity of
a doped crystal.   First, the experimental results on undoped crystals,
shown in Fig. 4, indicate that low sodium content favors the catalytic

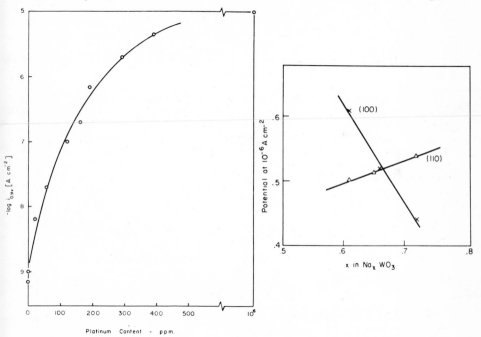

FIG. 3 *Activity (as log
[current density] at 0.9 volts
vs. RHE) of sodium tungsten
bronze for oxygen reduction in
0.1N H_2SO_4 as a function of
platinum content.*

FIG. 4 *Activity (as the poten-
tial where $i = 10^{-6}$ A/cm^{-2}) of pure
sodium tungsten bronze for oxygen
reduction in 0.1N H_2SO_4 as a function of composition for two crystal
faces [6].*

activity of a (100) face [6].   Second, the removal of sodium atoms
(which are known to act as donors [7]) is bound to alter the electronic
properties of the semiconducting layer, and significant effects upon
catalytic activity may be anticipated from theoretical considerations [8].

REFERENCES

[1]   D. B. Sepa, A. Damjanovic and J. O'M. Bockris, *Electrochim. Acta*,
      12, 746 (1967).

[2]   A. Damjanovic, D. Sepa and J. O'M. Bockris, *J. Res. Inst. Catal.*, 16, 1 (1968).

[3]   J. O'M. Bockris, A. Damjanovic and J. McHardy, *Proc. 3rd International Symposium on Fuel Cells*, Presses Académique Européenes, Brussels (1969) p. 15.

[4]   A. Damjanovic in *Modern Aspects of Electrochemistry*, J. O'M. Bockris and B. E. Conway (Eds.), Plenum Press, New York (1969), 5, p. 462.

[5]   J. McHardy, to be published.

[6]   R. A. Fredlein and J. McHardy, *Proc. 24th Annual Power Sources Conference*, Atlantic City (1970) p. 177.

[7]   W. McNeill and L. E. Conroy, *J. Chem. Phys.*, 36, 87 (1962).

[8]   J. O'M. Bockris, J. McHardy and R. Sen, these proceedings, p. 385.

ORGANIC CATALYSTS FOR OXYGEN REDUCTION

H. Alt, H. Binder, W. Lindner and G. Sandstede

Battelle-Institut e.V., Frankfurt am Main, Germany

ABSTRACT

It is only recently that organic compounds have been taken into
consideration as electrocatalysts. Jasinski has found that Co(II)
phthalocyanine is a fairly good catalyst for the cathodic reduction of
oxygen in alkaline electrolytes. Jahnke has investigated the performance of phthalocyanine cathodes in acid electrolytes. Extensive studies have shown that the catalytic activity depends critically upon the
metal ion, the substitution of the ligand, the nature of the support,
and the polymerization of the chelate. Our own investigations into
phthalocyanine catalysts have demonstrated that the support affects
not only the catalytic activity but also the stability of the chelate
in acid electrolytes. Only the Cu phthalocyanine turned out to be completely stable when adsorbed on the support. In the case of polymeric
Fe phthalocyanine free metal ions were identified in $2N$ H_2SO_4. Accordingly, the activity for oxygen reduction decreases gradually. Therefore, we looked for new organic cathode catalysts. As Pfeiffer complexes are known to act as reversible oxygen carriers, we have studied
the electrochemical behavior of monomeric and polymeric chelates of
this type. The Pfeiffer complexes, however, have only a very low activity for oxygen or no activity at all. Furthermore, they are not
acid-resistant. Another class of chelates we investigated is very
closely related to the natural oxygen carriers, the porphyrins. α, β,
γ, δ tetraarylporphyrins are stable in acid electrolytes and have a
surprisingly high catalytic activity. The performance is strongly affected by the metal ion, the structure of the aryl substituents, and
the preparation of the chelate-carbon catalyst. Co(II)tetra(p-methoxyphenyl)porphyrin is more active than polymeric Fe phthalocyanine,
the best phthalocyanine catalyst. The activity of the porphyrin catalysts has also been observed to decrease slowly, although the chelates
are completely resistant to the electrolyte. A tentative theoretical
interpretation of the influence of the metal ion and the ligand is
given on the basis of qualitative MO schemes.

INTRODUCTION

It has been well known for a long time that organic and organometallic compounds can act as catalysts in homogeneous and heterogeneous chemical reactions [1]. It is only recently that organic compounds have been taken into consideration as catalysts for electrochemical reactions. The starting point for investigations in this
field was the finding that particular metal chelates catalyze the oxidation of organic compounds by oxygen; for instance, the oxidation of
unsaturated fatty acids, cumene, olefins, ketones and esters [2]. Co(II)
complexes with phthalocyanine or octamethylporphyrazine as macrocyclic
ligands are among the most active catalysts. Uri [3] ascribes the

catalytic activity to the formation of a chelate-oxygen adduct in which
the oxygen molecule is chemically activated.

$$\text{Chelate} + O_2 \longrightarrow \overset{\delta+}{\text{Chelate}} \cdot \overset{\delta-}{O_2}$$

$$\overset{\delta+}{\text{Chelate}} \cdot \overset{\delta-}{O_2} + RH \longrightarrow \text{Chelate} + R\cdot + HO_2\cdot$$

The electron transfer from the metal chelate to the oxygen molecule
weakens the O-O bond, thus raising the reactivity of oxygen. Therefore,
bonding to a metal chelate may be expected to favor the cathodic reduction of oxygen, too.

A few years ago Jasinski [4] reported that Co phthalocyanine is a
fairly good catalyst for the cathodic reduction of oxygen in alkaline
electrolytes. Several years ago organic substances, including the
phthalocyanine and a porphyrin derivative, were also investigated as
catalyst candidates for the anode [5], but no pronounced activity has
been detected.

Recently, p-benzoquinone derivatives have been found to be suitable
cathode materials in acid electrolytes, but no direct reduction of oxygen is possible [6]. They can serve as redox systems and are not catalytically active.

PHTHALOCYANINE CATALYSTS FOR ALKALINE ELECTROLYTES

Looking for oxygen electrodes, Jasinski [4] investigated the electrocatalytic properties of phthalocyanine complexes in alkaline electrolytes. This class of metal complexes was chosen because it fulfills
the following requirements:

1) Chemical stability and insolubility in the electrolyte,
2) Resistance to hydrogen peroxide,
3) Stability up to potentials of about 1.3 V,
4) Capability for chemisorption of oxygen,
5) Simple and cheap synthesis.

The molecule is square and planar (Fig. 1), the metal atom being
symmetrically surrounded by four nitrogen atoms. The ligand is macrocyclic, and the π electrons are delocalized.

Jasinski found that Co(II) phthalocyanine is a satisfactory catalyst
for the cathodic reduction of oxygen in 35 percent KOH. In contrast to
the corresponding Fe complex, the Co chelate is stable in this electrolyte, even in the presence of H_2O_2. The Co phthalocyanine was blended
with acetylene black, and electrodes were prepared by bonding the mixture with PTFE to a nickel screen. The catalytic activity depends upon
the pretreatment of the electrode. High current densities have been
attained by polarizing the cathode to H_2 evolution potentials prior to
operation or by heating the electrode in H_2 at 370°C. The potential of
electrodes containing 14 mg/cm^2 Co phthalocyanine was about 800 mV at a
current density of 100 mA/cm^2 at 80°C.

Although the performance of the Co phthalocyanine electrode was
somewhat less satisfactory than that of a silver electrode, the use of
Co phthalocyanine is attractive in the light of relative cost. Furthermore, it is expected that the performance can be improved by modifying
the chelate and the support. Better results can be obtained, for

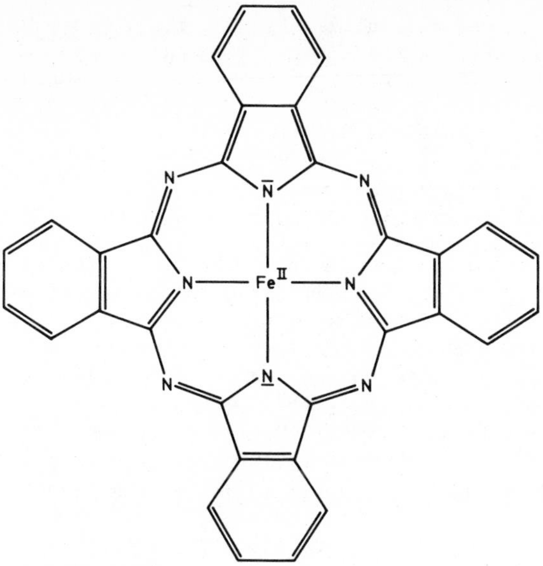

FIG. 1 *Fe phthalocyanine.*

instance, by using the Fe complex with a higher open circuit potential
[7]. According to Jasinski, however, the Fe compound is not sufficiently stable. As will be shown later, the stability of the chelate depends critically on the support.

PHTHALOCYANINE CATALYSTS FOR ACID ELECTROLYTE

In the last few years fuel cells with acid electrolyte have gained
increasing interest because they are insensitive to carbon dioxide.
Further development is stimulated by the recent discovery of tungsten
carbide as nonprecious metal catalyst for the anode [8,9]. For the
cathode of acid fuel cells, however, only platinum metals have so far
proved to be satisfactory catalysts, especially as far as their resistance to acid is concerned. Therefore, new cathode catalysts for acid
electrolytes have to be developed. In this context, Jahnke's results
are very important. He has found that phthalocyanines also catalyze
the electroreduction of oxygen in acid electrolyte [10]. This finding
has prompted several groups to undertake further studies into the performance of phthalocyanines in acid electrolyte. Only the most important results will be discussed here.

INFLUENCE OF THE METAL ION ON THE CATALYTIC ACTIVITY

For the preparation of the catalyst the chelate was dissolved in
concentrated H_2SO_4 and precipitated on acetylene black by addition of
water. Polyethylene and Na_2SO_4 were added, and the mixture was pressed
to give a mechanically stable electrode.

The central metal ion has a pronounced effect on the catalytic
activity of the phthalocyanine cathode. The best results are obtained
with Fe phthalocyanine. The order of decreasing catalytic activity is
Fe > Co > Cu [10]. The ligand without any metal ion has a very low
activity. The same order is observed in neutral electrolyte [11], Ni
phthalocyanine being slightly more active than the Cu compound. The

order (Fe > Co > Ni > Cu) holds also for the open circuit potentials, and Savy's [7] investigations have shown that this order is independent of the pH value. This finding is somewhat surprising because, according to Kozawa, Zilionis and Brodd [11], the reduction mechanism seems to be different in acid and alkaline media. In neutral and alkaline electrolyte, H_2O_2 is reduced at the Fe phthalocyanine electrode, whereas in acid electrolyte, it shows no activity for H_2O_2 reduction.

A tentative theoretical interpretation of the effect of the metal ion on the catalytic activity will be given at the end of this paper.

INFLUENCE OF POLYMERIZATION ON THE CATALYTIC ACTIVITY

Polymeric Fe phthalocyanine is the most active phthalocyanine catalyst. We have made some measurements with suspended catalysts in order to exclude the effect of the electrode structure. The apparatus used was similar to that described by Shlygin [12] and Gerischer [13]. The same method was also used by Jahnke [14].

The setup of the apparatus is schematically shown in Fig. 2. The catalyst suspended in the electrolyte is tossed against a gold gauze by vigorous stirring. Oxygen or nitrogen is bubbled into the solution. Hydrogen reference electrode, counter electrode and all the other equipment were used as in measurements with stationary electrodes.

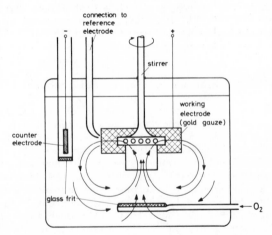

FIG. 2 *Diagrammatic view of the electrochemical measuring cell for catalyst suspensions.*

The measurements were carried out potentiodynamically at a voltage speed of 40 mV/min. Before measuring the curves, which are shown in the figures, the electrode was cycled approximately ten times between 400 and 800 mV until a temporarily nearly constant curve was obtained. Afterwards the activity of all catalysts investigated decreased slowly in the course of a few hours. This slow decrease seems to be due first of all to the measuring technique employed, but this effect may well be accompanied by a true decrease in catalytic activity.

Fig. 3 shows current-voltage plots of monomeric and polymeric Fe phthalocyanine catalysts. Only 50 mg catalyst were used. Ten percent by weight of the chelate was deposited on electrically conductive active carbon (Norit BRX). There is a marked difference between the monomeric and polymeric complex. At a potential of 600 mV, the polymeric Fe

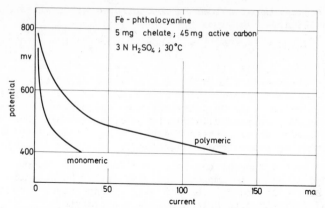

FIG. 3 *Potentiodynamic current-voltage curves for oxygen reduction at suspended phthalocyanine catalysts.*

phthalocyanine is about four times as active as the monomeric chelate.
This can be attributed to the higher electric conductivity of the
polymer [10] which differs from that of the monomeric compound by a
factor of 10^3 to 10^4.

When discussing the results, it should be noted that measurements
with suspended catalyst not only permit inferences to be made on catalytic activity, but also allow a more or less quantitative comparison
of different catalysts; however, the potentials in this case are lower
than those obtained with a solid porous electrode. According to Jahnke's
[15] results the potentials for bonded electrodes are 100 to 200 mV
higher.

INFLUENCE OF THE SUPPORT

Jahnke [14] reports that pure Fe phthalocyanine is not active at
all. On the other hand, acetylene black shows only a very poor activity for oxygen reduction. A suspension of the chelate and the support,
however, reduces oxygen when a gold gauze is used as an inert electrode.
The nature of the carbon support affects the performance of the catalyst. Jahnke claims that a specific interaction between the surface
oxides of the carbon and the complex is responsible for the synergistic
effect observed. From potentiodynamic measurements in the absence of
oxygen, he deduced that a current maximum observed with polymeric Fe
phthalocyanine at a potential of 160 mV is due to the complex-support
interaction. A correlation is found between the potential of this
maximum and the catalytic activity of the phthalocyanine catalysts [15].

In our own investigation on phthalocyanine catalysts, we have found
that the support affects not only the catalytic activity, but also the
stability of the catalyst in dilute H_2SO_4. When the catalyst was suspended in 2N H_2SO_4 for several days, it was possible to identify the
corresponding free metal ion in the solution in the case of Fe, Co and
Ni complexes. This must be due to decomposition of the complex into
the ligand and the central ion — the ligand remaining adsorbed on the
support. The decomposition rate depends on the nature of the support
and the temperature. If Fe phthalocyanine was precipitated on acetylene black, 3.2 percent of the total metal content was found in solution after four days at 25°C. This finding is unexpected because the

free chelates can be dissolved in concentrated H_2SO_4 without any sign of decomposition.

This negative effect of the carbon support on the acid resistance of the complex may be a general one. Presumably, there is a strong interaction between the ligand and the surface of the support which weakens the ligand-metal bond. Possibly the fine distribution of the chelate on the large surface area of the support may also play a role. In our experiments, only Cu phthalocyanine was completely stable on the support. The polymeric Co complex was more stable than the Fe compound. In agreement with these observations, the activity of the Fe phthalocyanine cathode decreased gradually.

Therefore, the support has to be modified or new supports have to be developed in order to improve the stability. Substitution in the periphery of the ligand by electron-withdrawing groups should also influence the resistance to attack by the electrolyte.

PFEIFFER COMPLEXES

Because of the instability of the phthalocyanine catalysts we looked for other chelates which might activate oxygen reduction. The ability to take up oxygen was chosen as a characteristic for the catalyst candidates.

The Pfeiffer complexes are known to act as reversible oxygen carriers [16]. Therefore, we studied the electrochemical behavior of monomeric and polymeric chelates of this type in acid electrolyte. The structures of some of the chelates synthesized are shown in Figs. 4 and 5.

The metal ion is complexed by two nitrogen atoms and two oxygen atoms. In contrast to the phthalocyanines, the ligand is not macrocyclic. The out-of-plane positions are again free to take up other ligands, for instance, amines and oxygen. As the complexes take up oxygen at the metal ion, we expected an activation of oxygen for the cathodic reduction. Unfortunately, however, only two monomeric Pfeiffer complexes gave electrodes which were slightly more active than the

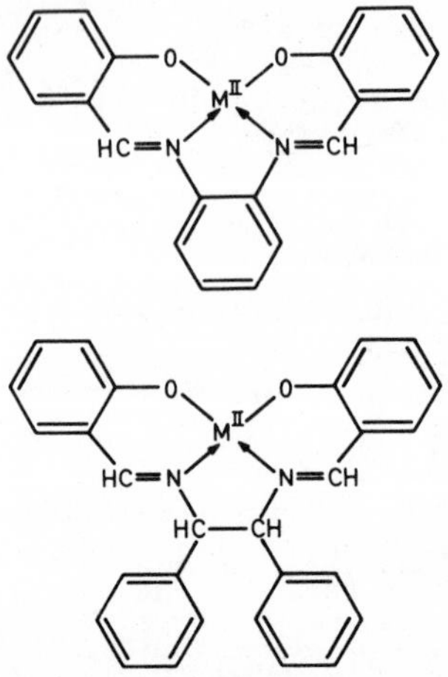

FIG. 4 *Monomeric Pfeiffer complexes (M-bis-(salicylaldehyde)o-phenylenediamine and M-bis-(salicylaldehyde)diphenyl ethylenediamine).*

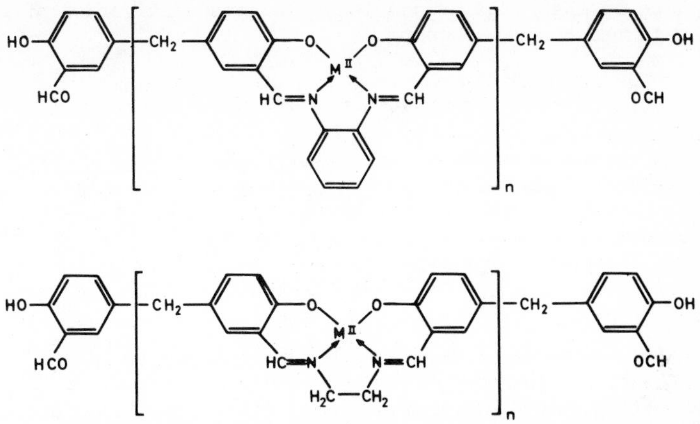

FIG. 5 *Polymeric Pfeiffer complexes (poly-M-bis(salicylaldehyde)-*
o-phenylenediamine and poly-M-bis(salicylaldehyde)-ethylenediamine).

carbon support itself. First, we thought that the preparation of the
catalyst, which was similar to the preparation of phthalocyanine catalysts, or the electrode structure might be the reason for the negative
result. We therefore made some measurements with suspended catalysts
using the apparatus already described. The experimental results were
in agreement with the measurements with porous electrodes; they showed
that monomeric Pfeiffer complexes have a very low activity and polymeric
complexes have no activity at all. Furthermore, the Pfeiffer complexes
were not stable in dilute H_2SO_4, especially at elevated temperatures.
The metal-ligand bonds were cleaved and the ligand was hydrolyzed. This
holds for all central metal ions investigated (Fe, Co, Cu). When electron-withdrawing groups, *e.g.*, NO_2 groups, were substituted, the ligand
was not hydrolyzed, but the metal-ligand bond was still broken by the
acid. From these experiments it may be concluded that the strong chelate effect of a macrocyclic ligand is necessary for a sufficient stability of the complex.

Our experiments with Pfeiffer complexes give clear-cut evidence
that a metal chelate taking up oxygen does not suffice to catalyze the
cathodic reduction of oxygen. It is possible that the nature of the
metal-oxygen bond plays an important role, especially the electron
transfer from the metal ion to the oxygen molecule. As far as the
Pfeiffer complexes are concerned, X-ray analysis has shown that in
solid state the oxygen is bound to two metal ions in the form of a
peroxo-bridge [17].

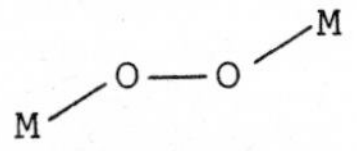

Of course, the oxygen is first bound to one metal ion:

$$M + O_2 \longrightarrow M - O_2 \cdot$$

$$M - O_2 \cdot + M \longrightarrow M - O_2 - M \quad .$$

The mononuclear complex can react with a second chelate. If this reaction also takes place in the catalyst, it could explain the inferior

performance of the Pfeiffer complexes, because activation of oxygen is unlikely in the binuclear complex.

PORPHYRIN DERIVATIVES

Another class of chelates we investigated is very closely related to the natural oxygen carriers, the porphyrins. These metal complexes are indispensable in life processes, notably in respiration. They are also involved in biochemical redox processes. Unfortunately, however, the porphyrins found in living organisms as well as most of the synthetic porphyrins are not stable in $2N$ H_2SO_4. Only a few derivatives are resistant to acid.

We have chosen the α, β, γ, δ tetraarylporphyrins because the ligand can be readily synthesized by refluxing a mixture of an aromatic aldehyde with pyrrole in propionic acid [18]. The structure of this type of porphyrin chelate can be seen from Fig. 6, where the formula of Co(II)tetra(p-methoxyphenyl)porphyrin is shown.

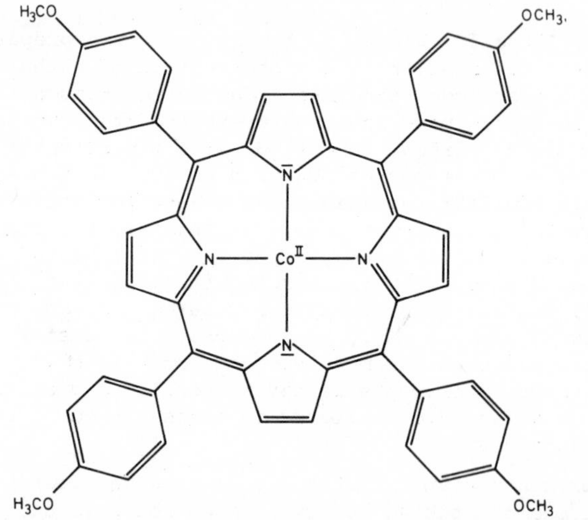

FIG. 6 *Co(II)α, β, γ, δ-tetra(p-methoxyphenyl)porphyrin.*

Again we tested the electrocatalytic activity of the chelates with suspended catalysts. The chelate was dissolved in organic solvents, *e.g.*, pyridine or dioxane. Active carbon (Norit BRX) was added as support, and the chelate was precipitated by addition of water. In other experiments it was possible to deposit the chelate from the gas phase because the tetraarylporphyrins are stable up to temperatures of about 400°C. The chelate support ratio was varied between 2 and 10 percent.

The catalytic activity of the porphyrin catalysts turned out to be surprisingly high [19]. The most important results obtained up to now are as follows. The metal ion has a pronounced effect on the catalytic activity, as can be seen from Fig. 7. The order of decreasing activity is Co(II) > Fe(III) >> Ni(II) $\approx$ Mn(III) $\approx$ metal-free $\geq$ Cu(II). Only with Co and Fe was the chelate-carbon catalyst distinctly more active than the support itself. The Cu complexes were not active at all.

The order of activity is very similar to that observed with phthalocyanine catalysts. It should be noted that in the porphyrin complex

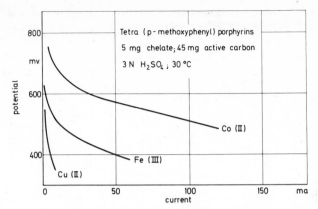

FIG. 7 *Potentiodynamic current-voltage curves for oxygen reduction at suspended porphyrin catalysts.*

Fe is in valence state III. This could explain the reverse order of Fe and Co, compared with the phthalocyanines.

The activity is also influenced by the nature of the aryl substituents. This will be discussed in more detail later on. Up to now Co(II)tetra(p-methoxyphenyl)porphyrin proved to be the most active chelate. Therefore, all the experiments discussed in the following were made with this compound.

The catalytic activity is affected not only by the metal and ligand, but also by the preparation and the pretreatment of the catalyst. If the chelate is precipitated on the support by adding water to an organic solvent, the solvent plays an important role. For instance, the performance at a potential of 600 mV is more than doubled if pyridine is used instead of dioxane. A strong effect on the activity of Co porphyrin is also observed in case of thermal pretreatment of the catalyst.

Because of the high activity of the Co porphyrin, we were of course interested in comparing the performances of Co porphyrin and polymeric Fe phthalocyanine, the best phthalocyanine catalyst. This was again done by using suspended catalysts (Fig. 8). It follows that the porphyrin derivative is more active than the Fe phthalocyanine catalyst.

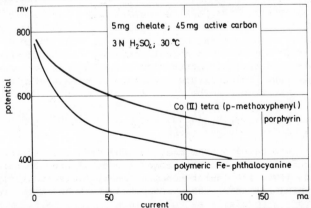

FIG. 8 *Potentiodynamic current-voltage curves for oxygen reduction at suspended porphyrin and phthalocyanine catalysts.*

This is worth noting because the porphyrin is monomeric and has a low electric conductivity, whereas the polymeric phthalocyanine is an organic semiconductor.

When we started our investigation on porphyrin catalysts, nothing was known about the capability of tetraarylporphyrins to take up oxygen. Very recently two ESR studies have been published which deal with the oxygenation of Co tetraphenylporphyrin [20,21]. When the Co porphyrin was exposed to oxygen, the original ESR signal decayed and a new signal developed. The spectral change was reversible in most cases, showing that this complex takes up oxygen reversibly. Furthermore, additional ligands, especially amines, affect the degree of oxygen complexing. This finding could explain our observation that the catalytic activity is influenced by the organic solvents used.

THEORETICAL ASPECTS OF CHELATE CATALYSIS

Phthalocyanines and porphyrins take up oxygen at the central metal ion, therefore, the metal-free compounds are catalytically inactive. In recent years X-ray analysis and ESR studies have given some insight into the bonding between the metal ion and the oxygen molecule. Oxygen can be bound in two ways to the metal ion:

Edge-on bonding has been proved to exist in Pfeiffer-type complexes, whereas side-on bonding is found in complexes of the Vaska type [22]. In the case of side-on bonding, the donor bond from the oxygen molecule to the metal is formed with the aid of π electrons from the double bond, as in olefin complexes of transition metals. Back-bonding takes place from occupied d orbitals into the antibonding π orbitals of the O_2 molecule, causing a shift in electron density from the metal ion to the oxygen molecule. In this way the O—O bond strength is reduced on coordination.

It can be assumed that a similar mechanism exists in edge-on bonding. For the donor bond, however, a lone pair of the oxygen molecule must be used.

The energy levels of the metal ion depend critically on the macrocyclic ligand. Fig. 9 gives a qualitative MO scheme of the interaction of the top-filled orbitals of the porphyrin ligand with the d orbitals of the metal ion. The orbitals of the porphyrins have been calculated using simple HMO theory. The splitting of the d orbitals of the central metal ion is caused by σ and π interaction with the top-filled orbitals of the porphyrin ligand, and has been estimated on the basis of simple symmetry considerations.

The $d_{x^2-y^2}$ orbital is the most destabilized orbital due to very strong σ bonding to the four nitrogen atoms. Since according to ESR measurements [23], in-plane π bonding can be neglected, only out-of-plane π bonding has to be taken into account. The latter raises the degenerate $d_{xz,yz}$ orbitals as well as the d_{z^2} orbitals. Thus, the ordering of the energy levels which are predominantly of d character

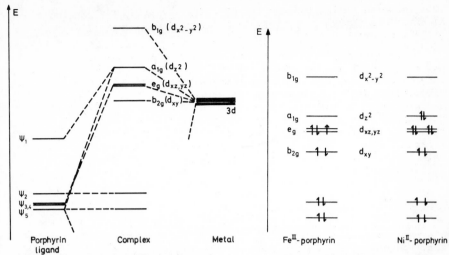

FIG. 9 *Qualitative MO scheme of the interaction of the top-filled orbitals of porphyrin with the d orbitals of the central metal ion.*

FIG. 10 *Electron configurations of the central metal ion of metalloporphyrins according to simple MO theory.*

is as follows [24]:

$$d_{x^2-y^2} > d_{z^2} > d_{xz,yz} > d_{xy} \quad .$$

Within a very simple model for the metal-oxygen bond, it is apparent that a vacant d_{z^2} orbital is necessary for the donor bond. Back-bonding requires filled $d_{xz,yz}$ orbitals. This holds both for side-on and for edge-on bonding. Whether these orbitals are filled or not depends upon the nature of the metal ion. The electronic population of the d orbitals in Fe^{III} and Ni^{II} porphyrin is shown in Fig. 10.

Assuming that chemisorption at the metal ion is a prerequisite for catalysis and that the degree of back-bonding governs the activation of oxygen, the following conclusions can be drawn: high catalytic activity is to be expected for complexes having a metal ion with less than 8 d electrons; the highest activity should be attained using d^6 and d^7 transition metals as central metal atoms of a macrocyclic complex, the basic requirement for maximum interaction between the ligand and the central atom is a planar structure of the ligand, which can be found in a macrocyclic compound.

These predictions should be valid for a number of macrocyclic ligands, in particular those with four nitrogen atoms surrounding the central metal atom. At present we are investigating several classes of macrocyclic compounds. The experimental results are in agreement with this very simple model, examples being the porphyrins and phthalocyanines. The Ni and the Cu chelates are almost inactive catalysts; the Co and Fe chelates are the most active.

Furthermore, the activity of the chelates should be enhanced by electron-releasing substituents because these substituents increase the electron density at the metal ion and thus favor the activation by back-bonding. The preliminary results obtained with Co porphyrin

support this conclusion. The activity is raised by introducing a
methoxy group in p-position into the benzene ring (Fig. 11).

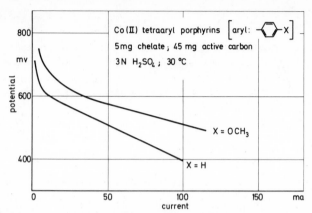

FIG. 11 *Potentiodynamic current-voltage curves for oxygen reduction at suspended Co tetraarylporphyrin catalysts.*

Another effect on the bond strength of the oxygen molecule, and
hence on the catalytic activity of the chelate, may result from the
support as a further ligand. This ligand occupies the 6th coordination
position and is opposite to the oxygen in the other axial position. If
it has the property of an electron donor, it can increase the electron
density at the central metal ion and thus cause a stronger polarization
of the metal oxygen bond. The result is a strengthening of the chemisorption by increased back-bonding.

Further experiments are necessary in order to improve and refine
the understanding of chelate catalysis, especially with respect to the
influence of polymerization on the activity and the role of the support.
Nevertheless, it can be expected that in the years to come substantial
progress will be made, including methods of quantum chemistry in a well-
planned research program. Thus a very interesting field of electrocatalysis is beginning to develop.

ACKNOWLEDGMENT

We thank Fräulein D. Berndt, Frau G. Kaufmann, and Drs. W. Klöpffer
and R. Reiner for the preparation of the substances and their analysis,
and Dr. W. H. Kuhn for valuable assistance.

This work was sponsored by the Bundesministerium der Verteidigung,
Bonn, under the guidance of Dr. R. Dieberg. The permission to publish
these results is gratefully acknowledged.

REFERENCES

[1] W. Langenbeck, *Fortschritte der chemischen Forschung 6*, Springer-
     Verlag, Berlin (1966).

[2] C. Paquot and F. Goursac, *Bull. Soc. Chim.*, 12, 450 (1945); C.
     Paquot, *ibid.*, 8, 695 (1941); *cf.* also [1].

[3] N. Uri, *Nature*, 177, 1177 (1956).

[4] R. Jasinski, *Nature*, 201, 1212 (1964); *J. Electrochem. Soc.*, 112, 526 (1965).

[5] J. O'M. Bockris and S. Srinivasan, *Fuel Cells: Their Electrochemistry*, McGraw-Hill Book Co., New York (1969) p. 346.

[6] H. Alt, H. Binder, A. Köhling and G. Sandstede, Extended Abstracts, *Comité International de Thermodynamique et de Cinétiques Electrochimiques*, Prague (September 1970) p. 319.

[7] M. Savy, Extended Abstracts, *Comité International de Thermodynamique et de Cinétiques Electrochimiques*, Prague (Sept. 1970) p.193.

[8] H. Böhm and F. A. Pohl, *Wiss. Ber. AEG-Telefunken*, 41, 46 (1968); *Comptes Rendus, Troisièmes Journées Internat. d'Etude des Piles à Combustible*, Presses Académiques Européennes, Brussels (1969) p. 180.

[9] H. Binder, A. Köhling, W. H. Kuhn, W. Lindner and G. Sandstede, *Nature*, 224, 1299 (1969); H. Binder, A. Köhling and G. Sandstede, *Preprints of papers, Amer. Chem. Soc., Meeting*, New York (1969), 13/3, 99; *Advan. Chem. Ser.*, in preparation.

[10] H. Jahnke, *Bunsentagung*, Augsburg (1968); abstract in: *Berich. Bunsenges. Phys. Chem.*, 72, 1053 (1968).

[11] A. Kozawa, V. E. Zilionis and R. J. Brodd, paper presented at the Fall Meeting of the Electrochemical Society (October 1969).

[12] J. A. Podwjaskin and A. J. Shlygin, *J. Phys. Chem. (Russian)*, 31, 1305 (1957).

[13] J. Held and H. Gerischer, *Berich. Bunsenges. Phys. Chem.*, 67, 921 (1963).

[14] H. Jahnke and M. Schönborn, *Comptes Rendus, Troisièmes Journées Internat. d'Etude des Piles à Combustible*, Presses Académiques Européennes, Brussels (1969) p. 60.

[15] H. Jahnke and M. Schönborn, *Bunsentagung*, Heidelberg (May 1970), abstract in *Berich. Bunsenges. Phys. Chem.*, 74, 944 (1970).

[16] M. Calvin, R. H. Bailes and W. K. Wilmarth, *J. Amer. Chem. Soc.*, 68, 2254, 2267 (1946).

[17] L. H. Vogt, H. M. Faigenbaum and S. E. Wibeley, *Chem. Rev.*, 63, 269 (1963).

[18] A. D. Adler, F. R. Long and J. D. Finarelli, *J. Org. Chem.*, 32, 476 (1967).

[19] G. Sandstede, *5th Symposium, Energie-Direkt-Umwandlung*, Essen (October 1970).

[20] F. W. Walker, *J. Amer. Chem. Soc.*, 92, 4235 (1970).

[21] K. Yamamoto and T. Kwan, *J. Catal.*, 18, 354 (1970).

[22] J. A. McGinnety, N. C. Payne and J. A. Ibers, *J. Amer. Chem. Soc.*, 91, 6301 (1969); A. L. Crumbliss and F. Basolo, *J. Amer. Chem. Soc.*, 92, 55 (1970).

[23] P. T. Manoharan and M. T. Rogers, in *ESR of Metal Complexes*, Teh Fu Yen (Ed.), Adam Hilger, London (1969), p. 143.

[24] M. Zerner and M. Gouterman, *Theor. Chim. Acta*, 4, 44 (1966).

SUMMARY OF DISCUSSION AND COMMENTS: SESSION CHAIRMAN II

C. E. Heath

Jersey Enterprises, Inc., New York, New York

The impasse reached in acid electrolyte fuel cell technology, the development of effective, however, utterly uneconomical, noble metal catalysts, has been often recognized in recent years. At least one organization has abandoned its efforts with the conclusion that the problem was not likely to be solved. The interesting work reported in this Session is thus most welcome. While at an early stage, this work has revealed that interstitials, oxides, and chelates possess some of the catalytic properties heretofore the unchallenged realm of platinum and its family. With these studies, carbonaceous fueled systems with invariant electrolytes again become possible.

Four papers were presented in the Session, one devoted to tungsten carbide catalysts, two to the interesting bronze compounds, and one to organometallic complexes.

The discussion followed several routes: the mechanism, how stability might be improved and surface area increased as well as the use of alternate substrates.

Centering on the assumption that d-band vacancies of the catalytic metal play a role in the electrode reactions, the discussion recognized the powerful tool we have in employing ligands to vary electron densities in metals, thus tailoring the surface for optimum adsorption of reactants and activation of intermediates. Further control can be exercised through the supporting material for the organometallic catalysts. The idea that the surface itself might be a ligand was contested by some.

The underlying carbides and bronzes which are doped with cocatalysts, are not necessarily stable, and the base may be dissolved under potential fields in strong electrolytes. However, it was suggested that where the catalyst consists of a stable species on the surface of a less stable material, that a "protective rind" be made to cover the less stable substrate.

Other improvements suggested included increasing the surface of the inorganic catalysts, although the current densities of the carbides reported were quite significant, considering their likely low surface area. No method for accomplishing this objective was proposed. Also, if one substrate can have a beneficial effect on non noble metal catalyst activation, why not use cosubstrates as applied in the petroleum industry?

127

SESSION III

FUEL CELLS WITH ALKALINE ELECTROLYTE

Chairman

D. P. Gregory

HYDROGEN AND METHANOL FUEL CELLS WITH AIR ELECTRODES IN ALKALINE ELECTROLYTE

H. Binder, A. Köhling, W. H. Kuhn, W. Lindner, and G. Sandstede

Battelle-Institut e.V., Frankfurt am Main, Germany

ABSTRACT

Electrodes of various structures and catalysts converting hydrogen, methanol, or oxygen (air) were investigated; all can be employed in fuel cells with alkaline electrolyte. The following catalysts were used predominantly: Raney silver-palladium for methanol, Raney nickel for hydrogen, and silver or active carbon for oxygen electrodes. Sintered metal bodies are suitable for gas electrodes if gases under pressure (0.5 atm) are used. They are generally too heavy as immersion electrodes and for gases at atmospheric pressure. The catalysts can be incorporated in plastic-bonded electrodes. They are satisfactory as immersion electrodes and also as gas electrodes if their gas side is made hydrophobic. Their thickness is less than 1 mm; their weight per unit area is extremely small (as low as 0.05 g/cm^2), especially if carbon is used as pore-forming substance or as catalyst (for oxygen).

1. INTRODUCTION

This paper is not intended to deal with the large field of hydrogen and methanol fuel cells as a whole, but to discuss some of our activities on alkaline fuel cells, the results of which have not yet been published.

Our investigations on electrodes for fuel cells generally covered two groups of problems: first, the activity of catalysts for distinct reactions; and second, the influence of electrode structures on these reactions and the best way to incorporate catalysts into electrodes. The catalysts for the cells under investigation — alkaline hydrogen and methanol cells — are of course specially adapted to the specific type of cell and the reagents, while many aspects concerning the electrode structures are also applicable to cells with acid electrolyte and other fuels.

For the construction of batteries, the basic principle of a filter-press arrangement with free liquid electrolyte appeared to be especially suitable, because heater, cooler or concentration controls can be built into such a system without any difficulty. This arrangement seems to be particularly suited for methanol batteries, since in these the electrolyte is consumed and has to be replaced continuously. The special advantages of the free liquid electrolyte are that the immersed methanol electrodes may have a simple pore structure [1].

For the investigation of both catalysts and suitable supporting structures, we used disc-shaped electrodes, which were examined in half-cell arrangement in the free liquid electrolyte (6.5N KOH). In some cases we fabricated workable but not completely developed battery models.

Our investigations were aimed at cells which operated without excess gas pressure. For this reason, we did not work on systems with immobilized electrolyte. These are suited particularly for cells with

131

pressurized gas because the electrolyte-filled matrix imparts additional mechanical stability to the electrode.

2. ELECTRODE STRUCTURES

2.1 COLD-PRESSED METAL SKELETON ELECTRODES

For investigations of the catalyst we often use gold skeleton electrodes. These are fabricated by pressing gold powder, catalyst and a water-soluble salt, *e.g.*, sodium sulfate, together without heating the mixture. A mixture of 30% by volume catalyst, 40% by volume gold and 30% by volume salt requires a pressure of about 2 tons/cm^2; the gold grains will then firmly adhere to each other, so that the catalyst is retained after the salt grains have been washed away with water, which yields correspondingly large cavities (pores). In the case of Raney catalysts, it is either possible to leach out the aluminum from the comminuted aluminum alloy by means of alkalis or acids and to produce electrodes from this "activated" catalyst, or to process the aluminum-containing alloy and to dissolve the aluminum from the pressed electrode. This leads to the formation of additional micropores.

These discs are suited for use as immersed electrodes or as bubbling gas electrodes because they are completely wettable (hydrophilic). Fabrication by cold pressing, moreover, is a very simple procedure and the catalyst is not attacked as a result of the ambient temperature.

2.2 SINTERED METAL ELECTRODES

For several experimental fuel cells we needed electrodes of larger dimensions and therefore used nickel skeleton electrodes. As is known, the sintering of powdered nickel produces stable, porous discs. When nickel carbonate is added and sintering is conducted in a reducing atmosphere, two advantages are gained: firstly, the sintering temperature can be kept lower than with a powdered nickel skeleton; and secondly, it is a simple way of increasing the porosity, especially the microporosity, of the disc.

Immersion Electrodes

This type of disc is absolutely wettable and is, therefore, highly suitable as an immersion electrode, which is needed for the methanol conversion. Catalysts such as palladium have to be built in to the skeleton. However, difficulties arose when a Raney alloy was used. When the Raney palladium or the Raney palladium alloy, serving as a methanol catalyst, was sintered together with the nickel, we observed that a reaction occurred between the nickel and the aluminum of the alloy — because of the comparatively low melting point of the Pd-Al-alloy. A part of the palladium alloy with reduced aluminum content was then no longer activatable (*i.e.*, it was impossible to leach out the aluminum) and the palladium was ineffective. On the other hand, the nickel-aluminum alloy, formed on the surface of the nickel grains, became fairly active on the dissolution of the aluminum, so that it could be oxidized and become nonconducting at the potential of the methanol oxidation. In this way the electrodes became ineffective or were destroyed. This effect can be prevented by admixing a metal with a high melting point, such as titanium, molybdenum or others, to the Raney alloy. Then the alloy remains far below its melting point at the sintering temperature of the nickel and a reaction during the sintering process is avoided.

Hydrophilic Gas Electrodes

The simple metal-sintered electrodes are not suitable for gas electrodes because one has to let the gas stream through the pores; *i.e.*, let it bubble in the electrolyte. But the production technique is — at least as regards the nickel skeletons — variable enough to enable the simultaneous sintering of several layers of different porosity, with coarse pores at the gas side and small pores at the electrolyte side. For this, however, one needs — contrary to our basic concept — gases under pressure. Such double or multilayer electrodes work under a gas pressure sufficient to blow the liquid out of the coarse pores, while the fine-pore layer remains impermeable to gas. Because of the high stability of such sintered bodies, gas overpressures of up to two atmospheres are allowed. Indeed such hand-made multilayer electrodes are fairly thick and heavy, but with improved mechanical methods it is possible to reduce the thickness significantly.

To produce silver skeleton bodies, as required for oxygen or air electrodes, we made discs from silver carbonate. The silver resulting from the reduction of the carbonate has a large surface area, is very active and therefore easy to sinter. However, because the large active surface area of the silver which one obtains first is then greatly reduced due to the sintering process, it is advisable to mix the silver carbonate with powdered nickel. This prevents the silver's being sintered and a mechanically stable skeleton is formed from the nickel. In this way one can produce electrodes of 100 mm or more in diameter. Due to their differing porosity, the electrode layers have differing degrees of shrinkage and are thus likely to lose their cohesion easily. One can compensate for this by altering the silver-nickel ratio.

2.3 HYDROPHOBIC GAS ELECTRODES

For multilayer gas electrodes, one needs an increased gas pressure in order to fill the coarse pores with gas. Since, however, our basic concept requires air-breathing cathodes, we preferred to use hydrophobic, plastic-bonded electrodes. Electrodes, made by different techniques, have been widely used in the last few years. Since nowadays most researchers into fuel cells have acquired experience in the field of plastic-bonded electrodes, we are only going to cover a few special points here:

Electric Conductivity

If a metal or other electrically conductive powder is admixed to a viscous liquid, for instance a polymer melt, the conductivity remains close to zero and increases only if there is a fairly high percentage of metal in the mixture (about 50 percent by volume). In contrast to this, the increase in conductivity of dry powder mixtures starts with a sharp break already at 10 percent to 30 percent depending on the sizes of the grains. If, in addition, pores are built in, by admixing soluble salts, the metal concentration at which conductivity begins again is a little higher (30% to 50%). This finding suggests that at the point of incipient macroscopic conductivity, not all metal grains are in conducting contact. Therefore, comparatively small changes in volume, such as arise when the pore-forming salt is leached out, may be followed by a substantial decrease in conductivity. For this reason one has to add more conductive material or reduce the proportion of nonconductive plastic. This, however, diminishes the mechanical stability of the material.

If a catalyst is admixed with such a body, it is very probable that a considerable part is not connected with the conducting skeleton of the electrode. Measurement of the electrode resistivity can only indicate whether the mixture concerned is suitable as regards conductivity; high macroscopic conductivity is a necessary but not a sufficient condition. It is therefore necessary to distribute the catalyst so that it has as close a contact as possible to the conductive particles. Thus volume percentage and grain size have to be optimised jointly.

In order to achieve sufficient porosity, the share of the volume occupied by the conducting substance must not be too high. For this reason it is usually necessary, particularly when only graphite is being used as the conducting substance, to build in a metal net or a graphite gauze so that the path is as short as possible between the point of reaction in the less conductive electrode mass and the gauze as current collector.

Porosity

The second point to be mentioned here is the porosity of the electrodes. The porosity may be furnished by the catalyst itself — as with Raney catalysts for instance — but in general, additional pore systems have to be built in. Admixing of salts leads to comparatively large pores (generally greater than 20 microns), desirable for the convection of a liquid, but normally not necessary for the transport of gas. For gas electrodes, a small-pore system is preferable, such as would be obtained from a porous carbon. Using electrically conducting porous carbon, an improvement in porosity and in conductivity can be achieved in one operation.

Wettability

There is a close connection between porosity and wettability of electrode bodies. Layers containing a catalyst and an electrically conducting substance are in most cases hydrophilic or completely wettable, even if the skeleton of the electrode consists of an extremely hydrophobic material, like PTFE. In this case, indeed, not all pores are filled with liquid, but numerous microareas remain unwetted and gas filled, and thus allow the transport of gaseous reactants.

In the case of immersed electrodes the remnants of the enclosed gas will be dissolved more or less rapidly by the surrounding electrolyte; we often observe an increase in activity over several hours, which is certainly due to an increase in wetted areas. With gas electrodes, however, this increase in wetting is undesirable and an additional barrier for the liquid must be applied to the gas side. Such a barrier may be provided by supplementary impregnation of the gas side, for instance with a film of PTFE spray or suspension. In our experiments, as a rule, such layers withstood the attack of hot alkali for only a few hours. A similar behaviour was observed for impregnation with paraffin solutions, especially in the case of low-melting paraffins.

Our experiments were in most cases directed toward the application of a layer of sintered PTFE powder of minimum thickness to the gas side of the electrode. The layers we obtained by this method were never under several tenths of a millimeter in thickness, *i.e.*, thick in comparison to spray films of the μ-order; but, on the other hand, these layers were fairly tough and made a welcome contribution to the stability of the electrode, especially in those cases where the active layer itself was fairly thin, too.

Undesired Reactions With PTFE

For gas electrodes PTFE is a very appropriate binder because of its
high hydrophobility; but, though PTFE is reported to be absolutely inert
to chemical attack, we observed in the case of powdered mixtures, different reactions at the elevated temperatures needed for sintering or
"baking" of electrode discs. With many materials in the pulverised
state — not only like nickel and other ignoble metals, but even with
compounds like calcium carbonate — vigorous reaction occurs with foaming,
glowing and total destruction of the electrode body. This is due to the
high heat of formation of most fluorides.

With Raney alloys the reaction between fluorine and aluminum can be
violent; it is therefore difficult to use such unactivated alloys for
the preparation of plastic-bonded electrodes with PTFE. On the other
hand, if a preactivated aluminum-free Raney catalyst is used, in most
cases a great part of the activity is destroyed at the required sintering temperature of about 350° to 370°C.

We found the solution to this problem to be the admixture of a
finely dispersed substance to the electrode mix, $e.g.$, carbon black or
a silver salt such as silver carbonate or oxalate of very small grain
size which is decomposed at the baking temperature. Thus the reaction
is prevented, because the grains of PTFE and those of the alloy are
separated from each other. The addition of silver salt has the advantage of resulting, after decomposition, in a very high conductivity of
the active layer; almost every catalyst grain is in contact with the
conducting silver skeleton.

2.4 PLASTIC BONDED IMMERSED ELECTRODES

In view of the heavy weight of the metal skeleton, we likewise use
a plastic bonding mostly for immersed electrodes for which the high
mechanical stability of the metal skeleton is not required. The
structure of the immersed electrodes causes less difficulty than that
of gas electrodes; only as regards electric conductivity, the same considerations apply as were described above for gas electrodes. With immersion electrodes, the pore structure must be optimized in such a way
that, on the one hand, many fine pores are present, so that the inner
surface area is as large as possible; and on the other, the electrode
is also traversed by a system of coarse pores, so that the dissolved
fuel can be conveyed quickly enough by diffusion and convection. Since
such an extremely hydrophobic bond as PTFE does not prevent the wetting
of the inside of the electrode — as mentioned above — but can cause
unnecessary impediments, we always use the easily processed polyethylene
as a binder. Although this material melts at the processing temperature
(approx. 130°C), it is sucked into the pores of the other ingredients to
a minor degree and only blocks the catalyst slightly.

3. CATALYSTS AND MEASURING RESULTS

3.1 METHANOL ANODE

Our investigations on methanol catalysts started from the well-known
high chemical activity of Raney catalysts, and a great part of our work
has been conducted in the field of Raney alloys. Since another paper in
these proceedings already reports on the behaviour of these catalysts

in general [2], our studies of Raney catalysts will only be discussed here insofar as they concern the aforementioned electrode structures.

In a screening program, we investigated the platinum metals and their alloys for methanol conversion in alkaline electrolyte. It was clearly observable that the Raney alloy catalysts were more active than the individual Raney metals. This means that after dissolving the aluminum the platinum metals are still present as an alloy and not as a mixture.

As palladium turned out to be nearly as active as platinum, and as we wanted to avoid platinum itself or alloys with platinum, we investigated palladium and alloys of palladium with metals such as silver, zirconium, molybdenum and others. The active Pd-phase is stabilized by the use of these additives, silver in particular, so that the activity remains the same even under a higher load. With pure palladium, the stable β-phase containing hydrogen is destroyed, because the hydrogen is dissolved by anodic reaction at high current density, $i.e.$, at elevated potential.

With the addition of nickel, however, one obtains palladium catalysts which, in the beginning, it is true, have a good effect on the methanol transformation, but which then become less active. At the potential which occurs in operation of above 200 mV, the nickel component will already be oxidized and thus will gradually render the catalyst ineffective. Fig. 1, curve 1, illustrates the behaviour of such an electrode; the dotted arrows indicate the increasing polarization of the palladium- and nickel-containing anode.

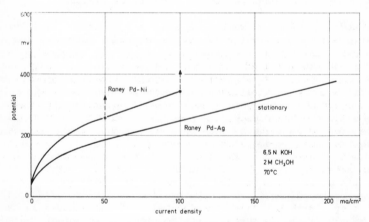

FIG. 1 *Galvanostatic current-voltage curves of polyethylene bonded methanol electrodes (25 mg Pd/cm²).*

We were, however, able with the aforementioned palladium-silver catalyst to produce PTFE- or PE-bonded electrodes whose activity did not decrease over a long period. The stabilization of the palladium can be achieved with as little as approximately 10 atomic percent silver — the optimal figure is around 40 percent; the activity naturally decreases, given a high percentage of silver, but not appreciably until after 50 atomic percent Ag.

Both the sintered electrodes of a 25-watt methanol-oxygen battery and a methanol-oxygen double cell were equipped with a palladium-silver catalyst; this was exhibited at the Achema Exhibition, Frankfurt (Main) in 1964 — as were the plastic-bonded electrodes of a 30-watt methanol-air battery, which we demonstrated at this exhibition in 1967.

The palladium-silver-aluminum alloys are very ductile and are accordingly difficult to grind to the desired small-grain size of less than 50 microns. If a small percentage of magnesium is admixed to the alloy, the reguli become brittle and, moreover, the catalyst is more active than a Raney catalyst without this magnesium admixture.

We stopped these investigations some time ago, but we are sure that the palladium content can be reduced significantly if the electrode structure is improved. Moreover, the use of a supported palladium-silver catalyst should lead to a substantial decrease in the amount of noble metal required.

3.2 HYDROGEN ANODE

Sintered Nickel-Silver Electrodes

For hydrogen conversion, we experimented with silver-containing, sintered nickel electrodes. They were produced from nickel and silver carbonate, as described in section 2.2. The silver content amounted to about 10%. Surprisingly, the carbonyl nickel became an active catalyst for hydrogen conversion because of the presence of the silver. It was important to use silver carbonate as a base material because the silver resulting from this compound is very finely distributed. Because nickel and silver do not alloy, probably a large surface area remains in the nickel-silver structure for the hydrogen conversion. Since such electrodes, however, are thick and heavy (up to 2 g/cm^2) we discontinued our experiments on them.

PTFE-Bonded Electrodes

For plastic-bonded structures we used nickel catalysts containing a small amount of palladium (5%) (Fig. 2, curve 1). These Raney nickel-palladium catalysts were — as were aluminum containing alloys — processed together with the plastic binder; the aluminum was then dissolved out of the finished electrode. The activity of the catalyst is constant over a long period and can, if diminished after storage of the electrode in air, be completely regained by reduction. Although the amount of palladium is only small, we did not continue with the experiments because of the noble metal needed in the alloys.

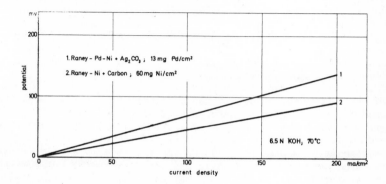

FIG. 2 *Galvanostatic current-voltage curves of PTFE bonded hydrophobic hydrogen electrodes.*

Plastic Electrodes with Raney Nickel

We developed hydrophobic plastic skeleton electrodes, consisting of polyethylene bonded, preactivated Raney catalyst and with a pore system produced by admixed carbon. The preactivated Raney nickel is not affect-ed by the preparation of the electrodes because 1) the compacting pressure applied is low, and 2) the active layer is "baked" at a fairly low temperature (150°C) and afterwards provided with a preformed porous PTFE layer. As the electrodes are used at atmospheric pressure, their mechanical stability need not be very high. Such electrodes have a thickness of less than one millimeter and their weight is below 0.1 g/cm^2. Their high activity is shown by curve 2 in Fig. 2.

If one does not pretreat the Raney nickel, its high activity does not remain stable for long. Especially at an elevated temperature (50° or 70°C) and high current densities (100 mA/cm^2 or more) the polarization increases slowly until after a few days a potential of about 160 to 180 mV is reached. Then the activity is more or less irreversibly affected.

Nonetheless, as shown by several researchers [3,4], one can stabilize the Raney catalyst by heat treatment after activation (400°C, hydrogen) without any loss in activity. We also found that tempering with argon brought successful results. In addition, the catalyst is in this case not pyrophoric; that is, it need not be carefully partially oxidized, as is the case with a Raney nickel containing hydrogen. The activity of this type of heat-treated catalyst remains constant.

3.3 OXYGEN AND AIR CATHODES

Silver Catalysts

Sinter Electrodes

As silver is known to be a very satisfactory oxygen catalyst, even in methanol-containing electrolyte, we prepared electrodes with Raney silver, according to the Justi method [5]. Our investigations then led us to silver catalysts generated from Ag_2CO_3, which are very active because reduction takes place under mild conditions (already at 200°C with hydrogen). The 1964 methanol cell mentioned in section 3.1 was provided with this type of catalyst in sintered air electrodes (for an overpressure of approx. 0.5 atm). The hydrophilic gas electrodes which were produced according to the method outlined in section 2.2 consisted of approx. 30 volume percent Ag_2CO_3 and 50 volume percent nickel; the remainder was a pore-forming salt. These electrodes were very active, as is indicated by Fig. 3, curves 1, but contained a lot of silver. The amount of silver can be reduced, if one keeps the fine-pore front layer of the electrode free of silver. This makes the fabrication more difficult, since the different compounds of the layers exhibit differing behaviour when sintered, which leads to warping and cracking during the cooling-down process. However, the high activity of the silver made from the carbonate was very encouraging.

PTFE Bonded Electrodes

Consequently we made attempts to use silver carbonate in plastic-bonded structures and succeeded in developing very satisfactory hydrophobic air electrodes suitable for methanol cells as well as for hydrogen cells, and working with the surrounding air (Fig. 3, curves 2). When working with air, their polarization is often only 50 mV greater than with oxygen (at 100 mA/cm^2). They remained nonwettable for a

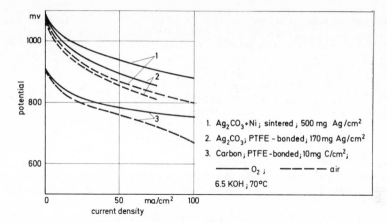

FIG. 3 *Galvanostatic current-voltage curves of oxygen electrodes.*

prolonged period of use both under load and at open circuit. At Robert
Bosch GmbH in Stuttgart, a methanol battery was built with similar
electrodes and has been working with interruptions for several years
[6,7]. Our 30-watt methanol battery, referred to in section 3.1 and
made in 1967, was also fitted out with this type of electrode [7]. The
battery was constructed from double cells in such a way that the air
electrodes of two adjacent cells were placed towards each other. A
space remains between them through which the ambient air is led by convection.

The amount of 170 mg/cm^2 of silver still seems too high, but it is
necessary in order to render the electrode mass conductive. It can still
be reduced if one uses a current collector made out of, for example,
nickel gauze.

Carbon Catalysts

It is well known that certain types of active carbon, and especially
mixtures of active carbon with carbon black, are fairly active oxygen
catalysts [8]. Curves 3 of Fig. 3 show the results obtained with one
of our plastic-bonded carbon electrodes; the potential is not as high
as in the case of silver already at open circuit, but the current-
voltage curve for pure oxygen is clearly less steep. (The steeper course
of curve 3 for air can be explained by the transport inhibitions on the
layer facing the gas side, which is not yet optimally developed.) A
decrease in the amount of carbon per unit area brings about only a very
small decrease in activity. At the present experimental stage, the
cathodes contain only 10 mg carbon/cm^2; this means that, given a current
density of 100 mA/cm^2, a true current density of about 10 amps per gram
is obtained.

Permeation of Electrolyte

Still, the use of carbon cathodes presents a severe problem. Our
electrodes have a composition similar in principle to that of silver
electrodes: one layer of plastic-bonded catalyst (here a mixture of
several carbons or carbon blacks) and a second layer of porous, sintered
PTFE. After a few hours under load (at *e.g.*, 100 mA/cm^2 and about
700 mV), droplets of electrolyte are always observed on the gas side. As
the hydrophobic layer itself is able to withstand a liquid pressure of
more than 0.1 atm, it is obvious that a pressure above this figure is

built up at flowing current in the interior of the electrode. The
phenomenon is well known to many investigators, but as we hardly ever
observed it in silver electrodes, it seems that its severity is due to
a special property of the carbon catalyst or of the carbon layer. There
are two possible interpretations for the occurrence of this phenomenon:
1) according to Wong [9] transport of material to the cathode always
takes place if the cations are obviously taking an active part in the
electrolytic conduction. Whether an accumulation of matter (*i.e.*, an over
pressure) can arise or remain in existence is dependent on the speed
that the exchange in matter by convection can take place in the porous
electrode material. If a sufficient number of wide pores are available
through which the matter, transported in the narrow pores to the place
of reaction, can stream back again to the electrolytic area, no pressure
will arise which will press the electrolyte through the back layer of th
electrode.

      2) An electroosmotic transport in the porous electrode can take
place if a difference in potential exists between the front and back of
the active layer, and if the wall matter of the pores is of such a kind
that a diffuse, movable electric double layer is formed which is pushed
through the electrode because of the difference in potential. Thus the
movement of the liquid must disappear in this case if no motive electric
field exists; that is to say, if the electrode body is so conductive tha
no potential difference can last between the front and back. The transport may be altered by utilizing other electrode matter because the
charge of the diffuse double layer is modified (different ζ-potential).
The fact that the above-mentioned cathodes with metallic, conducting,
silver skeleton and with metallic pore walls did not show the passage
of electrolyte, indicates that an electroosmotic transport is taking
place with the carbon cathodes. Naturally coarse pores, even in the
case of electroosmotic transport, would prevent the formation of a pressure. However, usually the conductivity and the activity of the electrodes is considerably reduced by the presence of coarse pores.

      Since the phenomenon occurs particularly at higher temperatures, one
could consider running a battery at as low a temperature as possible.
Fig. 4 shows the temperature dependence of the potential of air cathodes
(upper) and hydrogen anodes (lower) in an experimental cell. The curves
indicate the considerable loss of performance when the cell works at
unnecessarily low temperature.

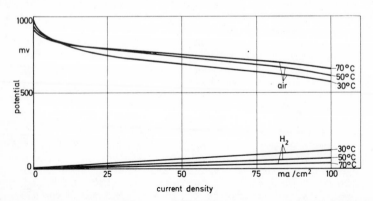

FIG. 4 *Temperature dependence of the polarization of hydrogen and
air electrodes in 6.5N KOH.*

We are, therefore, carrying on with the attempt to diminish the penetration of the electrolyte by changing the structure and conductivity, *e.g.*, by admixing conductive substances with differing grain sizes. Apart from this, we will alter the nature of the walls of the micropores by impregnating the bodies with various hydrophobic substances (paraffin, PTFE) or with wetting agents. We do not want to dispense with carbon as a cathode catalyst on account of its high activity, low weight and its relatively low price.

ACKNOWLEDGEMENT

Some of these results were acquired while working on a contract for Robert Bosch GmbH, Stuttgart, Germany. We wish to express our thanks to them for giving us their kind permission to publish results of these tests.

REFERENCES

[1] H. Binder, A. Köhling, W. H. Kuhn, W. Lindner and G. Sandstede, *Chemie-Ing.-Technik*, 40, 171 (1968).

[2] H. Binder, A. Köhling and G. Sandstede, "Raney Catalysts", these proceedings, p. 15.

[3] F. v. Sturm, H. Cnobloch, H. Nischik and M. Marchetto, *Abh. Sächs. Akad. d. Wiss. Leipzig*, 49, 165 (1968).

[4] F. P. Dousek, J. Jansta and J. Řiha, *Coll. Czechosl. Chem. Commun.*, 31, 457 (1966).

[5] E. Justi and A. Winsel, *Kalte Verbrennung — Fuel Cells*, Franz Steiner Verlag, Wiesbaden (1962).

[6] K. Brill, *Proceedings of the Third Internat. Symposium on Fuel Cells*, Presses Académiques Européennes, Brussels (1969) p. 39.

[7] W. Haecker, H. Jahnke, M. Schönborn and G. Zimmermann, *Metalloberfläche*, 24, 185-228, especially p. 193 (1970).

[8] K. V. Kordesch, these proceedings, p. 157.

[9] W. S. Wong, *AICE-Symposium: Transport in Electrode Processes*, Paper No. 61c, Washington, D. C. (November 1969).

STATE OF DEVELOPMENT OF HYDRAZINE FUEL CELLS, THEIR PROBLEMS AND POTENTIAL TECHNICAL APPLICATIONS

M. A. Gutjahr

Institut Battelle, Geneve, Switzerland

ABSTRACT

Hydrazine is one of the most interesting fuels used in fuel cells, because of its high energy density with respect to volume and weight, and its remarkable electrochemical activity under the conditions of anodic oxidation. Thus, there is no need to use noble metals as anode catalyst. Alkaline electrolyte, and in particular KOH, is to be preferred both in order to obtain good energy efficiency and to avoid corrosion problems. Hydrazine oxidation is always accompanied by gas evolution which occurs even at the rest potential, thus indicating a fuel consumption by chemical decomposition or by a mixed potential mechanism. Another influence decreasing the coulombic efficiency is the direct fuel oxidation at the cathode by a chemical short-circuit. To avoid these losses, special cell design and fuel concentration control are required. As oxidants, air, hydrogen peroxide and pure oxygen are used, examples of hydrazine batteries employing these depolarizers will be considered. At present, the most promising system seems to be the couple N_2H_4/air. Power units of this type have been built from a few watts up to several kilowatts with good efficiency and reliable operation. Despite the favorable properties of these constructions, it is doubtful that fuel cells using hydrazine could gain acceptance in the most important civil market of the future: the electrotraction, because of the high price of the fuel.

INTRODUCTION

Hydrazine is one of the most attractive fuels used in fuel cells. Three properties are mainly responsible for this dominating position:

High energy density in relation to volume and weight.
Good electrochemical activity under anodic oxidation.
Reaction products that do not contaminate the electrolyte.

In an alkaline solution, the overall reaction for converting energy by anodic oxidation can be described as follows:

$$N_2H_{4_{aqu.}} + 4OH^- \longrightarrow N_2 + 4H_2O + 4e^- \quad E_0 = -1.16V$$

$$O_2 + 2H_2O + 4e^- \longrightarrow 4OH^- \quad E_0 = +0.4\ V$$

$$N_2H_{4_{aqu.}} + O_2 \longrightarrow N_2 + 2H_2O \quad \Delta E_0 = 1.56\ V$$

(E_0 — theoretical equilibrium potential, calculated from the value of free energy.)

The potential of the hydrazine electrode should be at about 300 mV more negative than the potential of a reversible hydrogen electrode in the same electrolyte. Using an oxygen electrode as cathode, the theoretical E.M.F. should be 1.56 volt.

The available cell voltage is lower than this theoretical E.M.F. because both the air electrode and the hydrazine electrode deviate considerably from the thermodynamic equilibrium potential. The reasons for the potential shift of the air electrode are well known and need not be discussed here. There exist several possible explanations of the hydrazine electrode shift.

First of all, for every hydrazine catalyst with low hydrogen overpotential, a mixed potential can be expected due to the partial currents of the hydrazine-oxidation and electrolyte reduction:

Mixed Potential Mechanism

Anodic Reaction:

$$N_2H_4 + 4OH^- \longrightarrow N_2 + 4H_2O$$

Cathodic Reaction:

$$4H_2O + 4e^- \longrightarrow 2H_2 + 4OH^-$$

On the other hand, it may be assumed that the open circuit potential is determined by hydrogen formed in a purely chemical decomposition, according to:

Chemical Decomposition Mechanism

$$N_2H_4 \xrightarrow{\;\;\text{Electrode}\;\;} N_2 + 4H_{ad}$$
$$\text{Surface}$$

$$2H_2 \rightleftharpoons 4H_{ad} + 4OH^- \rightleftharpoons 4H_2O + 4e^-$$

When the adsorbed hydrogen is electrochemically oxidized by anodic current flow, the decomposition rate may be increased.

As yet, it is not certain which of the mentioned possibilities is applicable for this system, though extensive research has been carried out to investigate this problem [1-6].

Both mechanisms mentioned agree with a gas evolution at any potential of the electrode. The rate of the gas evolution and its composition depends upon the potential value as shown in Fig. 1. At the OCV the rate reaches a minimal value. Here the composition corresponds to the formula of the hydrazine (33 1/3 vol. % N_2, 66 2/3 vol. % H_2). Shifting the potential to a more positive position increases the reaction but lowers the hydrogen concentration in the gas. At high cathodic overpotential, the reaction rate is also accelerated whereby the exhaust gas contains more hydrogen the more negative the chosen potential.

The gas evolution at open-circuit potential is controlled by several parameters, the most important of which are the hydrazine concentration in the electrolyte and the kind of material used as electrode catalyst.

The decomposition rate is proportional to the fuel concentration and the reaction is accelerated with increasing concentration. The influence of several different catalysts on the hydrazine decomposition at the open-circuit potential is illustrated in the following table:

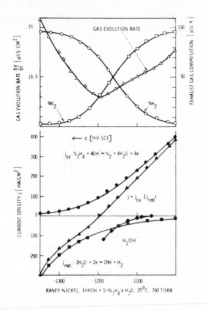

FIG. 1 *Influence of potential on the current density, gas evolution rate and composition.*

In this table the hydrazine decomposition was measured by following the gas evolution rate, which was then converted into current density. The decomposition rate is very high for Raney Nickel; lower, but not negligible for both the platinized electrodes. To keep such losses low it is necessary to maintain the fuel concentration as low as possible according to the current requirements.

Hydrazine is electrochemically active in both basic and acid solutions. This is shown by means of the triangular sweep potential diagrams measured on smooth platinum (Fig. 2). The potential velocity for these two measurements was 1 Volt/sec.

TABLE I

Hydrazine Decomposition at Rest Potential

Metal	Decomposition Rate (mA/cm^2)	Potential $(mV:H_{\frac{1}{2}}/OH^-)$
Raney-Nickel	56	- 148
Platinized Rhodium	12	- 28
Platinized Platinum	8	+ 50
5nKOH, 20°C, $1mN_2H_4$		

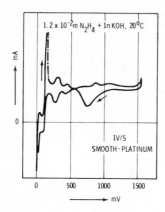 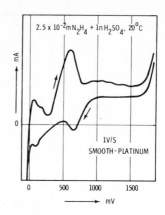

FIG. 2 *Triangular potential sweep characteristics of hydrazine in*
alkaline and acid electrolyte (after Grüneberg, Vielstich, Gutjahr [7,8]).

These current-potential diagrams show that the oxidation of the fuel
takes place at lower potentials in acid solutions than in alkaline. A
basic electrolyte should therefore be preferred. Another important
reason for this preference is the superior corrosion resistance of most
of the electrodes in basic solutions. The best known basic electrolyte
is potassium hydroxide.

The electrochemical activity of several transition metals often
applied as catalysts in fuel cells is shown in Fig. 3. Here a 1N
potassium hydroxide electrolyte was used and the reversible hydrogen
potential in the same electrolyte was taken as reference potential.
The diagram shows that at each metal tested the anodic hydrazine oxidation is possible. There are, however, quite important differences
depending upon the starting potentials of the reaction. The most interesting materials are obviously the two active nickel electrodes. The
anodic branch of the current-potential diagram of these electrodes
begins at potentials about 100-140 mV negative to the potential of the
reversible hydrogen electrode in the same electrolyte.

It is important to note that even gold and silver show a noticeable
activity at higher potentials. Silver is frequently used as a catalyst
for the air electrode, and an unwanted N_2H_4/O_2 mixed potential might
shift the potential of the positive electrode to a more negative position
and reduce the Ah-efficiency of the N_2H_4.

To avoid these losses it is necessary either to separate the positive
electrode by a diaphragm, or to design a special anode to which the
hydrazine is supplied from the back. Both techniques have been applied
successfully.

It is advisable to add no more N_2H_4 to the electrolyte than necessary
according to the current needed. If the N_2H_4 quantity becomes too small,
irreversible damages to the catalyst can be caused by corrosion, especially when nonnoble metal catalysts are applied.

For these reasons the fuel concentration should be controlled, for
instance by means of a potential measurement technique. A different
method uses the limiting current to check the concentration. At high
overvoltage and in a certain range of the concentration, the hydrazine

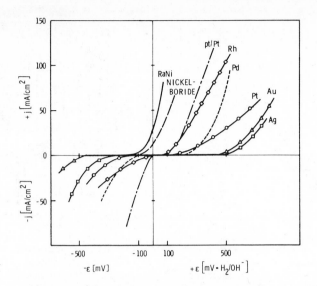

FIG. 3 *Electrochemical activity of several metals with respect to the hydrazine oxidation.*

oxidation is limited mainly by the fuel diffusion to the anode. In this case the corresponding current becomes proportional to the concentration of hydrazine [10,11].

   Since the proportion between current density and hydrazine concentration is restricted by other kinetic parameters, concentration control using this technique is only possible in the range up to approximately one molar. At higher concentrations no defined limiting current is observed. Furthermore, this method can only be used when all the parameters influencing the diffusion process are reproducible over a long period; these parameters are diffusion coefficient, thickness of diffusion layer and activity of the electrode. To assure the constancy of these parameters, laminar flow conditions and constant temperature are unalterable. The construction of the measuring device must therefore be chosen with great care. Recently, a description of such a system was published by Dreher, Schwindke and Grüneberg [11]. Employing the limiting current at gold electrodes, their device allows the control of fuel concentrations up to 0.6 molar at temperatures between 20-40°C. The potential of the test anode is kept at 1200-1300 mV (with respect to the hydrogen potential in the same electrolyte). By means of a potentiostat, constant diffusion conditions are guaranteed using a special system design consisting of a membrane pump, gas separator, gas chamber and flow controller. An NTC resistance in the electrolyte circuit serves to scan the temperature and, connected with an electronic device, to compensate the influence of variations on the diffusion coefficient. This compensation is especially necessary at higher fuel concentration as illustrated in Fig. 4. A cross section of the whole control device is given in Fig. 5.

   As mentioned before, the system described allows concentration control only up to 0.6 molar. Other authors, employing the same limiting current technique but a different cell design, reached concentration control just up to 1.2 molar [12].

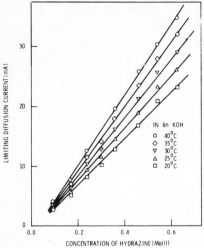

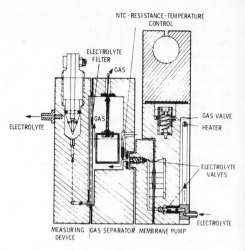

FIG. 4 *Limiting current depen-*
dence from the hydrazine concentration
at different electrolyte temperatures
(after Dreher [11]).

FIG. 5 *Cross section of the*
concentration measuring system
proposed by Dreher, Schwindke
and Grünberg [11].

Good control and exact supply measurements enable the construction
of a hydrazine cell with a current efficiency of more than 90% corresponding to 90-100 Ah/mol fuel [9]. Furthermore, the formation of
ammonia, which can be an undesirable side reaction especially at the
positive electrode, is reduced when the hydrazine concentration is no
higher than that which is necessary to feed the current requirements.

COMPLETE CELLS

Since the development of hydrazine cells in the 1950's, there has
been considerable progress in this field. The principle problems of
hydrazine fuel cells, such as the choice of catalyst, electrolyte, etc.
have been successfully solved. Recent research, therefore, concentrates
almost entirely on the improvement of the automation and the auxiliary
aggregates.

Today nearly all highly perfected cells work with a basic electrolyte
(KOH with concentrations between 5 and 12 molar), whereas hydrazine concentrations up to 1.5 molar are used (with the exception of the UCC-Bi-
cell primary battery).

As catalyst for hydrazine oxidation, skeleton nickel (like Raney
nickel, nickel boride), palladium-doped nickel and platinum on nickel
are used. With these catalysts, current densities up to some hundreds
of milliamperes per square centimeter can be drawn from the electrodes.
Electrolyte temperature of hydrazine fuel cells varies in a large range;
most of these power plants work between 40 and 80°C. To facilitate the
removal of the reaction products (water and nitrogen) and the temperature
control of the system, the electrolyte is normally conducted in a circuit
on which devices are affiliated such as a water evaporator, gas separator,
cooler and the systems needed for maintaining the fuel concentration.
Sometimes the circulations is not driven by pumps but by use of the gas
lift due to the nitrogen bubbles generated by the hydrazine oxidation.

As oxidants, oxygen, air and hydrogen peroxide are available. The
application of nitric acid, proposed in the early sixties by the Monsanto
Corporation, could not succeed. This failure is caused by corrosion
problems in the acid-cathode compartment (the anode compartment contained 10 molar sodium hydroxide as electrolyte). H_2O_2 is used both
directly as liquid oxidizer and indirectly as an oxygen donator which
is decomposed to feed the electrode with gaseous oxygen. Though H_2O_2
already reacts electrochemically with high current densities at low
temperatures, and though the problems of the construction of electrodes
with a three-phase boundary are abolished by the use of H_2O_2, there are
many disadvantages to its use. The principal disadvantage is its price
in comparison with air and compressed oxygen. Another disadvantage is
the large quantity of water associated with the low concentration of
commercial H_2O_2 solutions.

Finally it is necessary to separate the anodic from the cathodic
compartment and this implies the occurrence of two separate electrolyte
circuits. The high decomposition rate at elevated electrolyte temperatures with a permissible range of 40°C-45°C is also disadvantageous.
Summing up, the combination N_2H_4/H_2O_2 is useful for special applications
only — for example, when high performances are wanted immediately at low
temperatures.

In the following, the principle of hydrazine fuel-cell systems will
be presented by means of some typical examples.

OXYGEN AS OXIDANT

The application of gaseous oxidants necessitates the construction
of a three-phase boundary cathode, where the gas is fed to the electrode
from behind. To obtain optimal electrochemical performances and a satisfactory durability, a well-defined pore structure and careful hydro-
phobisation of the gas side of the electrode are required. As mentioned
before, it is also useful to supply the anode from behind. The principal
of such a cell construction is illustrated in Fig. 6. This arrangement,
proposed by Tomter and Antony [13] (Allis-Chalmers Manufacturing Co.,
Milwaukee) has been such a success that it is used today in many other
laboratories. Anode and cathode are separated by an asbestos diaphragm
containing electrolyte. The grooved walls of the cell press the electrodes onto the diaphragm. Fuel and oxygen are supplied to the porous
electrodes from behind. Thus, neither oxygen nor fuel can in practice
reach the opposite electrode. The losses by unwanted "chemical short
circuit" are practically eliminated and current efficiency reaches
values between 75-90%. The load performances of this type of power unit
are shown in Fig. 7. This battery was constructed by Grüneberg [14].
In the diagram, the characteristics for two different operating conditions are specified.

The curves A give the operation at maximum yield, while the curves
B were measured after the cell was adjusted to optimum fuel efficiency
(economic operation).

To get maximum power load it is necessary to increase the pressure
of oxygen and to work with an excess of fuel. Under these working
conditions the terminal voltage is a function of the current load.

In the economic operation, however, the hydrazine concentration is
controlled in such a manner that a terminal voltage of exactly 24 volts
results. In this case, the terminal voltage is independent of load.

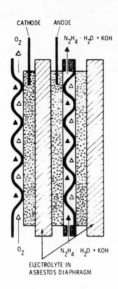

FIG. 6 *Principal arrangement of the N_2H_4/O_2 fuel cell.*

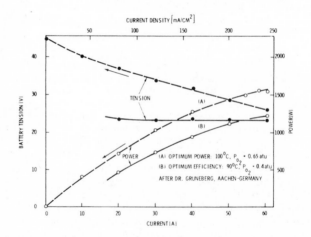

FIG. 7 *Load characteristic of 1.5 kW hydrazine/oxygen power unit constructed by Arbeitsgruppe Dr. Grüneberg, Aachen [14].*

To enable concentration control, the anode potential of each cell is measured using mercury oxide reference electrodes. By means of these potentials, the hydrazine concentration is determined and the fuel-supply device operated so that the Faraday efficiency in economic operation reaches more than 80%. Grüneberg takes flame-sprayed Raney nickel as anode catalyst and silver-plated sinter nickel as cathode catalyst. (The geometrical surface of the electrodes is about 250 cm^2). The whole system consists of 4 modules, each containing 10 single cells in series. The weight of the module (without electrolyte) is 6.300 kp, the volume 2.17 litres. Electrolyte circulation is done by gas lift and thermosyphon principle. Operation temperature is rather high: 90-100°C. The removal of reaction water by evaporation serves at the same

time to cool the electrolyte. To clean the exhausted gas, cartridges
containing excelsior saturated with phosphoric acid are applied.

The feasibility of N_2H_4/oxygen power plants as an energy supply in
electrical vehicles was demonstrated spectacularly by Allis-Chalmers.
Examples include the well-known golf car (3 KW), the fork-lift truck
(6 KW with fluid oxygen storage) and the one-man submarine driven by
750 W battery.

AIR AS OXIDANT

Among the oxidants suitable for hydrazine cells special attention
should be paid to the air-oxygen. Using air, it is possible to keep
the volume and weight of the cell low. In general, only a slight overpressure of air is applied. Therefore, the energy needed to drive the
air ventilation is not very high. There is no problem involved in
purifying the air of CO_2, which can poison the basic electrolyte. In
many cases, mainly for batteries of small performance, this purification
is not even necessary.

Cell constructions are similar to the Antony-Tomter arrangements
already mentioned. Also, the same electrode catalysts are applied.

Today, hydrazine/air electrodes cover the range from several watts,
for use as a transportable energy source, to several KW which may be used
as a power source in electrical vehicles.

As an example of such a transportable power set, the 300 W system
developed by Union Carbide [15] will be outlined because of its
simplicity in service and reliability in operation. Porous nickel
electrodes activated with palladium (1 mg/cm^2) are used as anodes. The
cathodes consist of special nickel/active-carbon catalysts (metal composite) which are free of noble metals. Compared with the Allis-Chalmers
cell previously mentioned, this battery does not contain a diaphragm.
The mixture of electrolyte/fuel is pumped across between both electrodes.
In order to avoid a breakdown at the cathode, the N_2H_4 concentration
applied must be comparatively low. This concentration is controlled by
an indicator electrode (or sensory device) which is incorporated into
the electrolyte circuit. 100% N_2H_4 hydrate from a storage tank can be
injected to the circulating electrolyte. The fuel concentration is 0.3%
at the entrance of the cell and 0.2% at the exit.

At current densities about 60 mA/cm^2 the cell voltage reaches 0.7
volts. The efficiency for the conversion varies from 65% to 75%: the
working temperature being 55°-65°C. The oxygen electrode is supplied
with four- to six-fold excess of air. As the CO_2 is not removed from
the air, the electrolyte must be changed after 100 hours of working.

The energy density of this system is about 230 Wh/kg or 145 Wh/L,
the corresponding power density being 20 W/kg or 12 W/L. These figures
relate to the complete system, including a fuel storage for 12 hours.

The development of this power unit was made under a U. S. Army
contract. A possible civil application was shown by Dr. Kordesch: he
replaced the engine of a motorcycle by an electric drive. The primary
battery consisted of two 400 W hydrazine/air modules as described before.
The hydrazine concentration was raised to 0.5% and the air was roughly
cleaned by "sodalime". A storage battery of Ni/Cd type could be connected parallel to assist in acceleration. The vehicle reached a cruis-
ing speed of about 60 km/h with a consumption of 1.5L $N_2H_4 \cdot H_2O$ for 100
km.

Greater power units have been developed by the Monsanto Research Corporation [16] for application as energy sources in electrically driven cars. The construction of these batteries corresponds to the Tomter-Antony type with an asbestos diaphragm between the electrodes. The nickel anode is dotted with palladium, while the cathode consists of carbon, activated with platinum. So far, several modules have been constructed with a load performance of up to 5 kW at a current of up to 180 A and a voltage of 28 V.

Four of these 5 kW modules have been assembled to produce a power supply of 20 kW. This unit was used as an energy source in an electrically powered jeep of the U. S. Army. The total weight of this block was 240 kg, and thus a performance density of 12 kg/kW was reached.

Recently the development of a new kind of N_2H_4/air battery was reported [17,21], which corresponds to the chemical rechargeable primary battery type. These batteries possess neither an external electrolyte circulation nor an air-purification system. The electrodes do not contain noble metal catalysts. In comparison with other hydrazine batteries, the fuel concentration is strongly increased. The electrolyte consists of equal parts of 12 molar potassium hydroxide and 64% hydrazine. To prevent the fuel decomposition at the cathode, a new cell construction with barrier layer between anode and cathode was demanded (Fig. 8). This barrier should have high diffusion resistance to hydrazine, but on the other hand, a good hydroxyl-ion conductivity. As yet, not all of the problems of this layer could be solved. Nevertheless, the present state of development of these batteries is already promising. The current rate is sufficient whereas the efficiency of energy conversion is lower than in other hydrazine cells. The high concentration strongly promotes the self-decomposition. Durability can reach more than 100 chemical loads.

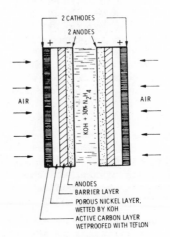

FIG. 8 *Dual sandwich construction for chemical rechargeable hydrazine/air primary battery (Kordesch [17]).*

H_2O_2 AS OXIDANT

The problems arising from the application of H_2O_2 have already been briefly discussed. As mentioned before, batteries with this oxidant are only efficient for special applications; for example, if only a small air quantity is available (submarine, space vehicle) or if high efficiencies are wanted immediately after starting power operation. Contrary to the systems described previously, the N_2H_4/H_2O_2-cell systems need a double electrolyte circulation.

Fig. 9 describes the principle of such a N_2H_4/H_2O_2 cell, designed by Grüneberg [18]. Anode and cathode are separated by an asbestos diaphragm. The walls of the cell consist of special formed soldered nickel sheets. Its structure is built up of a system of internal tubes which are connected to a separate water cooling circulation. In addition, the profiles of the sheet provide for the correct spacing of the electrodes and current collectors. They also connect the cathode of the k-th cell with the anode of the k + 1st cell, resulting in a series connection for the cells of each unit. Catholyte and anolyte are supplied to the electrodes from behind. The cathodes are made of silver-plated nickel tissue (70-200 mg Ag/cm^2), the anodes of flame-sprayed Raney nickel on the same support. The electrolyte consists of 6N KOH to which 0.5 Mole N_2H_4 is added for the anolyte and 0.8-1N H_2O_2 for the catholyte.

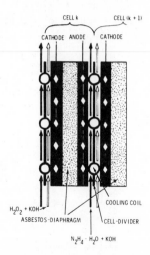

FIG. 9 *Cell arrangement of the N_2H_4/H_2O_2 battery with two electrolyte circuits (after Grüneberg [8]).*

In Fig. 10 the polarization characteristic of the described cell at 40°C operation temperature is presented. The feeble polarization of the cathode at current densities of the order of 800 mA/cm^2 should be underlined. Such current densities are obtainable only when using H_2O_2 as oxidant.

As anode catalyst, different materials have been tested and their polarization characteristic plotted in the diagram. It is clearly shown that the Raney-nickel catalyst without noble metal doping is much more interesting than the platinized or palladinized one. Even when the Raney nickel was oxidized by air (curve c) the polarization curve is positioned in a more cathodic range than the noble metals. It is possible to regenerate such an air-oxidized electrode by a treatment with hydrazine in potassium hydroxide at elevated temperature: the difference in activity between original (curve a) and regenerated (curve b) Raney nickel is very small.

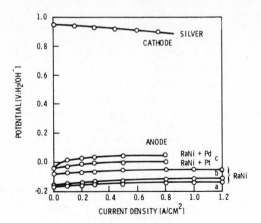

FIG. 10 *Polarization characteristics of the electrodes applicable in N_2H_4/H_2O_2 batteries (after Grüneberg [8]).*

The load performances of a 850 W module, consisting of 13 cells, are shown in Fig. 11. Under standard operation (850 W) this power unit delivers 77 amperes at a tension of 11 volts.

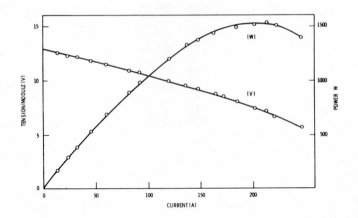

FIG. 11 *Load characteristics of a 1.5 kW N_2H_4/H_2O_2 power unit (after Grüneberg [8,14]).*

The maximum power is 1.5 KW at an EMF of 7 volts and a current of 210 A. At nominal load the battery uses 90% of the N_2H_4 capacity and 40% of the H_2O_2 capacity. The working temperature is 30°C at the entrance and between 40°-45°C at the exit. For higher loads, the Farada efficiency decreases because of the increasing temperature of the electrolyte, which is causing a more rapid H_2O_2 decomposition.

Under steady working conditions, the systems described reach a specific power output of 150 W/Kg, respectively 200 W/litre.

H_2O_2/N_2H_4 systems have also been developed in other laboratories with principles of cell construction similar to those already described.

Sometimes other catalysts were used for the anode, for example Raney Cobalt in the Alsthom cell [19,20].

CONCLUSION

This short survey on the present state of the development of N_2H_4 fuel cells shows that the electrochemical problems of these systems are practically solved. It is possible today to build power units from a few watts up to several kilowatts with good efficiency and reliability in operation.

The most interesting of these batteries use air as oxidant, and do not need noble metals to catalyze the electrochemical reactions.

A lifetime of several thousand hours is no longer utopic. Despite these favorable properties, it is doubtful that the hydrazine cell is the battery of the future, because of the high price of approximately $2/kg. As a result of this, the price of 1 kw hour for the best developed system with the best efficiency is about $1.50. This price does not include the cost of material for the cell construction. If the hydrazine cells should gain acceptance in the best and widest-spread potential market of the future — namely electrotraction — the price of the fuel would have to be reduced to $0.15 to $0.20/kg.

The future of small hydrazine bicells can be viewed more optimistically. This chemical rechargeable battery will gain acceptance mainly for those applications where small, transportable units demand high performance, for instance in the field of television receivers.

ACKNOWLEDGEMENT

A substantial amount of the information discussed in this paper was contributed by Dr. G. Grüneberg, *et al.* (Arbeitsgruppe, Aachen), Dr. K. V. Kordesch (UCC, Parma, Cleveland/Ohio) and Prof. W. Vielstich (Universität Bonn). Dr. R. Dieberg (BMVtg, Bonn) was kind enough to permit the publication of some research work sponsored by the German Ministry of Defense.

REFERENCES

[1] T. O. Pavela, *Suomen Kemistilehti*, 30 B, 240 (1957).

[2] G. Susbielles and O. Bloch, *Compt. Rend. Ac.d.Sciences*, 255, 685.

[3] A. J. Bard, *Analyt. Chem.*, 35, 1602 (1962).

[4] S. Szpak, P. Stonehart and T. Katan, *Electrochim. Acta*, 10, 563 (1965).

[5] B. P. Nesteroy and N. W. Korovin, *Electro. Chimia.*, 11, 1296 (1966).

[6] M. Gutjahr and W. Vielstich, *Chem. Ing. Tech.*, 40, 180 (1968).

[7] M. Gutjahr, *Diplomarbeit*, Bonn (1966).

[8] G. Grüneberg, private communication.

[9] W. Vielstich, *Brennstoffelemente*, Verlag Chemie, Weinheim/Bergstr. (1965).

[10]   J. Hölzel, thesis, Technische Hochschule Aachen (1968).

[11]   J. Dreher and P. Schwindke, *Messtechnik*, 78, 242 (1970).

[12]   H. Kohlmüller, H. Cnobloch and F. v. Sturm, paper presented at the Power Sources Conference, Brighton (1970).

[13]   S. S. Tomter and A. P. Antony, *Fuel Cells*, A.I.Ch.E., New York (1963).

[14]   G. Grüneberg, Arbeitsgruppe der Fraunhofer Gesellschaft Aachen, with kind permission of Dr. Dieberg, BMVtg Bonn, Germany.

[15]   K. Kordesch, *Elektrotechnik und Maschinenbau*, 86, 451 (1969).

[16]   E. A. Gillis, *Proc. 20th Ann. Power Sources Conf.*, Atlantic City (1966) p. 41.

[17]   K. V. Kordesch and M. B. Clark, *Proc. 24th Ann. Power Sources Conf.*, Atlantic City (1970) p. 207.

[18]   G. Grüneberg, private communication.

[19]   B. Warzawski, *Entropie*, 14, (1967).

[20]   W. Vielstich, *Fuel Cells*, D. J. G. Ives (trans.), Wiley-Interscience, London (1970).

[21]   J. Perry, *Proc. 22nd Ann. Power Sources Conf.*, Atlantic City (1968

OUTLOOK FOR ALKALINE FUEL CELL BATTERIES

K. V. Kordesch

Union Carbide Corporation, Parma, Ohio

ABSTRACT

Of all the types of fuel cells, alkaline cells have received the greatest attention in past research and development work because of their advantageous power output and relatively low cost. However, serious doubts have been voiced as to whether they may ever become practical batteries because they cannot operate directly on hydrocarbon fuels; even for air operation, a CO_2 scrubber is needed. The discovery of catalysts containing little or no noble metals and operating in non-alkaline electrolytes now provides real competition for some alkaline systems, but as with any comparative judgment, the type of cell, kind of fuel, and cost ratios must be properly appraised. A picture of the present state of the art, efforts, and trends may be obtained by studying published military contracts and industrial work during the years 1969-70. It will be shown that CO_2 removal from the air is easy; the resulting improvements warrant the expense, and once more make the alkaline system a preferred choice. This is particularly true when one considers a simple means for shutting down and starting up the fuel cell systems so that very long life expectancies are possible, even on intermittent-duty cycles. Applications ranging from portable batteries to vehicle power systems are discussed.

I. INTRODUCTION

Alkaline fuel cells are technologically the most developed fuel cell type today to a large extent because of their early selection for space and military batteries. However, this choice was based on real advantages when compared with other systems:

High power output at ambient temperatures and low pressures;

Use of conventional materials (carbon, nickel, silver, plastics);

Available catalysts (noble metals, Raney alloys);

Easily oxidizable fuels such as H_2, N_2H_4 can be used;

Air operation with no precious metal catalyst is possible.

All of these advantages are still recognized, but alkaline cells also possess the following disadvantages:

Hydrocarbons cannot be used directly in fuels, and

Air operation requires a CO_2-scrubbing system.

It is also frequently pointed out — first as an advantage, then later as a disadvantage — that hydrocarbons, alcohols, or ammonia can be (or must be) reformed in order to be utilized in alkaline-hydrogen cells.

II. RECENT PROGRESS AND TREND OF FUEL CELL ACTIVITIES

ALKALINE CELLS

The perfecting of electrodes requiring little or no noble metals was probably the most promising progressive step in alkaline-type fuel cell manufacture during the last few years. It projected wider applications for fuel cells. For alkaline electrolytes, carbon-air electrodes are known to operate at high current densities without noble metals [1]. Carbon-hydrogen electrodes perform satisfactorily with as little as $0.1 \ mg/cm^2$ noble metal content. Recent work also indicates that different types of thin porous metal anodes using Raney alloys can match the carbon anode performance with respect to output-per-weight figures [2,3]. The life expectancy of a low-cost combination cell employing a carbon cathode and metal anode looks especially good.

ACIDIC CELLS

Competition for the alkaline cells was greatly increased when the noble metal content was lowered to a few mg/cm^2 for hydrogen and hydrocarbon anodes operating in acidic electrolytes, and special promoters also increased life considerably [4]. Actually, such catalysts as tungsten carbide promise to eliminate the need altogether for Pt metals as anode catalysts [5], but oxygen electrodes in acidic electrolytes still require precious metals at high-level loadings. Organic catalysts such as phthalocyanine [6] and the more recent quinone types [7] make the future look even more promising. Clearly, acid-type, low-temperature fuel cells are progressing at a high rate.

This trend is also confirmed by the only truly large commercial fuel cell endeavor remaining in the United States; $i.e.$, the TARGET Program [8]. In this project the all-gas home is the goal by using impure (reformed) hydrogen as the fuel and air as the oxidant. The platinum cost problem has not yet been solved, but once the level is under $1 \ mg/cm^2$, a large field of applications may open up, especially when lifetimes in the 10,000's of hours are being proclaimed.

The involvement of industry in fuel cell work is guided largely by Government procurement. This is understandable when one considers that research and development work in this field is expensive, and until a larger production level is achieved, the product cost will remain high. It is interesting, therefore, to study the U. S. Government expenditures for 1969 and 1970 in this respect. Table 1 lists the contractual work done for the Army, Navy, Air Force, and for NASA in 1969-70, as reported by Power Information Center, University of Pennsylvania, and summarized by the author.

The selectivity of the contracts when related to the different Government agencies using these fuel cells is obvious. The Army has shifted almost completely to (acidic) hydrocarbon fuel cell work, and has supported alkaline systems only in connection with metal-air cells and work on hydrazine cells. The Navy has sustained only work on fuel cell designs for the Deep Submergence Search Vehicle (DSSV) System. The Air Force has spent its money on H_2-O_2 (alkaline) cells (regenerative type and high-current output studies). NASA is also exclusively concerned with H_2-O_2 (alkaline) cells for work primarily related to the Manned Orbiting Laboratory (MOL) and space-shuttle projects, plus studies of the capillary-matrix system, with the latter now being finished.

TABLE I

Government Expenditures on Fuel Cells 1969-70
(Amounts in $1,000)

1969 (1 year)	Army	Navy	Air-Force	NASA	Total
H$_2$-O$_2$ (Air)	---	556	634	161	1351
Hydrazine	75			21	96
Metal-Air	230				230
Hydrocarbon	40				40
H$_2$ & Hydrocarbon Acidic	461				461
Totals	806	556	634	182	2178

1970 (10 mos.)	Army	Navy	Air-Force	NASA	Total
H$_2$-O$_2$ (Air)	50	---*	226	889	1165
Hydrazine	110				110
Metal-Air	189				189
Hydrocarbon	44				44
H$_2$ & Hydrocarbon Acidic	997				997
Totals	1390	---*	226	889	2505

Government sponsored Contracts on Fuel Cells and Metal Air Cells, listing based on Project Briefs from the Power Information Center, University of Pennsylvania (through October 1970).

*DSSV effort for 1970 not yet listed.

However, the total amount of effort is rather small when compared with expenditures during the peak period of fuel cell research and development work, as can be seen from Table II. This table lists the contractual work of the U. S. Government during 1962-65, as compiled by the Battelle Memorial Institute (Columbus, Ohio) in their "Final Report on the Advancement of Fuel Cell Technology," in 1965. When comparing Tables I and II, it should be noted that the section headed "Hydrocarbon" contains work in which emphasis during the earlier years was on alkaline systems with reformers. This effort was gradually shifted to acidic cells employing thermal crackers. The emphasis on metal-air cells (zinc-air rechargeable and replaceable anode types) appeared only in the last few years as an outgrowth of fuel cell cathode work. In the earlier compilations, zinc-air cells are listed under "Other Fuels."

III. DISCUSSION OF ALKALINE FUEL CELL SYSTEMS

The H$_2$-O$_2$ system is favored for space-oriented projects. Since fuel and oxidants are specialized for this application, and performance

TABLE II

*Government-Sponsored Contracts on Fuel Cells**

	Estimated Value of Contracts, million dollars						
Totals	Hydrogen	Hydrocarbon	HTFC	Biocells	Other Fuels	Publi-cations	Totals
1963	10.82	2.61	0.89	0.93	0.60	----	15.85
1964	11.75	2.45	0.43	0.20	0.74	0.03	15.60
1965	9.44	3.17	0.22	0.20	0.43	0.03	13.49

	Estimated Value of Contracts, million dollars, by Government Agency							
Totals	Army	Navy	Air Force	NASA	Department of Interior	AEC	ARPA	Totals
1961	1.3	1.8	2.4	0.1	---	---	---	5.6
1962	2.7	0.7	1.5	8.9	1.0	---	0.8	15.6
1963	4.0	1.5	1.0	8.5	0.5	0.2	0.2	15.9
1964	1.6	0.2	0.4	9.6	0.5	---	3.3	15.6
1965	1.6	0.4	0.3	7.8	---	---	3.3	13.5

Based on Project Briefs from the Power Information Center, University of Pennsylvania, as referenced by Battelle Memorial Institute of Columbus, Ohio.

is expected at a premium price, the outlook for alkaline cells is pre-dictably good in this field [9].

Similar considerations are due for undersea explorations [10]. An exchange of hydrogen for hydrazine, for example, for better adaptation to the environment may be expected [11]. For applications on earth, however, operation on air is all-important, and many firms have con-structed large hydrogen-air systems [12]. Factually, alkaline electro-lytes are sensitive to CO_2 in the air, but this disadvantage is far more serious for such small units as zinc-air batteries which have no means of protection against carbonate formation once they are exposed to the atmosphere. While cathode arrangement, electrode thickness, current density, and electrolyte composition can change the time scale, as yet no satisfactory solution has been found for consumer demands. Con-ceivably, only a highly selective membrane with good O_2-transport features can remedy the situation for batteries intended for intermitten use; *e.g.*, in flashlights.

Larger sized batteries can employ a CO_2-scrubbing means (externally or internally), depending on their intended use. The best (low-cost) external scrubber is probably soda lime with exhaustion indicator. One kg of soda lime can clean 1,000 m^3/hr of air from 0.03% to 0.001% CO_2.

In some systems the caustic electrolyte itself can be used as an air cleaner, but KOH must be considered as an extra expenditure. For instance, it is sufficient to let the intake air pass through a labyrint of caustic-wet, deflecting plates to remove most of the CO_2 before the

air reaches the battery cathodes. Dissolved carbonate in the electrolyte does not affect electrode performance, but any carbonate deposited
in the pores of the electrodes reduces their "breathing" [13,14].

In one version of a hydrazine battery, the air electrodes had a
small portion of carbon-free, KOH-wet, porous nickel exposed to the in-
coming air: a 3/4-inch-wide strip cleaned CO_2 from the air sufficiently
to prevent damage to other portions of the electrodes [15]. Fig. 1
pictures a cut-through view of a hydrazine-air cell showing the arrangement of the CO_2-removal strip at the top edge of the electrode.

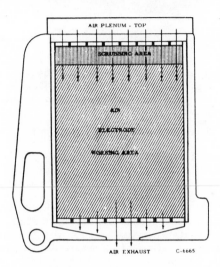

FIG. 1 *Cut-through section of a hydrazine-air cell showing cathode
with top-edge CO_2 scrubbing area.*

The need for fuel cells to operate with hydrocarbon fuels is undis-
puted for certain applications and for cost reasons. Compact, steam-
reforming units have been constructed [16]. Very effective CO_2 removers
operating on the circulating principle may be used to clean the reformer
gas. The use of Pd or Ag diffusers is technically simple but costly.
Acid cells have the advantage of being able to operate on less purified
gas.

The competition between alkaline and acidic fuel cells is also
related to current density. If the alkaline cell can produce twice the
current at the same battery cost, more expensive reformer and cleaning
systems can be afforded. Fuel cost is small when compared with the
installment expense — at least in medium-sized units. Large power
plants — if any are ever constructed with fuel cells — will have a better
chance with high-temperature cells of the solid-electrolyte type.

A stronghold for alkaline cells is to be found in the field of liquid
fuel cells. The use of such cells in parallel with secondary batteries
is probably the best way to combine large capacity (Wh/lb) with high
current density (W/lb). Good performance has been experienced with the
parallel operation of simplified hydrazine-air batteries with Ni-Cd
batteries to power communication equipment [17].

Technologically, the alkaline systems are still ahead: A hybrid
system consisting of a 6-kW hydrogen-air battery and a 4-kWh lead-acid
battery has been built into a four-passenger automobile [18]. Fig. 2

FIG. 2 *Electric car powered by a hydrogen-air/lead battery hybrid power system.*

shows the picture of this car (a converted Austin). Such a system must be powerful and reliable, require reasonably low maintenance, and be simple to operate. A life expectancy of several years will be needed to make such an endeavor practical.

Surprisingly, the shut-down operation appears to affect most the life of the fuel cell. When electrolyte is let out of the cells into a reservoir, no parasitic currents flow in the 90-volt battery, and the hydrogen supply can also be shut down and replaced by nitrogen (for purging to switch from hydrogen over to air, and vice versa). The electrodes do not seem to deteriorate at all during such a stand-by condition. Operating conditions can be established within a minute or two, with the help of the secondary battery. Also, very high polarization of the electrode is impossible since the PbO_2 cells serve as a protection against overload. This feature is shown in Fig. 3, picturing polarization curves measured during actual driving conditions.

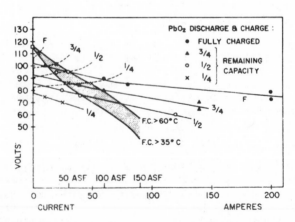

FIG. 3 *Test polarization curves of H_2-air/lead battery hybrid syste*

The peak output of the combined system is 20 kW (top speed: 55 mph) At an average speed of 40 mph, the 2,000-lb. car has an operating range of over 200 miles, and its lightweight hydrogen cylinders can be refueled (from high-pressure tanks) in about three minutes. The air is supplied by means of a blower, and soda lime (20 lb.) is used to remove the CO_2. A change of the air cleaner is expected to be routine maintenance every 1,000 miles. The use of an ammonia-dissociator unit would be a further improvement of this system (weightwise) [19]. In this application, of course, an acidic fuel cell developed to a similarly high standard might do better.

IV. OUTLOOK FOR FUEL CELLS

The outlook for alkaline cells is a part of the total fuel cell picture. Unfortunately, no extensive commercial applications have materialized despite the technical achievements of so many workers in this field. To a certain extent, the collapse of contract financing by U.S. Government agencies is responsible for the abrupt halt in fuel cell activities in this country, at a time when many of the systems face their highest development costs; *i.e.*, during the transition periods from pilot plant to large-scale production.

Only one commercial group has been able to pick up the enormous costs of a program which certainly will not bring profits in the near future, and in this case (TARGET Program) the financial load was distributed over a large number of firms. As far as automobile companies are concerned, it is far more promising for them to study means for reducing the pollution caused by combustion engines rather than to support battery development work on a large scale.

Therefore, the conclusion to which one must come is rather pessimistic. At the present time, fuel cell work will not be first-in-mind in battery development laboratories. However, in research laboratories, studies are going on and the trend here is undoubtedly in the direction of nonalkaline cells. Hydrocarbon-consuming electrodes for acid electrolytes need far more work, whereas alkaline cells of the H_2-KOH type have already passed the research stage.

REFERENCES

[1]   K. V. Kordesch, "Thin Carbon Electrodes," in *Proceedings of the 21st Power Sources Conference*, PSC Publ. Comm., Red Bank, N.J. (1967) p. 14.

[2]   H. Cnobloch, M. Marchetto, H. Nischik, G. Richter, and F. v. Sturm, *Proceedings of the Third International Symposium on Fuel Cells*, Presses Académiques Européennes, Brussels (1969) p. 203.

[3]   A. Winsel, *Electrochimica Acta,* 14, 961 (1969).

[4]   U. S. Army Mobility Equipment R & D Center, contracts with Engelhard Ind., DAAK-02-70-C-0517 (J. G. Cohn), DAAK-02-67-C-0219 (O. J. Adlhard) and Tyco Inc., DA44-009-AMC-1408(T), (S. B. Brummer).

[5]   K. Mund, G. Richter, and F. v. Sturm, *Extended Abstracts, 21st Meeting of CITCE*, Prague (1970) p. 347; and J. Heidemeyer, D. Baresel, W. Gellert, and P. Scharner, *Extended Abstracts, 21st Meeting of CITCE*, p. 335; See also: H. Böhm and F. A. Pohl, *Proceedings of the Third International Symp. on Fuel Cells*, Presses Académiques Européennes, Brussels (1969) p. 180; H. Binder, A. Köhling, W. Kuhn, W. Lindner, and G. Sandstede, *Nature,* 224, 1299 (1969); and *Energy Conv.*, 10, 25 (1970).

[6]   R. J. Jasinski, *J. Electrochem. Soc.*, 112, 526 (1965); and also H. Jahnke, and M. Schönborn, *Proceedings of the Third International*

Symp. on Fuel Cells, Presses Académiques Européennes, Brussels (1969) p. 60.

[7]   H. Alt, H. Binder, A. Köhling, and G. Sandstede, *Extended Abstract 21st Meeting of CITCE,* Prague (1970) p. 319.

[8]   M. V. Burlingame, "The Target Project" (Team to Advance Research for Gas Energy Transformation, Inc.) in *Fuel Cell Systems II, Proceedings of the 5th Biennial Fuel Cell Symposium,* Am. Chem. Soc., Washington D.C. (1969) p. 377.

[9]   National Aeronautics and Space Agency (NASA) contracts with Allis-Chalmers Mfg. Co., NAS-9-10443 and NAS-8-25609.

[10] P. E. Grevstad, *Proceedings 4th Intersociety Energy Conv. Eng. Conf.,* paper 17b, Washington D.C. (1969).

[11] R. J. Bowen, H. B. Urbach, D. C. Icenhower, D. R. Gormley and R. E. Smith, *ibid.,* p. 845.

[12] Large Hydrogen-Air Systems (Alkaline), Energy Conv. Ltd., England, *ibid.,* p. 1057, Generale d'Electricite & Cie, France, in *Power Sources Conf.,* by D. H. Collins (Ed.), (1970) and Shell Research, *5th Intersociety Energy Conv. Eng. Conf.,* Las Vegas (1970).

[13] K. V. Kordesch in *Hydrocarbon Fuel Cell Technology,* by B. S. Baker (Ed.), Academic Press, New York (1965) p. 17.

[14] H. R. Kunz, and M. Katz, *Extended Abstract No. 7, Electrochem. Soc. Meeting,* Atlantic City (Oct. 1970).

[15] G. Fee, and E. Storto, *Proceedings of the 23rd Annual Power Source Conf.,* PSC Publ. Comm., Red Bank, N.J. (1969) p. 8.

[16] H. I. Henkel, C. Koch, H. Mentschel, H. Stamm, and E. V. Szabo, *Proceedings of the Third International Symp. on Fuel Cells,* Presse Académiques Européennes, Brussels (1969) p. 274.

[17] K. V. Kordesch, *Extended Abstract No. 273; Electrochem. Soc. Meeting,* Los Angeles (May 1970); also *Proceedings of the 24th Power Sources Symp.,* PSC Publ. Comm., Red Bank, N. J. (1970) p. 207.

[18] K. V. Kordesch, *Extended Abstract No. 10, Electrochem. Soc. Meeting,* Atlantic City, N. J. (October 1970).

[19] O. J. Adlhard and P. L. Terry, *Fourth Intersociety Energy Conv. Eng. Conf.,* Washington, D. C. (September 1969).

SUMMARY OF DISCUSSION AND COMMENTS: SESSION CHAIRMAN III

D. P. Gregory

Institute of Gas Technology, Chicago, Illinois

The discussion on Dr. Kuhn's paper was confined to the cathode phenomenon which he described, in which electrolyte is "pumped" out through the air electrode pores when current is drawn. There were several different opinions as to the cause of this, and possible remedies, but it was agreed that the effect could not be allowed to occur freely in a practical system.

The discussion on Dr. Gutjahr's paper concerned the fuel cost of a hydrazine-air cell, and the possibilities of the basic cost of hydrazine being reduced. Dr. Bockris stated that he had been given to understand that a reduction in present costs by a factor of five was possible if the demand increased. It was generally agreed that the chemical industry had not been given sufficient incentive yet to develop cheap hydrazine, and the fuel cell industry would not pour large development funds into hydrazine cells until fuel cost problems had been resolved. Although toxicity was not such a severe problem as might be thought, the question was raised of what kind of limits the air pollution control people might place upon its use as a general fuel. The point was made that, even at present hydrazine costs, primary hydrazine-air cells were much cheaper than conventional Léchanché batteries.

The discussion on Dr. Kordesch's paper led into general discussion on alkaline fuel cell systems. There were many questions about the operating and building costs of the hydrogen-air automobile that he had described, together with its safety of operation. Many people raised misgivings about regulations which may already exist to prevent compressed gas or liquid hydrogen being used in private automobiles. Doubts were raised about the capability to operate alkaline systems on hydrogen from reformed hydrocarbons without expensive purification.

FUEL CELLS WITH ACID ELECTROLYTE

Chairman

E. J. Cairns

THE ELECTROCHEMISTRY PROGRAM AT USAECOM

S. Gilman

U. S. Army Electronics Command, Fort Monmouth, New Jersey

ABSTRACT

*USAECOM is active in research, development and advanced develop-
ment of a wide spectrum of battery devices. A few examples of our
activities are discussed in the paper. High energy batteries utilizing organic solvents, organic cathode depolarizers and fused
salts are in the research and development stage. Air-depolarized
batteries with hydrazine, methanol, zinc anodes are reaching the
stage of final development. Magnesium dry cells have been successful
for some time but are undergoing further refinement.*

INTRODUCTION

The U. S. Army Electronics Command has primary responsibility for
research and development relevant to all types of battery devices for
communication, surveillance, munitions and other military applications.

Past successes and future expectations are illustrated in Fig. 1.
The figure plots energy density, both initial and after storage, as
it has varied or is expected to vary with both time and the emergence
of new devices. In particular, a fourfold improvement over post-World
War II zinc batteries was achieved with the introduction of the magnesium dry cell into field use in 1968. Additional improvement in the
performance of that cell is expected over the next few years and a
large advance is expected upon introduction of organic electrolyte
cells into the field by 1974.

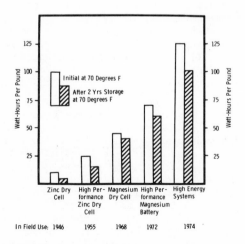

FIG. 1 *Improvement of primary battery performance.*

169

ADVANCED DEVELOPMENT

MAGNESIUM DRY CELL

The magnesium dry cell battery has been a key development. It was introduced into the field approximately three years ago and was highly successful in use with a small radio transmitter and receiver set and other portable equipment. The specific advantages of the magnesium battery over the more conventional zinc battery are depicted by Fig. 2. Even more significant than its double capacity is the magnesium battery's good storageability at normal and high temperatures. This feature has eliminated the need for refrigeration of dry cell batteries, and has vastly simplified the handling of batteries in the field and increased their cost effectiveness.

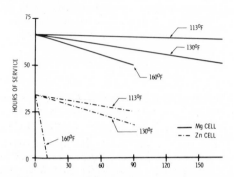

FIG. 2 *The magnesium dry cell (its advantages over the zinc dry cell). Note: Temperatures shown are storage temperatures.*

ZINC-AIR BATTERIES

Until recent development of high current density air electrodes for fuel cells, the zinc-air battery was a very low-drain device, unsuitable for military applications. Now this system offers the advantages of high energy density (over 100 Wh/lb), good power density, long shelf life and full environmental capability. Further, it has been possible to construct a device [1] with replaceable zinc anodes (mechanically rechargeable) so that the advantages of a secondary cell may be enjoyed without access to a power generator in the field.

An entire family of rechargeable zinc-air batteries was first announced in 1967 [2]. Some of the characteristics of these devices are summarized in Table I. All are designed around a battery box of standard length 12.2 inches and standard width 4.0 inches (see Fig. 3). Typically, 20 cells are used in a 24-cell system; each cell is rated at 1.2 V. The battery is activated by pouring water in each cell and inserting a zinc electrode-electrolyte composite. Air transport through the cell stack is accomplished through convection. A cell of the zinc-air battery is diagrammed in Fig. 4. It consists of a porous zinc-KOH composite electrode, a KOH solution as the electrolyte, and a catalyzed gas diffusion electrode as the cathode. Two cathode electrodes are connected in parallel and cemented to a frame to form a watertight, gas-permeable "cathode box". To activate the cell, the cathode box is filled with water. The zinc-KOH composite electrode, wrapped in suitable separators, is then inserted. A

TABLE I

Characteristics of Standard Family Mechanically Rechargeable Zinc-Air Batteries

Nomenclature	BA-525()/U	BA-526()/U	BA-527()/U	BA-530()/U	BA-528()/U
Size	20 Ah	32 Ah	48 Ah	75 Ah	150 Ah
Voltage	12/24	12/24	12/24	12/24	12/24
Dimensions (in)					
Length	12.2	12.2	12.2	12.2	12.2
Width	4.0	4.0	4.0	6.7	6.7
Height	4.8	7.0	9.0	6.0	10.0
Weight (Lbs) (Without Cover)	7	10	15	21	35
Rated Capacity (Watthours)	480	768	1152	1800	3000
Energy Density (Wh/Lb)	69	77	77	86	86

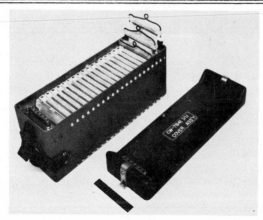

FIG. 3 *Zinc-air battery.*

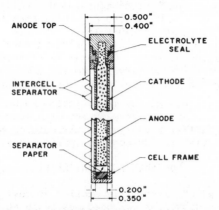

FIG. 4 *Cell of zinc-air battery.*

platinum-catalyzed cathode will last for 50-100 anode replacements. Various ignoble catalysts, including carbon, have been examined for suitability. Thus far they have provided cathodes of very limited cycle life. It has not yet been established whether the degradation is due to a decline in mechanical or catalytic properties of the electrode. We are continuing the search for catalysts which are both inexpensive and capable of providing improved low temperature performance.

The performance of these batteries has been determined over a wide range of temperature and load conditions [3]. Optimum performance is obtained over the 5 to 50 hour rate range. The performance falls off at high discharge rates, due, in large part, to slow oxygen reduction kinetics. Another source of performance loss at all discharge rates is water loss and resulting cell dry-out. This is aggravated by high temperatures at fast discharge and by lengthy exposure to evaporation at slow discharge.

The weight and service advantages of the zinc-air battery for one particular application, the new radio-transceiver Radio Set AN/PRC-70, are depicted in Table II. The receiver requires about 3 watts for operation; the transmitter about 125 watts. The table shows the weight

TABLE II

Weight and Service Life of Batteries Powering Radio Set AN/PRC-70

Battery Type	Weight, Lbs.	Hours of Service
Magnesium dry cell (BA-4840/U)	12	18
Nickel-cadmium (BA-655/U)	12	9
Zinc-silver oxide (BB-534/U)	7	12
Zinc-air (BA-525/U)	7	30

and hours of service for several battery systems operating on a 9:1 receive/transmit duty cycle. The magnesium dry cell battery, itself vastly superior to the standard zinc dry cell, cannot effectively handle the relatively heavy transmit load. The zinc-air battery has a 2.5:1 advantage over the zinc-silver oxide battery and almost 6:1 over the nickel-cadmium battery. Of equal importance is the capability of recharging the zinc-air battery in the field with a 3-lb. replacement of zinc anodes and sufficient water for activation.

Serious commercial use of the mechanically rechargeable zinc-air battery will have to await the development of a less expensive electrocatalyst. An inexpensive, rechargeable air cathode might make the zinc-air cell attractive for automotive applications.

PORTABLE FUEL BATTERIES

The term "fuel cell" is commonly applied to devices utilizing a continuous feed of hydrogen or carbonaceous fuels at the anode. By contrast the interest at USAECOM is confined mainly to portable "fuel battery" devices, where fuel is added by a batch process. The highly soluble and relatively reactive substances, hydrazine and methanol, lend themselves well to such usage, as does solid lithium hydride (acting as a source of hydrogen).

Several versions of the portable fuel cell, utilizing hydrazine or hydrogen (by reaction of lithium hydride with water) as the fuel, are reaching the state of final development [4]. A methanol system is in earlier stages of development.

The "fuel battery" is inherently a high-energy, low power device. When combined with a rechargeable battery (*e.g.*, nickel-cadmium), the resulting fuel battery-secondary battery hybrid is a high energy source capable of supporting moderate steady loads and heavy transient loads. Such an electrical source is highly attractive for powering certain communications equipments. For example, a 28 volt, 60 watt fuel battery operating on liquid hydrazine monohydrate and mated to a nickel-cadmium battery has been used [4] to power a military radio set, as illustrated in Fig. 5.

FIG. 5 *Hybrid power supply for radio sets.*

The hydrazine fuel cell system comprises three main replaceable components: the fuel cell module, the hydrazine fuel tank, and the power conditioning module. The fuel cell module contains two 20-cell stacks and controls circuitry for the stacks to maintain proper mass and heat balance and operating temperature.

Table III reveals the comparative advantage of the hydrazine-fueled hybrid fuel cell for use with Radio Set AN/VRC-12, requiring 40 watts

TABLE III

Radio Set AN/VRC-12 Power Options

12 Hours Service	
Nickel-cadmium batteries	102 lbs
Two nickel-cadmium batteries, engine-generator, fuel	94 lbs
Fuel cell, nickel-cadmium battery hybrid	30 lbs
Support for Additional 12 Hours of Service	
Engine-generator and fuel	60 lbs
Fuel, alone, for fuel cell	2 lbs

of power on receive and 240 watts during transmit.

Fig. 6 illustrates a manpack fuel cell, utilizing gaseous hydrogen.
The hydrogen is generated in an internal Kipp generator using lithium
hydride and water. An energy density of 1000 Wh/lb of hydride is
obtained with a one-quarter pound pellet of hydride providing 8 hours
of service.

   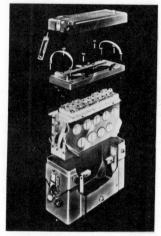

FIG. 6 *Metal-hydride fuel*   FIG. 7 *Battery construction for*
cell.   *interchangeability.*

ALKALINE SECONDARY BATTERIES

Programs, underway for several years [5], have been directed
at packaging the common alkaline secondary cells in configurations
that would allow interchangeability and compatibility with various
electronic equipments and with portable fuel cells (for hybrid use).

Fig. 7 illustrates battery construction. To achieve battery
interchangeability, only the height of the cells and of the battery
case is varied to accommodate to the characteristics of the battery
and to the requirements of the mission. Work has been completed on
Ni-Cd, Ag-Cd and Ag-Zn batteries. Work is progressing on Ni-Zn and
Ni-Fe cells. Interchangeability allows the optimum battery to be
used for an application after a trade-off is made for performance
under the specific environmental and service conditions involved.

RESEARCH AND DEVELOPMENT

ORGANIC ELECTROLYTE CELLS

Where reserve batteries are needed for portable communication and
electronic surveillance equipments and the environmental temperature
is not lower than approximately 0°F, it appears that the zinc-air
cell will be the usual choice over the next several years. The zinc-
air battery will be available in standard line, water-activated, 60 to
80 Wh/lb, throw-away configuration. For field usage, where stabilized
battery temperatures may reach -40°F or lower, the lower energy (30-40
Wh/lb) magnesium perchlorate battery will be employed in the reserve
configuration. It is clear that the high-energy capability and wide

operating temperature range of organic electrolyte cells would be highly
valuable particularly at the lower range of operating temperatures.
Poor wet stand life is perhaps one of the greatest difficulties encountered with state-of-the-art cells. Fig. 8 is the polarization curve
for a cell with copper chloride cathode discharging at a simulated
transmit-receive regime, immediately after activation [6]. The performance was deemed acceptable in spite of the power conditioning
difficulties introduced by distinct two step discharge voltage. In
Fig. 9, a similar cell was discharged after six days activated stand

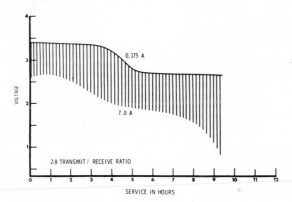

FIG. 8 *Li/CuCl$_2$ organic electrolyte cell discharge after six day
activated stand at 75°F.*

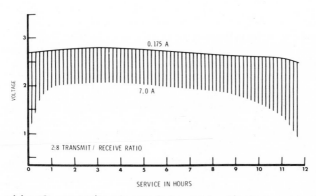

FIG. 9 *Li/CuCl$_2$ organic electrolyte cell, discharge at -20°F.*

at 75°F. The performance was then markedly curtailed. The deterioration during activated stand is attributable to solubility of the cathode
material. Other organic electrolyte systems are being evaluated at
USAECOM. One approach proposed at USAECOM [7], for eliminating the
cathode solubility problem, involves the use of graphite-fluorine
intercalation compounds. These compounds are nonstoichiometric and
may be represented by the formula C_xF when x usually varies between
3.5 and 4.5. In addition to its low solubility, C_xF is a potentially
attractive cathode material because of its electronic conductivity and
chemical stability. The theoretical capacity of C_xF is plotted as a
function of x in Fig. 10. For 3.5 < x < 4.5 the theoretical capacity
is 179 > Ah/lb (C_xF + Li) > 152. This places C_xF at approximately

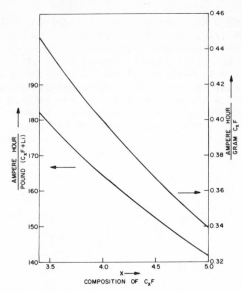

FIG. 10 *Theoretical capacity of Li/C_xF-cell as a function of the C_xF composition assumed cell reaction: $Li + C_xF \longrightarrow xC + LiF$.*

the level of $CuCl_2$ on Table IV. Actual capacities are close to 100 percent of the theoretical when the cathode was discharged to 0 volts.

TABLE IV

Cell Characteristics of Various Lithium Systems

Cell Reaction	Theoretical Capacity Amp.hrs. lb active material	Operating voltage*	Energy Density, Watthours/lb Theoretical (based on weight of active material)	Practical* (based on cell weight	Reference
$C_4F+Li \rightarrow 4C+LiF$	165	<3.0	553	70**	ECOM
$CuF_2+2Li \rightarrow Cu+2LiF$	211	>3.0	749	140	ESB
$CuCl_2+2Li \rightarrow Cu+2LiCl$	164	2.5	504	> 70	Electro-Chimica
$CuCl_2+2Li \rightarrow Cu+2LiCl$	164	2.5	504	109	SAFT
$CuS+2Li \rightarrow Cu+Li_2S$	222	<1.8	499	100	SAFT
$Ni_3S_2+4Li \rightarrow 3Ni+2Li_2S$	182	<1.8	388	<100**	TYCO
$CdF_2+2Li \rightarrow Cd+2LiF$	148	1.5	385	64	Whittaker
$AgO+2Li \rightarrow Ag+Li_2O$	177	2.0	520	106	ESB
$AgCl+Li \rightarrow Ag+LiCl$	81	2.5	230	30	Lockheed

*Data based on discharge at room temperature, approximately at
   1 mA/cm^2
**Estimated Value

Open circuit voltages were found to be in the range 3.2 to 3.5V at
25°C, depending on conditions of preparation. The theoretical energy
density, based on this observation, is 526 to 576 Wh/lb of active material, which places it towards the top of Table IV. The practical
energy densities are unfortunately considerably lower than the theoretical as a consequence of declining cathode potential during discharge
(See Fig. 11). The cause of electrode polarization is not yet understood. With $LiClO_4$-PC as electrolyte, and a $C_{4.43}F$ cathode, 200 Wh/lb
of active material could be obtained. Assuming a design efficiency of
0.35, practical energy densities of 49 to 105 Wh/lb may be anticipated
for complete batteries. This would compare favorably with some of the
entries on Table IV, but the absence of a steady voltage is a serious
problem. Further research and development effort is required to make
this a practical system.

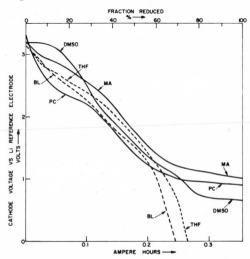

FIG. 11 *Reduction of 0.8 gm $C_{3.47}F$ at 1 mA/cm^2 (20 mA), 25°C
(Compound vacuum stored over CaO).*

META-DINITROBENZENE DRY CELLS

It has already been stated above that the magnesium reserve cell
(with magnesium perchlorate electrolyte and MnO_2 cathode depolarizer)
is presently the prime choice for low-temperature, high-energy primary
applications. However, the theoretical capacity of MnO_2 is only 127
Ah/lb of active material as opposed to 172 Wh/lb for AgO. Replacement
of MnO_2 by m-dinitrobenzene would, on the other hand, afford a theoretical charge density of 465 Ah/lb. This depolarizer has received some
attention at USAECOM [8]. In battery discharge tests, it was determined
that cell voltages are lower for the organic than for the inorganic
depolarizer, but that the voltage is more constant during discharge
(See Fig. 12). Service life is a more sensitive function of temperature
(See Fig. 13). In spite of the shortcoming, the organic depolarizer
provides an energy density 40 percent greater than that of the inorganic
depolarizer at the 100-hour discharge rate (Table V). Many of the
shortcomings of the organic depolarizer are attributable to its incom-
plete reduction and to formation of dimeric compounds. Fig. 14 presents
proposed reaction sequences for reduction of m-dinitrobenzene. Complete
and efficient reduction to the phenylenediamine (IX) is known only in

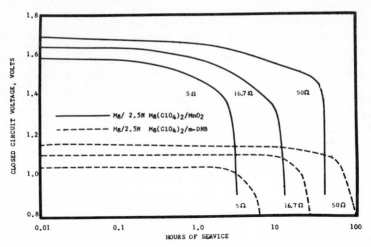

FIG. 12 *Comparison of discharge curves at 70°F for "A" cells.*

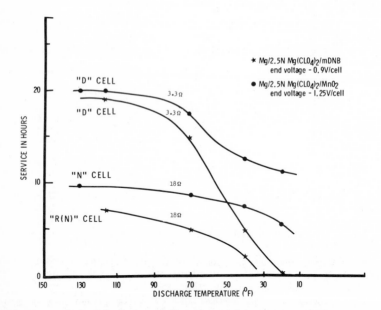

FIG. 13 *Comparison of service at various temperatures.*

acid solutions. In order to learn how to optimize m-dinitrobenzene
reduction, several lines of research are being followed at USAECOM.
A number of the proposed intermediates on Fig. 14 have been synthesized. Coulometric and polarographic measurements have been conducted
on the intermediates in aqueous solutions in an attempt to trace
reaction path indirectly. More direct studies of intermediates are
being performed in nonaqueous media, using infrared, visible and ESR
spectroscopic techniques. The kinetics of formation of radical intermediates is under investigation by rapid-scan internal reflectance
techniques.

TABLE V

Energy Density Comparisons Between
$Mg/Mg(ClO_4)_2/m\text{-}DNB$ and $Mg/Mg(ClO_4)_2/MnO_2$
at the 25- and 100-Hour Rate

Electrochemical System	Cell Size	Hourly Rate	Watt-Hr/Lb	Watt-Hr/In^3
$Mg/Mg(ClO_4)_2/m\text{-}DNB$	R(N)	25	44.4	2.2
	A	25	58.2	2.9
	D	25	40.4	1.9
	R(N)	100	58.3	2.9
	A	100	76.0	3.8
	D	100	71.6	3.4
$Mg/Mg(ClO_4)_2/MnO_2*$	A	25	45.3	3.0
	D	25	36.2	2.1
	A	100	50.2	3.4
	D	100	52.5	3.1

All cells taken to end-voltage of 0.9 Volt/cell

*Synthetic MnO_2

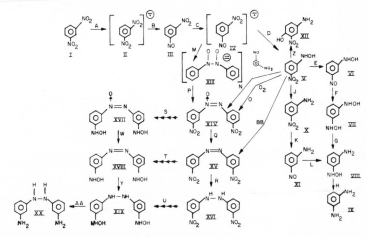

FIG. 14 *Reaction sequence for the electrochemical reduction of m-Dinitrobenzene.*

REFERENCES

[1] H. R. Knapp, C. A. Nordell and D. Linden, "Advanced Primary Zinc-Air Batteries," *Proc. 23rd Annual Power Sources Conference*, Atlantic City (May 1969).

[2] D. Linden and H. R. Knapp, "Metal-Air Standard Family," *Proc. 21st Annual Power Sources Conference*, Atlantic City (May 1967).

[3] H. R. Knapp, "New Designs for Zinc-Air Batteries," *USAECOM Technical Report ECOM-3182* (October 1969).

[4] D. Linden, "New Concepts for Recharging Electrochemical Batteries,"
     Proc. of Design Engineering Conference and Show, Chicago (May 1970).

[5] E. J. Settembre, "Standard Alkaline Cells and Secondary Batteries,"
     USAECOM Technical Report ECOM-2929 (March 1968).

[6] N. T. Wilburn, "Engineering Evaluation of High Rate Organic Electro-
     lyte Reserve Batteries," *USAECOM Technical Report ECOM-3278*
     (May 1970).

[7] K. Braeuer, "Feasibility Study of the Lithium/CₓF Primary Cell,"
     USAECOM Technical Report ECOM-3322 (August 1970).

[8] J. B. Doe, "Characteristics of Meta-Dinitrobenzene Dry Cells,"
     USAECOM Technical Report ECOM-3089 (January 1969).

COMMENT ON PAPER OF S. GILMAN
BY J. O'M. BOCKRIS

The outstanding problem in respect to fuel cell and battery
research at present remains the oxygen electrode; its improvement
would have great significance. In our discussions, we do not seem to
look very much to the bronzes or other oxide conductors. These seem
to be the best hitherto tried. Although this research was begun at
the University of Pennsylvania, it is being continued in other parts
of the world, *cf.* the publications of Tseung [1] and also Meadowcraft
[2].

The Tseung work seems to give some relation to magnetic properties, which should be an important criteria in choosing structures.

The theory of the bronzes is being developed by J. McHardy and
R. Fredlein at the University of Pennsylvania. What seems to be
emerging is a general formula for a suitable oxygen electrode. It
should be a conducting oxide (*i.e.*, chemical attack possibility
removed; anodic dissolution improbable). However, it should contain a
thin surface layer of semiconducting properties. This is then subject
to <u>doping</u>. It is this which shifts the Fermi level to an extent that
can give rise to good catalysis (the Fermi level <u>does</u> have an effect
on the reaction rate in semiconductors, but not in metals). As with
the Cahan concept for high-power oxygen electrodes with a low weight
of platinum, the field and the concept seem open for considerable
development.

The organic alternatives along the lines of the porphyrin-metallic
type of compound are attractive too, although in performance they do
not approach, as yet, the tungsten bronzes, doped with a few hundred
ppm. of noble metals.

REFERENCES

[1] A. C. Tseung and H. L. Bevan, *Electrochem. Soc., 138th Meeting*,
     Atlantic City (October 1970).

[2] D. V. Meadowcraft, *Nature*, **226**, 847 (1970).

THE AIR-COOLED MATRIX-TYPE PHOSPHORIC ACID CELL

O. J. Adlhart

Engelhard Minerals and Chemicals Corporation, Newark, New Jersey

ABSTRACT

Systems considerations have stimulated a substantial research and development effort on fuel cells with phosphoric acid electrolyte. Although materials problems are severe in this medium-temperature cell and precious metal activation is a necessity, the advantages in respect to systems simplicity and operational reliability are remarkable.

To date the development of the phosphoric acid cell — a matrix-type construction is commonly used — is well advanced and multi-KW units have been fabricated and evaluated.

The air-cooled version of the matrix-type cell is described below. This mode of cooling is of interest for devices with medium or lower power ratings. Questions are discussed relating to the mass and energy balance in this cell type, as are features of cell construction aimed at maintaining temperature uniformity.

INTRODUCTION

Increasingly, operational reliability combined with comparatively maintenance-free performance is recognized as a decisive feature in favor of fuel cells. Unique in this respect is the phosphoric acid cell. For one thing, this cell type lends itself readily to the preferred matrix-type construction. Furthermore, a simple mass and energy balance system can be devised because of the low water content of the electrolyte and the elevated operating temperature. Thus, the cathodic air stream can be relied upon for heat and water removal. This adds considerably to the simplicity of the overall system.

Carbon monoxide tolerance of the phosphoric acid cell has often been emphasized since it opens a more direct route to the utilization of carbonaceous fuels, a long standing goal in fuel cell development. Thermal cracking or steam reforming of hydrocarbons are examples of such approaches.

Equally, of course, dilute hydrogen streams such as cracked ammonia can be readily utilized [1]. Direct oxidation of such fuels as methanol or propane is also possible. Power densities, however, are small in the latter cases and the utility of such cells is likely to be confined to low-wattage devices.

Much information has been accumulated in recent years on this type of fuel cell, including in particular the question of the precious metal requirements [2].

The discussion below is confined to the mass and energy balance in the air-cooled phosphoric acid cell and to certain related features of cell construction.

181

MASS AND ENERGY BALANCE

As mentioned above, cathodic air cooling is an attractive mode
for mass and energy balance in the matrix-type phosphoric acid cell.
In general, air is admitted to the system at ambient temperature and
leaves the cell stack hot, removing waste heat and moisture while
supplying oxygen to the cathode.

A prerequisite for this mode of operation is the use of a concentrated electrolyte with low water vapor pressure. With the removal
of the reaction water, the electrolyte volume changes little over a
wide temperature range, and most important, it is largely independent
of air flow rates. This is the case with the phosphoric acid cell.
Another advantage of this electrolyte is the elevated operating
temperature, which favors cathodic cooling. The preferred temperature
is 125°C, which provides ample ΔT for heat removal.

The choice of this temperature is a compromise between cell performance and stability of various components.

A point of interest in the cathodically cooled cell is the specific
flow requirements for oxygen supply, water and heat removal. These
values are dependent on the structure and thermal characteristics of
the specific cell components. They have been experimentally determined
where necessary. Fig. 1 illustrates flow requirements as a function
of ambient temperature and current density. The shaded area to the

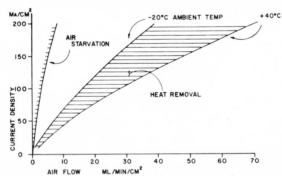

FIG. 1 *Air flow requirements for mass and energy balance.*

right shows the air flow needed for heat removal, assuming waste heat
removal exclusively by the cathodic air stream. The shaded area to
the left indicates the region of air starvation. Flow rates of 1.3-2x
stoichiometric are required to sustain optimum cathode performance.
Interestingly, water removal at the point of beginning air starvation
is still complete. The data obviously point towards a system which is
solely controlled by the stack temperature, since flow requirements
for heat removal exceed considerably those for water removal. This
finding has led to the development of a cell controlled by a variable
speed air blower which increases or decreases the air flow depending
on ambient temperature and stack heat load.

CONSTRUCTION FEATURES OF THE CELL MODULE

The basic cell module shown in Fig. 2 comprises all the components
required for current generation, collection and reactant distribution.
The modules are preferably stacked in a biopolar mode. Individual

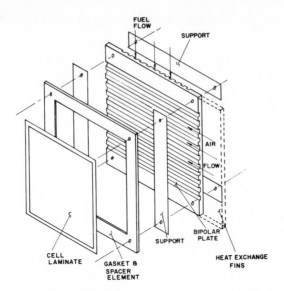

FIG. 2 *Cell module design.*

components are the electrode and electrolyte membrane assembly,
a one-mm-thick laminate described earlier [3]. The laminate is
contained by an EPR rubber gasket with dimensional elements to limit
the degree of compression in the stack. A critical component is
the current collector and reagent distribution plate. This plate
consists of aluminum or carbon three mm in thickness. Aluminum is
gold plated for protection; a gold flash is applied to carbon to
minimize contact resistance. The configuration of the module is
usually rectangular with the dimension in the direction of air flow
small to minimize air pressure drop. Fuel channels are perpendicular
to the air channels and both reactants are manifolded externally to
the stack.

UNIFORMITY OF CELL TEMPERATURE

In the air-cooled matrix-type phosphoric acid cell, as in most
fuel cells, temperature uniformity over the active cell area is a
critical requirement. Although heat and water management may not be
affected by gradients, temperature uniformity is needed for good cell
life and fuel utilization when dilute hydrogen streams are used.

Admitting the air directly to the cathode cavities leads to a
thermal gradient which is only counteracted by the thermal conductivity within the cell module. Various options are available to
achieve uniformity. One is to recirculate the air through the cell
to minimize the temperature difference between the air entering and
leaving the cell. This requires high air-flow rates and considerable
blower capacity. Below are described two approaches that utilize
the thermal conductivity of the collector plates primarily to achieve
temperature uniformity.

FINNED CELL CONSTRUCTION

One type of cell construction relies on the horizontal thermal
conduction in the current-collector plate. Waste heat is conducted

to a fin where heat exchange with air takes place [4]. This construction is shown schematically in Fig. 3.

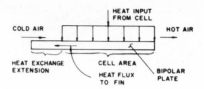

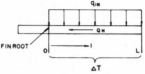

FIG. 3 *Finned cell heat transfer.*

The plates extend into the reactant air stream. Sufficient
heat-exchange area is provided to transfer the entire waste heat to
the air on this fin. The temperature differential over the remainder
of the plate is then determined by cell geometry and plate configuration as follows:

$$\Delta T = \frac{q_{in}}{K_h \cdot A_s} \int_0^L (L-1)dl$$

or assuming in a first approximation a constant q_{in},

$$\Delta T = 1/2 \, \frac{Q_h \cdot L}{K_h \cdot A_s} \quad .$$

ALTERNATE FLOW AIR COOLING

The finned cell heat removal concept is attractive for small cells
or cells operating at comparatively low power densities. It offers a
simple solution from the viewpoint of cell construction. However, for
larger cells or high power density operation, collector plates of large
cross sectional area are required to minimize thermal gradients. Equally, one is confined for the manufacture of the collector plates to
materials of high thermal conductivity.

The alternate flow air cooling offers an alternative to the finned
cell cooling concept. It relies on the thermal conductivity in the
horizontal and vertical direction in the stack. Collector plates of
lower thermal conductivity can be used even at high power density
operation. The basic concept of the alternate flow cooling is shown
in Fig. 4. Air is admitted to single cells or groups of cells from
opposing sides. The effect is that the cold cell region where air
enters is adjoined by hot ends of bipolar cells where the air leaves
the cell. Thermal conductivity in the vertical direction provides a
short path for heat transfer between what would have been the hot end
of one cell to the cold end of the other.

With reference to Fig. 4, the following equations are of primary
consideration in the thermal analysis of the counterflow heat removal.

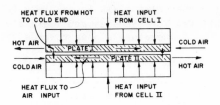

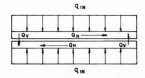

FIG. 4 *Alternate flow cooling.*

<u>Heat transfer plate to air:</u> $Q_s = U\,A_c\,T_{1\bar{m}}$

<u>Horizontal heat flux in plate:</u> $Q_h = \dfrac{K_h \cdot A_s}{1}$

<u>Vertical heat flux in stack:</u> $Q_v = K_v\,A_c\,n\,T_c$

    A typical temperature distribution obtained with the alternate flow air cooling is shown in Fig. 5. The cell has a plate length of

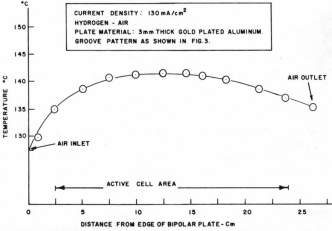

FIG. 5 *Temperature profile over bipolar plate.*

26 cm and an active cell area of 22.5 × 22.5 cm. Operated at a current density of 130 mA/cm^2 a maximum temperature differential of 70°C developed over the active region of the cell. Over a broad region the cell temperature is uniform, although towards the end and particularly at the air inlet, a considerable slope is evident. The shape of the temperature distribution is due to the fact that heat transfer from plate to air (Q_s) and the vertical thermal flux (Q_v) are large compared with the waste heat generated (q_{in}) at the entrance section. Thus much of the total waste heat is removed in the first

quarter of the cell. This also leads to some cooling of the adjoining cell, and as a result the hot air does not leave the cell at maximum temperature. With the temperature slopes towards the cell ends, thermal conductivity in the horizontal direction (K_h) does aid in evening out temperature differential between air inlet and cell center.

The alternate flow air cooling was employed in fuel cells with larger power ratings such as shown in Fig. 6. This particular unit is a 750-Watt power plant with the cover removed. The channels along side the stack serve to manifold the air as indicated schematically in Fig. 4.

FIG. 6   *28 volt air cooled phosphoric acid fuel cell.*

LIST OF SYMBOLS

ΔT = Max. temperature differential over cell area in °C.

Q_h = Total heat flux in horizontal direction counter to the air flow in watts.

q_{in} = Heat input per unit cell area in watts.

K_h = Horizontal thermal conductivity of plate in watts/°C/cm.

A_s = Unit cross sectional area of bipolar plate. This area is 0.178 cm^2/cm for the plate shown in Fig. 2.

L = Total cell width in cm.

l = Length of plate in cm.

Q_v = Thermal flux in vertical direction in watts.

K_v = Thermal conductivity in cell perpendicular to the electrode plane in watts/°C/cm^2. K_v was experimentally determined for the configuration shown in Fig. 2 at 0.098 watts/°C/cm^2.

U = Surface heat transfer coefficient for forced convection. Experimentally determined at 0.006 watts/cm^2/°C for configuration given in Fig. 2.

A_c = Geometric cell area in cm^2.

$T_{l\bar{m}}$ = Log mean temperature difference over surface in °C.

Q_s = Heat flux in watts from plate to air.

n   = Number of cells.

ΔT_c = Temperature difference in °C between adjoining cells.

REFERENCES

[1]   O. J. Adlhart and P. Terry, "Ammonia-Air Fuel Cell System,"
      Intersociety Energy Conversion Engineering Conference, Washington,
      D. C. (September 1969).

[2]   W. M. Vogel and J. T. Lundquist, "Reduction of Oxygen on Teflon
      Bonded Platinum Electrodes," *J. Electrochem. Soc.*, 117, 1511
      (1970).

[3]   O. J. Adlhart and A. Hartner, "Plastic Bonded Electrodes,"
      Proceedings of the 21st Annual Power Sources Conference, Atlantic
      City (May 1967) p. 4.

[4]   O. J. Adlhart, "Design and Performance Characteristics of a
      Matrix Type Phosphoric Acid Hydrogen-Air Cell," *Proceedings of
      the Third International Symposium on Fuel Cells*, Brussels
      (June 1969) p. 227.

THERMO-CATALYTIC HYDROGEN GENERATION
FROM HYDROCARBON FUELS

M. A. Callahan

U. S. Army Mobility Equiment Research & Development Center
Fort Belvoir, Virginia

ABSTRACT

As a result of the U. S. Army research and development efforts in the fuel cell field, an open-cycle, indirect hydrocarbon-air, acid electrolyte fuel cell system has been designed and feasibility tested at USAMERDC. The system consists of two subsystems, namely the cell stack and the fuel conditioner. The fuel conditioner, with which this paper deals, is a hydrogen generator that utilizes a high temperature (850°C to 1200°C) catalyzed pyrolysis reaction to strip hydrogen from a fluid hydrocarbon fuel. Using nickel as a catalyst, the fuel conditioner has proved to be both simple and versatile; the reactor has no moving parts and is capable of operation over a fairly wide temperature range on a variety of liquid or gaseous hydrocarbons. The process is a very efficient one; at fuel space velocities on the order of 230 cc's of fuel per 1000 cc's of reactor volume per hour, it is capable of conversions in excess of 95 percent, with stream purity of over 90 percent hydrogen.

The reactor was tested using three basic catalyst configurations: 1) an uncatalyzed reaction using alumina or zirconia packing material, 2) a metal catalyst coated on the alumina, and 3) the uncoated packing material mixed with pure metal catalyst. The first configuration withstood the extreme environment well, but the efficiency was mediocre. The second case showed good efficiency, but material degradation was severe. The third configuration showed both good endurance and efficiency.

The testing showed several influencing factors upon the reaction: 1) temperature, 2) fuel space velocity, 3) oxidation state of catalyst, and 4) air flow rate. The influence of pressure was not investigated.

The data indicate that two reactors, with internal volumes of 675 cc's each, could produce about 12 cubic ft./hr. of hydrogen from 0.5 lb. of a typical hydrocarbon fuel, such as combat gasoline. This would be sufficient to run a 500-watt fuel cell stack for the same length of time.

INTRODUCTION

The fuel cell's potential as a superior electrical power source has been recognized for some time. This has led the Department of Defense to sponsor fuel cell research and development, both in-house and through contracts with industrial and academic institutions. Part of this effort has been directed toward the development of a family of fuel cell power sources which could eventually replace the existing Military Standard spark ignition engine-driven generator family in the supply

system. The requirements for such a family of power sources include
high reliability, low operating noise levels, light weight, and an
ability to operate economically on a variety of logistically available
hydrocarbon fuels. At the U.S. Army Mobility Equipment Research and
Development Center (USAMERDC), a fuel cell system that appears to meet
all of the above requirements has been conceived and feasibility tested.
The system, termed the Open Cycle Fuel Cell Power Source, is an indirect
hydrocarbon-air system comprised of two subsystems: 1) a phosphoric acid
fuel cell stack which operates on impure hydrogen and air, and 2) a hydrogen producing fuel conditioner; the latter is the subject of this
paper.

The overall system, described by Gillis [1], is illustrated schematically in Fig. 1. The phosphoric acid stack subsystem can use any
of the state-of-the-art stacks which operate on impure H_2 fuel [2,3,4].
The stack can tolerate considerable amounts of impurities such as N_2,
CO_2 or CH_4, the adverse effect being chiefly one of dilution of H_2, and
in addition can tolerate up to several volume percent CO. The effect
of excess CO on the cell seems to be one of catalyst poisoning, but the
effects are reversible if the exposure is not too severe or for too
great a duration.

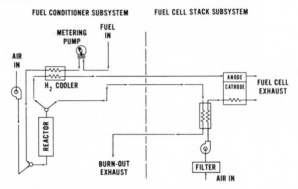

FIG. 1 *Simplified schematic of open cycle fuel cell system.*

The major purpose of the fuel conditioning subsystem is to convert
the combined hydrogen tied up in a liquid hydrocarbon fuel, such as
octane, to the free hydrogen which can be consumed by the fuel cell
stack. The liquid fuel is converted to hydrogen by a catalyzed pyrolysis reaction in which the hydrocarbon is ultimately degraded to its
element, carbon and hydrogen. The reaction is carried out in a reactor
packed with catalyst and inert packing material at temperatures in the
range of about 850°C to 1200°C. During the course of this reaction,
the carbon formed is deposited on surfaces in the reactor. After a
suitable period of time, the fuel flow is interrupted and air is introduced into the reactor to oxidize the carbon to CO and CO_2, thus re-
heating and regenerating the clean reaction bed. The process then is
cyclic; half the cycle is a thermo-catalytic cracking of fuel to hydrogen (hereafter called the "cracking cycle" or "fuel cycle"), and the
other half cycle (hereafter called the "burnoff cycle" or "air cycle"),
a burnoff of the carbon deposited during the previous fuel cycle. For
continuous hydrogen generation, two or more reactors are used so that
one is always producing hydrogen while the other is being regenerated.

In addition, this subsystem acts as a filter for lead and sulfur
and other additives or impurities which are present in most fuels in

small quantities and have a potentially harmful effect upon fuel cell
stack performance. Substances such as tetraethyl lead are unstable at
the reaction temperature, and the lead is deposited upon the surfaces
in the reactor. Sulfur compounds likewise break up and the sulfur forms
sulfides on the catalyst; but evidence indicates this sulfur is at least
partially burned off during the air cycle and is vented from the system
as SO_2.

Several major problems remain to be solved at this time. Many of
these are engineering considerations in the development of the cracker
as an integrated part of the overall fuel cell system. Other problems
are concerned strictly with the process itself or the materials degradation caused by the extreme environment. The purpose of the work presented here was to investigate the chemical aspects of the thermo-catalytic cracking process and some of the effects of the chemical and
thermal environment on several materials.

EXPERIMENTAL

EXPERIMENTAL SETUP AND METHOD

Fig. 2 is a schematic diagram of the test setup (Fig. 3) used to
study the reaction. The air and fuel enter the reactor alternately
through the same tube, the flow being switched with solenoid valves.
The reactor, which sits in a propane-fired furnace, is a 12 in. length
of Inconel 600, 2 in. schedule 40 pipe, with 1/2 in. stainless steel
plates welded into both ends. The internal volume is 675 cc. The inlet end plate is tapped for a 1/4 in. pipe to allow the fuel and air
to enter. The exit end of the reactor is tapped for 3/4 in. pipe and
also serves as the means for filling the pipe with catalyst and packing
material.

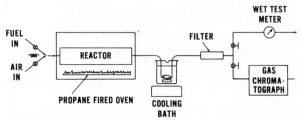

FIG. 2. *Schematic of test setup.*

During the course of the study, several types of packing material
were investigated in several configurations. Among those tested were
berl saddles* and several sizes and shapes of alumina and zirconia
packing.** These were tried in three basic configurations: 1) uncatalyzed, 2) coated with a catalyst, and 3) uncatalyzed but mixed with a
catalyst. The endurance of all the packing materials was poor when
coated with catalyst. The berl saddles were of marginal endurance,
even uncoated, and were discarded as unacceptable. The alumina materials showed excellent endurance when uncoated. After preliminary
findings, two studies were made: one used 3 to 5 percent Ni on porous

*Berl saddles were obtained from Fisher Scientific and U.S. Stone-
ware, the latter holding up considerably better under test conditions.

**Norton Company, Akron, Ohio.

FIG. 3a *Reactor oven during operation.*

FIG. 3b *Oven with top removed, showing reactor.*

zirconia spheres (Norcat 100); the other employed the mixed catalyst concept using alumina rings (SA5118 — 1/8" x 5/16" x 5/16" (Norton Co.)) mixed with pure nickel Raschig rings (1/4" 99+% nickel (Metallo Gasket Co., New Brunswick, N. J.).

ANALYTICAL EQUIPMENT

The basic analytical instrument was a modified Hewlett-Packard 5750B Research Gas Chromatograph. The chromatograph as modified has three columns, and is essentially described as a dual-column instrument and a single-column instrument sharing a common oven. In this way, the chromatograph has the versatility of two chromatographs, especially in its ability to utilize two different carrier gases, while having the convenience of a single unit. A schematic of the internal flow is included (Fig. 4). The dual-column system was used with a helium carrier gas to analyze for N_2, O_2, CO, CO_2, CH_4, C_2H_6, SO_2, H_2S, H_2O, COS, CS_2, etc. The single-column system was used to monitor hydrogen concentrations

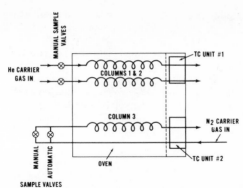

FIG. 4 *Chromatograph internal flow schematic.*

solely, employing high purity nitrogen as the carrier gas. The chromatograph was calibrated regularly using certified gas mixtures [5].
Table I is a summary of the pertinent chromatograph parameters.

TABLE I

Col. No.	Material	Carrier Gas	Gases Analyzed	Time Required Between Samples
1	5X Mole Sieve - 6 ft.	He	O_2, N_2, CH_4, CO	2 minutes
2	Poropack Q - 4 ft.	He	CO_2, H_2O	20 seconds
3	5X Mole Sieve - 6 ft.	N_2	H_2	7.5 seconds

Column carrier flow rates: He, 125 cc/min; N_2, 550 cc/min.
TC Bridge amperage: TC1 (for Cols. 1, 2) 150 mamps; TC2 (for Col. 3)
                     100 mamps.

Temperatures

Column Oven:	94°C	AUX1	150°C
TC1	220°C	AUX2	165°C
TC2	190°C		

The single-column system was fitted with an automatic sampling
valve* which allowed the hydrogen to be sampled every 7.5 seconds. The
dual-column system allowed sampling of CO_2 concentrations every 15 seconds on one column, while the second column provided a slower analysis
of some of the components not separated on the first column. At first
the cycles were analyzed using a slow procedure to obtain good accuracy.
This consisted of sampling at a specific point in the cycle (say, 30
seconds after cycle switch) and, allowing time for the peaks to come
out, then sampling again (say, at 90 seconds after cycle switch). By
sampling at different points in the cycle and averaging the results over
several runs, graphic representations of the cycle could be made, such
as those seen in Figs. 5 and 6. However, after several of the runs were
completely analyzed and represented graphically, it became possible to
read qualitatively the analysis pattern of Column 1 directly in terms

Chromatronix Co., Berkeley, California, Model SSVA-8031.

of most of the components. Fig. 7a is a representation of the Column 1
recorder readout, with peaks every 15 seconds. Fig. 7b shows the interpretation of the pattern in light of Fig. 5. This method of esti-
mating the component concentrations, when used for qualitative or semi-
quantitative observations, has proved to be a valuable tool, as it
provides quick and continuous information concerning the effluent gas
composition. The time delay between the actual reaction and the readout pattern seen on the chromatograph recorder is on the order of a
minute; thus the effects of any imposed condition changes can be
quickly seen.

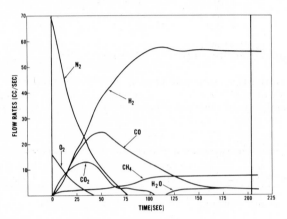

FIG. 5 *Flow data for fuel cycle.*

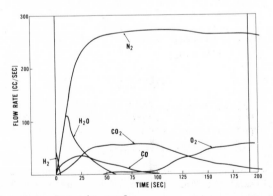

FIG. 6 *Flow data for air cycle.*

The reaction temperatures were read as the reactor skin temperature,
which was taken using an optical pyrometer. A wet test meter was used
to read the volume of effluent gas. Qualitative analysis of the coke
and similar residues for metallic contents was performed at USAMERDC
using X-ray fluorescence techniques. Quantitative coke and gasoline
analysis for carbon, hydrogen, and sulfur was performed by Aldridge
Associates of Washington, D. C.

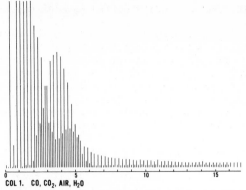

FIG. 7a   *Representation of typical recorder readout.*

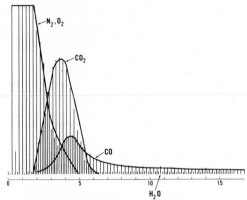

FIG. 7b   *Interpretation of Fig. 7a.*

RESULTS

UNCATALYZED PYROLYSIS

The first tests performed were on uncatalyzed packing material. The purpose was to provide a base line for future work and to screen out any packing which would not stand up to the stringent thermal environment. These in-house tests showed a good correlation with the work performed by Fleming, *et al.* [6], under contract to USAMERDC. In the temperature range of 1088°C to 1371°C, they found hydrogen effluent concentrations in the range of about 70 to 85 percent, the remainder being chiefly CH_4 and CO. Their proposed mechanism for the pyrolysis reaction is a fast breakdown of the fuel to C_2 fragments, followed by slower cracking of the C_2 fragments to carbon and hydrogen. It is during the slow step that the methane forms. The in-house work showed hydrogen concentrations in the same range, with the same major contaminants, CH_4 and CO. Fleming suggests that in the noncatalytic cracking, kinetics plays the major role in the determination of reaction products, while it is the system equilibria that control the stream composition in the catalytic process. The work with catalytic cracking which follows shows that although the equilibria play a much more significant role, kinetics, especially the kinetics of side reactions, still prove to be a major factor in the composition of the effluent stream.

CATALYTIC PYROLYSIS USING COATED ALUMINA AND ZIRCONIA

The catalyzation of the pyrolysis reaction would yield a more pure product if a suitable catalyst-substrate combination could be found. Both alumina and zirconia were investigated as substrates, while Ni, Co, Cr, and Fe were used as catalyst. Nickel and cobalt were decidedly better than the others in performance, and nickel was chosen for further work because of its availability. Several forms of alumina and zirconia were tested with catalyst coatings deposited by electroless methods and with nitrate solutions. The substrates invariably cracked, spalled, crumbled, etc., after several hours of running. The presence of the catalyst resulted in this behavior, since the uncoated forms were able, in most cases, to withstand the environment with little degradation. Although the study of catalyst-coated materials resulted in preliminary comparisons between the catalyzed and uncatalyzed reactions, the substrates proved to be so unstable when subjected to the experimental conditions that an in-depth study of the reaction became extremely difficult. One catalyst-substrate combination that withstood the environment long enough for evaluation was Norcat 100. The catalyst is, strictly speaking, not a coated one, since the nickel is an integral part of the makeup of the material. The performance of this catalyst was good (Fig. 8) but the endurance was poor. After 100 hours of run time the catalyst was examined, with severe degradation noted (see Table II).

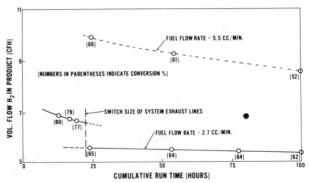

FIG. 8 *Norcat 100 catalyst performance vs. time.*

TABLE II

Breakdown of Zirconia Spheres after 100 Hours

Start: 580 grams   After 100 Hrs: 611 grams   Difference: +31 grams

Condition of zirconia spheres catalyst:

Intact Spheres	137 g	22.5%
Large Fragments	189 g	30.9%
Small Fragments:		
> 10 mesh	95 g	15.6%
10 — 20 mesh	66 g	10.8%
20 — 40 mesh	77 g	12.7%
40 — 60 mesh	25 g	4.1%
60 — 80 mesh	10 g	1.6%
80-140 mesh	6 g	1.0%
<140 mesh	5.5 g	0.9%

An increase in the size of the system exhaust lines after 20 hours
resulted in a lessening of the back pressure on the reactor. A slight
loss of reaction efficiency resulted, possibly indicating a direct relationship between pressure and efficiency. In purely thermal cracking,
the relationship is an inverse one [6], but at lower temperatures the
"purely" catalytic reaction does show this direct relationship [7].
This would indicate a strong catalytic influence on the pyrolysis reaction. A definite inverse relationship between fuel flow rate and
efficiency is exhibited (Fig. 9), which is as expected.

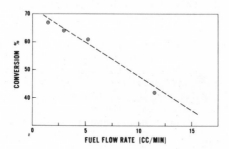

FIG. 9 *Fuel flow vs. efficiency for Norcat 100 run.*

The volume percentages of H_2 were of the order of those obtained by
pyrolysis alone, *i.e.*, 70 to 80 percent. However, the comparison of
the two sets of volume percentage data can be somewhat misleading.
First, the steady decline of the efficiency of the catalyzed test can
be at least partially attributed to the physical decay of the catalyst
support. Because of the horizontal orientation of the reactor,
crumbling of the support can lead to settling and consequently to gas
flow channeling, thus limiting the effective area of the catalyst. Had
the support retained its integrity, it is felt that the percentage
would have remained above 80 percent for the 100 hours. Secondly, the
average temperature of the catalytic reaction was about 200°C less than
the thermal reaction. Thirdly, the space velocity for the catalyzed
test was three times that for the thermal reaction. So even though the
same resulting volume percentage of hydrogen was found in both the thermal and the catalytic studies, the conditions of the catalytic study
were less favorable to the reaction.

A more generally reliable guide to the efficiency of the reaction
than the volume percentage H_2 reading is the percent conversion. This
is defined as:

$$\% \text{ conversion} = \frac{H_2 \text{ recovered in effluent}}{H_2 \text{ available in liquid fuel}} = \text{reaction efficiency}$$

The terms "efficiency" and "percent conversion" will hereinafter be
used interchangeably. The efficiency of the reaction with Norcat 100,
as shown in Fig. 8, decays as time progresses, but again, this long-
term decay can be at least partially traced to catalyst-support dis-
integration.

CATALYTIC PYROLYSIS USING MIXED BED

After encountering poor environmental resistance of coated (including Ni-zirconia) catalysts, it was proposed that a pure metal such as
nickel be used as both catalyst and packing material. Consequently, a

supply of nickel Raschig rings were obtained and tested in the reactor.
It became obvious after just a very few hours that the pure Ni arrangement would not be suitable. The nickel was a good catalyst, but so
much so that upon entering the reactor, the fuel immediately cracked
almost completely to hydrogen and carbon, thus causing severe carbon
plugging at the reactor entrance. Equal mixtures of nickel and alumina
rings were tried in an attempt to reduce the plugging, but this mixture
still caused some stoppage due to the reaction taking place in the entrance of the reactor. In addition, severe (75°C) thermal gradients
were evident from the entrance to the exhaust ends of the reactor on
burnout cycle. In order to combat these problems, a graded bed was
devised, and is illustrated in Fig. 10. The fuel enters a zone where

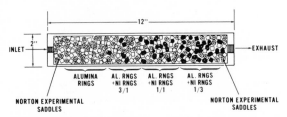

FIG. 10 *Cutaway view of reactor showing layered catalyst structure.*

pure thermal cracking can take place, after which the gas moves to a
progressively more heavily catalyzed zone in which the catalytic cracking takes place. Results using this bed show: 1) almost no plugging
in the reactor, 2) the disappearance of severe longitudinal thermal
gradients, and 3) a better product.

    In observing this graded bed, a study was made of the basic pyroly-
sis reaction, with some consideration being given to the side reactions
caused by imposed engineering considerations, such as regenerative cycling. Qualitative and some quantitative data were first taken over
the complete cycle, and then an attempt was made to isolate the basic
cracking reaction and to study it under laboratory conditions. The
endurance of these graded beds was considered good, with lifetimes running several hundred hours. Final failure was not usually due to breakdown of the bed itself, but rather to failures resulting from very
strenuous or unusual test conditions imposed.

    In working for an extended period of time with the graded bed,
several general trends became evident. There are at least four variables which have a significant effect on the quality and quantity of
product:

          1) Oxidation state of catalyst.
          2) Fuel flow rate.
          3) Temperature.
          4) Rate of carbon burnoff during air cycle.

    The effect that the oxidation state of the catalyst has on the
product is illustrated in the product stream flow charts (Figs. 5-6).
It is evident that an alarming amount of oxygen shows up as oxides of
carbon during the first part of the fuel cycle. Fig. 11 is a representation of the amounts of oxygen, tied up as CO, CO_2 and H_2O as well
as molecular O_2, which appear during the fuel cycle. As an indication
of the origin of this oxygen, Fig. 12 is a plot of the N_2/O_2 ratio during the burnoff cycle. In Fig. 12 it is seen that during the hottest
part of the air cycle, the effluent is depleted in oxygen. This, then,

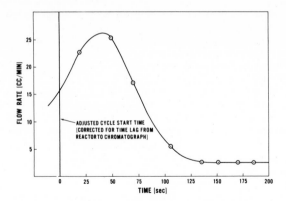

FIG. 11   *Oxygen flow (as O_2 from CO, CO_2 and H_2O) during fuel cycle.*

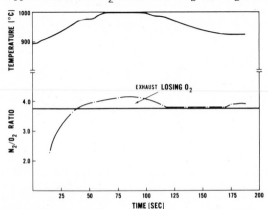

FIG. 12   *N_2/O_2 ratios during air cycle shown with corresponding*
reactor temperatures.

should be the source of oxygen for the oxides present in the fuel cycle.
The substances which are holding the oxygen could be 1) the alumina
packing, which could possibly trap air to be released later, or 2) the
nickel catalyst and/or the Inconel reactor walls.

The metals (M) could react as follows:

$$M + O_2 \xrightarrow{\text{burnout}} MO \xrightarrow[\text{(C + H$_2$)}]{\text{Fuel}} M + CO, CO_2, H_2O \quad .$$

Since the substitution of nonporous alumina rings has an insignificant effect on the product composition, the most likely suspect for
oxygen holdup is the reaction with a metal, especially nickel. Nickel
undergoes oxidation to NiO in air at high temperatures and obeys a parabolic rate law [8]:

$$(\text{rate})^2 = Kt + K' \quad ,$$

where K and K' are constants and t is the exposure time. There is a
considerable difference in the values of K which are found in the
literature [8-10]. The rate data from Gulbransen and Andrew [9] and
Baur, *et al.*, [10] indicate the amount of oxygen found in the fuel
cycle could come strictly from an oxidation-reduction cycle of the

nickel itself. As a check of the hypothesis that the oxygen in the
fuel cycle was a result of a redox phenomenon, the system was prepared
as follows. First, the carbon was burned off the catalyst until no CO
or CO_2 was seen in the exhaust. Then, the oxidized nickel was reduced
by passing CO over it for about one hour, after which time the CO_2 content in the effluent was stable at less than one percent. Finally, the
CO was swept out by purging for five minutes with N_2, after which time
only N_2 was seen in the exhaust. Fuel was then introduced into the
reactor. The results are shown in Fig. 13; this should be compared
with Fig. 7, which shows the same conditions without prereduction of

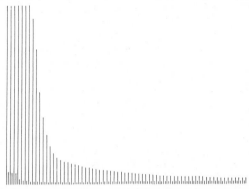

FIG. 13 *Run under conditions like those in Fig. 7 but after CO
reduction of catalyst.*

the catalyst. Rather dramatically, after the prereduction, the N_2 concentration falls off rapidly with no corresponding large rise in CO or
CO_2. Consequently, the H_2 concentration (not shown) rapidly jumps to
higher values. It becomes more readily apparent, then, that the CO and
CO_2 in the fuel cycle effluent is a problem of an insolable side reaction, and not a drawback of the pyrolysis reactions. With this
technique, study of the cracking reaction was separated from the side
reaction imposed by the systemic cycling. A series of data was then
taken to indicate the basic performance of the reactor under various
temperatures, fuels, and fuel flow rates. The data, which are summarized in Table III, illustrate the reactor's ability to perform well on
a wide variety of fuels, ranging from diesel fuel to commercial gasoline.* The reactor is expected to run routinely on almost any liquid
hydrocarbon fuel from diesel to super premium gasolines. The 23 runs
which provided the data in Table III were made on a single bed, and
are meant to illustrate typical data over a wide range of conditions.

The CO and H_2O concentrations seen in the data are for the most
part a function of technique, *i.e.*, it was difficult to reduce the
catalyst to the same point before each run, so the CO varies according
to how well this was done. Any correlation which may appear in the
data for the CO or H_2O concentrations is an artifact of technique.
Also, the accuracy of the percent conversion figures are about ±5%,
the accuracy of the fuel flow meter.

The effect of fuel space velocity on the product is very significant and rather clearly illustrated by the data. Table IV gives the
compositions of the products in Table III averaged over all the fuels

*The Sun Oil Co., Philadelphia, Pennsylvania, provided a sample
barrel of Sunoco 190.*

TABLE III

Summary of Data for Various Conditions

Run No.	Fuel Used	Temp. °C	Fuel Flow cc/min.	H_2 Flow in (CFM)	Product Analysis (%)				H_2 Flow Out (CFM)	% Conversion
					H_2	CO	CH_4	H_2O		
1	Combat	921	2.6	.100	91.5	5.0	1.0	2.5	.099	99%
2	190	935	2.6	.117	95.6	2.2	1.7	0.5	.113	97%
3	190	1025	2.6	.117	95.0	2.5	2.0	0.5	.113	97%
4	Combat	1037	2.6	.100	93.5	5.0	---	1.5	.096	96%
5	Diesel	1075	2.8	.113	94.7	1.8	3.4	0.1	.105	92%
6	JP4	1112	2.6	.114	97.8	1.9	0.1	0.2	.111	97%
7	Combat	1125	2.6	.100	98.5	1.5	---	---	.096	96%
8	190	1130	2.6	.117	95.7	2.1	2.2	---	.112	96%
9	190	935	4.8	.218	90.2	3.1	6.5	0.2	.198	91%
10	190	1025	4.9	.218	91.8	1.8	6.1	0.3	.192	88%
11	Diesel	1075	4.5	.182	93.4	4.8	1.8	---	.168	92%
12	JP4	1112	4.9	.213	94.5	5.4	---	0.1	.181	85%
13	190	1130	4.9	.218	92.9	3.1	3.9	0.1	.200	92%
14	190	935	7.7	.346	87.2	1.2	11.4	0.2	.280	81%
15	190	1025	7.7	.346	89.7	4.0	6.0	0.3	.296	85%
16	Diesel	1075	7.0	.284	88.7	2.1	9.2	---	.250	88%
17	JP4	1112	7.7	.337	94.9	2.3	2.7	0.1	.270	80%
18	190	1130	7.7	.346	85.2	0.9	13.9	---	.259	75%
19	190	935	12.0	.540	86.6	0.6	12.7	0.1	.296	55%
20	190	1025	12.0	.540	90.6	0.9	8.3	0.2	.391	72%
21	Diesel	1075	10.4	.422	81.9	1.2	16.7	0.2	.302	72%
22	JP4	1112	12.0	.526	87.7	1.2	10.9	0.2	.316	60%
23	190	1130	12.0	.540	84.8	3.8	11.4	---	.331	61%

FUELS

Combat: Combat gasoline, approximate composition $C_{7.36}H_{13.5}$

190: Sunoco 190, approximate composition $C_{7.32}H_{15.15}$

JP4: Jet fuel, approximate composition $C_{8.53}H_{17.59}$

Diesel: Diesel fuel, approximate composition $C_{15.10}H_{24.76}$

TABLE IV

Average Values for Analysis Data

Fuel Flow Rate (cc/min)	%H_2	%CO	%CH_4	%H_2O
2.6	95.3	2.8	1.3	0.7
4.8	92.6	3.6	3.7	0.1
7.7	89.3	2.1	8.5	0.2
12.0	87.4	1.6	10.8	0.1

and grouped according to fuel space velocities. As Fig. 14 shows, the
trend is clearly that space velocity and H_2/CH_4 ratio are related by
some inverse function. Since there is a direct relation between H_2/CH_4
ratio and efficiency, then the space velocity and efficiency are also
inversely related, which is what was also seen for Norcat 100 (Fig. 9).

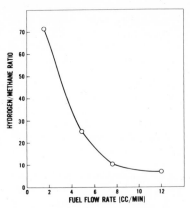

FIG. 14 *Effect of fuel flow rate on H_2/CH_4 ratio.*

The effect of temperature on the reaction products is not clearly
evident from the data in Table III. Temperature is a much less dominating influence on the reaction than fuel space velocity once the
cracking range has been achieved (850°C to 1200°C). The review of a
great deal of data at different temperatures shows a direct relationship between reactor efficiency and temperature. Temperature rises
generally cause the methane concentrations to go down and the H_2 to
go up. Temperature, however, has a strong effect on the nickel oxidation and reduction rate. The temperature, of course, governs the
equilibria of the system, but in the actual process, the physical influences on the system kinetics appear to be at least as important.
The rate of carbon burnoff during the air cycle must be optimized for
each system. There is an optimum value for the air flow rate, corresponding to an oxygen-carbon ratio of between one and two, depending on
fuel space velocity, cycle time, system design, etc. If the system is
run at or reasonably near this optimum value, the system performance
is good. If the air flow rate is substantially lower, more methane
will appear during the fuel cycle, lowering efficiency. If still lower
air rates are used, the reactor will eventually plug up with carbon. If
substantially higher rates than the optimum are used, the carbon will
burn off faster and there will be more oxidation of the catalyst and
subsequently more CO in the fuel stream. At very high flow rates, the
carbon particles will blow out of the reactor, clogging up valves and
filters, and the hot zone of the burnout cycle will extend beyond the
end of the reactor, thus putting severe thermal strain on parts not
designed for this.

The effects of pressure have not been fully investigated as yet,
nor completely understood, but there appears to be a slight favorable
effect on the reaction efficiency with pressure increase.

The data indicate that two reactors with internal volumes of 675 cc
each could produce about 12 cubic feet per hour of hydrogen from 0.5 lb.
of a typical hydrocarbon fuel. This would be sufficient to run a 500-
watt fuel cell stack.

FUEL CONTAMINANTS AND ADDITIVES

Several other reactions occur during the process which, while minor from a material standpoint, are of great importance to the compatibility of the fuel conditioning system with the fuel cell stack. There are many additives and impurities in fuels which contain substances such as sulfur, lead, bromine, chlorine, etc., which are or may be harmful to fuel cell electrodes. Through various analytical methods, some of these elements (Pb, S) were traced so that their fate during the process is fairly clear. Others (*e.g.*, Br, Cl) are in such a small concentration that they may have an insignificant effect on the fuel cell. The product stream was analyzed chromatographically for such sulfur compounds as H_2S, SO_2, COS, and CS_2, with detectability limits less than 100 ppm, and even on the most sour fuel (diesel fuel, sulfur 0.55% w/w), negative results were seen. Traces of SO_2 were seen in the exhaust of the burnout cycle. The sulfur, then, is temporarily retained on the nickel surface as sulfides during the fuel cycle. The nickel surface is a catalyst also for the dissociative adsorption of H_2S [11], which keeps the H_2 from picking up the sulfur during this part of the cycle. During the burnout cycle the sulfur is at least partially stripped off and passes out of the system as SO_2. Tetraethyl lead and similar compounds also decompose, with the lead being deposited on the nickel. Analysis of the used nickel rings indicates the presence of the lead either as a coating or as an undetermined alloy. Traces of chlorides and bromides have been found in the coke samples analyzed. These elements are present in very small quantities, however, and their effect on fuel cells is questionable.

OUTLOOK

The catalyzed pyrolysis of liquid hydrocarbon fuels has been shown to be a feasible method for high-purity hydrogen generation. The advantages of this method over other hydrogen production systems (*cf.* steam reforming) are notably its simplicity, its tolerance to fuel additives, and the purity of the product. The cracking reaction can be made to be close to 100 percent efficient in H_2 yield under laboratory conditions, with a stream purity of greater than 95 percent hydrogen. The problems associated with this method of hydrogen generation are the high temperatures needed — and consequently the related engineering problems — the undesired side reactions such as the nickel redox cycle, and the careful heat-transfer balance necessary for the reaction to be self-sustaining when cycling. The engineering-related problems are being pursued both in-house and through contractual efforts [12,13]. The search for better high-temperature materials and attempts to minimize the side reactions are presently continuing in-house.

ACKNOWLEDGMENT

The author wishes to thank his coworkers on the in-house fuel conditioning team, especially Mr. Edward Gillis and Mr. Robert Trader, for their help and timely suggestions. The author is indebted to Dr. James Huff and Dr. J. B. O'Sullivan for their helpful comments in the preparation of this manuscript. All Figs. are released U. S. Army photographs.

REFERENCES

[1]   E. A. Gillis, "Open Cycle Hydrocarbon-Air Fuel Cell Power Plant,"
      5th Intersociety Energy Conversion Engineering Conference, Las
      Vegas (1970).

[2]   O. J. Adlhart and P. L. Terry, "Phosphoric Acid Fuel Cell Stacks,"
      Final Technical Report on U.S. Army Contract No. DAAK02-68-C-0407,
      Engelhard Industries, Newark, N. J. (May 1969).

[3]   Pratt & Whitney Aircraft, under U.S. Army Contract DAAK02-70-C-
      0518.

[4]   USAMERDC In-house design fuel cell stack.

[5]   Air Products and Chemicals, Inc., Washington, D. C.

[6]   D. Fleming, *et al.*, "Research, Development, and Design of a
      Breadboard Fuel Conditioner Sized for a 1.5KW Fuel Cell System,"
      Final Report, U.S. Army Contract No. DAAK02-69-C-0452, Institute
      of Gas Technology, Chicago, Illinois (February 1970).

[7]   A. Voorhies, *Ind. and Eng. Chem.*, **37**, 318 (1945).

[8]   L. Berry and J. Paidassi, *Compt. Rend., Ser. C*, **262** (18), 1353
      (1966).

[9]   E. A. Gulbransen and K. F. Andrew, *J. Electrochem. Soc.*, **104**,
      451 (1957).

[10] J. P. Baur, R. W. Bartlett, J. N. Ong, Jr., and W. M. Fassell, Jr.
      J. Electrochem. Soc., **110**, 185 (1963).

[11] M. W. Roberts, *First Internat. Congr. Metallic Corrosion.*,
      London (1961), p. 84.

[12] U.S. Army Contract No. DAAK02-70-C-0517, Engelhard Ind.

[13] U.S. Army Contract No. DAAK02-70-C-0518, Pratt & Whitney Aircraft.

PROBLEMS OF THE ELECTROCHEMICAL OXIDATION OF HYDROCARBONS IN LOW TEMPERATURE FUEL CELLS[*]

D. Doniat

Institut Battelle, Geneve, Switzerland

ABSTRACT

The practical application of fuel cells employing hydrocarbons is limited by the fact that the kinetics of the overall electrochemical oxidation is very slow. In order to improve the kinetics, a better knowledge is required of the reaction mechanism occurring at the electrode. In particular, further knowledge is required of the adsorption step which appears to be the rate determinant. The majority of methods generally used for investigating this problem have various disadvantages. A new method has been employed to study these adsorption phenomena. Basic equations have been established for the analysis of the adsorption kinetics. Certain kinetic constants have been determined without resort to consideration of mass transfer or charge transfer phenomena.

INTRODUCTION

Several studies have been made in recent years in order to elucidate the mechanism of the direct electrochemical oxidation of hydrocarbons. Their use as fuel in low-temperature fuel cells is, in effect, of great economical interest.

However, the realization of such fuel cells is fraught with considerable difficulties. The problems posed are related to the fact that the only known catalysts giving an acceptable rate of reaction are essentially platinum or other noble-metal-based compounds. Cheaper catalysts always give an extremely feeble overall reaction rate and thereby their use is excluded for the construction of electrical generators of sufficient specific power.

Thus, it is necessary both to increase the reaction rate of the oxidation of hydrocarbons on known catalysts and to select new and cheaper catalysts.

One of the approaches used to attain these objectives is to procede with an empirical screening of the properties of the greatest possible number of metals and alloys without the choice being, *a priori*, particularly motivated. This type of study has been carried out, without great success, in a certain number of laboratories, and it would now seem that the solution of the problem lies rather in a better understanding of the reaction mechanisms at the electrodes. In effect, only this will point to the rate-determining steps and will lead to the understanding of the reasons why platinum is such an active catalyst. The fundamental data collected will help in the choice of other catalyst formulae.

[*]This work has been carried out in the Laboratoire d'Électrolyse C.N.R.S. — Bellevue, France.

Nevertheless, the study of the reaction mechanism of the electro-chemical oxidation of hydrocarbons is complex and must take into account several different processes:

Mass transport.
Reactant fixation at the electrode (adsorption......).
Charge transfer and associated heterogeneous chemical reactions.

The electrochemical methods employed until now have not allowed the independent study of these diverse processes and their role in the overall process. Nevertheless the knowledge of the rate-determining steps of the reaction should permit its control and the improvement of the kinetics.

In particular, it seems that it is indispensable to analyze selectively the processes governing the fixing of the reactants on the electrode so as to improve our knowledge of the nature of the species from which the charge transfer process takes place.

ADSORPTION PHENOMENA STUDY METHODS

The study of adsorption phenomena is usually carried out via the analysis of isotherms, that is to say the curves representing the variation of electrode coverage as a function of the concentration of the species in the medium under consideration. This method is not very appropriate at very reduced concentrations where the isotherms are almost linear; this is the case for hydrocarbons whose solubilities in aqueous solutions are very low.

However, the analysis of the curves of the surface coverage as a function of adsorption time allows us, even for low concentrations, to differentiate among the various postulated adsorption mechanisms; this is because the shape of the curves is strongly representative of the theoretical kinetic function.

Several authors have based their studies of adsorption kinetics on the analysis of the curves of surface coverage as a function of time. However, in all cases, the classical approach presents a certain number of characteristic, fundamental drawbacks. The most important of these is due to the fact that the electrode is polarized and thus can give rise to a nonnegligible current during the period in which the adsorption is supposed to take place. Under these conditions, the overall adsorption kinetics, *i.e.*, the evolution of the electrode coverage, depends not only on the characteristic constants of the elementary adsorption processes, but also on those of the charge transfer process. Thus, the shapes of the experimental curves obtained by these methods for the variation of surface coverage against the time of adsorption are functions of the value of the overall charge transfer terms, which themselves depend upon the applied overvoltage. The differences observed between the curves measured at several adsorption potentials are thus difficult to exploit. Also the existence of a faradic current, always nonnegligible, results in the formation of different reaction products whose electrode adsorption and reactivity with the adsorbate could considerably modify the kinetics of the adsorption under study.

A second inconvenience is related to the fact that in the methods normally used, the kinetics of the mass transport phenomena, in particular of diffusion, cannot always be considered as being rapid compared with the phenomenon under study. Thus the analysis of curves

to yield information on the adsorption process implies that a diffusion
term should be taken into account. This correction is often difficult
or even impossible to estimate.

Thus we felt it essential to develop a method for the selective
analysis of the processes involved in the fixation of reactants on the
electrode. The principle and the experimental equipment relative to
this method have been discussed in several publications [1,2] and
are not developed here.

ANALYSIS OF THE ADSORPTION KINETICS

In most heterogeneous catalysis studies, an empirical equation is
proposed for the representation of the adsorption kinetics of diverse
gases on metals. This equation can be represented as:

$$\frac{d\theta}{dt} = V_0 \exp(- a\theta) \tag{1}$$

where V_0 is the initial rate of adsorption ($\theta = 0$).

Most authors explain such a relationship, named after Elovich [3],
by a linear variation of adsorption activation energy with surface
coverage. Several hypotheses have been put forward to explain this
linear variation. Both a model of heterogeneous active sites with
noninteracting adsorbed molecules and a model of homogeneous sites and
interacting molecules have been proposed. In the latter case the
parameter "a" of equation (1) is related to a coefficient of interaction.

Such considerations have also been adopted to explain adsorption
in an electrolytic medium. However, in certain cases, even when the
kinetics assumes an Elovich form, it is unrealistic to assume marked
site heterogeneity or interactions between the adsorbed molecules.
For this reason, we felt that it was absolutely necessary to examine
whether the explanations that would not imply a relationship between
the activation energy and the surface coverage could lead to functions
of the same shape.

Let us consider first the cause of nondissociative adsorption of
a compound on a surface so homogeneous that the reaction surface is
the same as the apparent surface.

The general equation for the kinetics of adsorption is:

$$\frac{d\theta}{dt} = \frac{kT}{h} C (1-\theta) \exp\left(- \frac{W}{RT}\right) - \frac{kT}{h} \theta \exp\left(- \frac{W'}{RT}\right) \tag{2}$$

in which k, T, h and R have their usual meanings, C is the partial
pressure of the active substance, and W and W' are respectively the
activation energies of adsorption and desorption.

For low coverages, it is possible to disregard the desorption term
$\frac{kT}{h} \theta \exp\left(- \frac{W'}{RT}\right)$.

We then have:

$$\frac{d\theta}{dt} = \frac{kT}{h} C (1-\theta) \exp\left(- \frac{W}{RT}\right) \tag{3}$$

The term $(1-\theta)$ for $\theta \ll 1$, can be identified at the start of the development in series of $(\exp - \theta)$, which means that equation (3) is equivalent under these circumstances to:

$$\frac{d\theta}{dt} = \frac{kT}{h} \ C \ \exp\left(-\frac{W}{RT}\right) \ \exp(-\theta) \qquad . \tag{4}$$

Consequently, we see in this case that the kinetics observed can be described by a law or formula analogous to equation (1); the term "a" then is equal to 1.

When, for a reaction of the same type, the reacting surface is composed of two fractions, X_0 and Y_0, characterized by sites having different activation energies, W_1 and W_2, we can write expressions analogous to equation (3) for X_0 and Y_0.

Assuming:

$$\theta_x + \theta_y = \theta \quad \text{and} \quad \frac{d\theta_x}{dt} + \frac{d\theta_y}{dt} = \frac{d\theta}{dt}$$

we obtain by means of a development in limited series a function $\frac{d\theta}{dt} = f(t)$ of the form:

$$\frac{d\theta}{dt} = mX_0 + nY_0 - t(X_0 m^2 + Y_0 n^2) \tag{5}$$

with: $m = \frac{kT}{h} \ C \ \exp\left(-\frac{W_1}{RT}\right)$ and: $n = \frac{kT}{h} \ C \ \exp\left(-\frac{W_2}{RT}\right)$.

Now, equation (1) expressed in the form $\frac{d\theta}{dt} = f(t)$ leads, with the same approximations, to:

$$\frac{d\theta}{dt} = V_0 - V_0^2 at \qquad . \tag{6}$$

Equation (5) can therefore be identified with equation (6) in which the term "a" has the following value:

$$a = \frac{X_0 m^2 + Y_0 n^2}{(X_0 m + Y_0 n)^2} \qquad .$$

Similar treatment carried out for a surface that can be decomposed into n fractions also gives us, for $\frac{d\theta}{dt} = f(t)$, a formula that can be identified with equation (6).

It is easy to show that when $(W_2-W_1) < 50$ cal., "a" tends toward 1, and the system acts substantially the same as in the case of a homogeneous surface. On the other hand, when $(W_2-W_1) > 200$ cal., surface Y_0 can be considered inactive and in that case "a" is equal to $\frac{1}{x}$, that is to say, to the ratio of the apparent surface to the active surface. Let us now consider the case of a dissociative adsorption produced on a surface the active fraction of which is X_0.

If S is the number of sites occupied after adsorption of a molecule of reagent, the kinetics of the reaction can be written, neglecting desorption and assuming $\frac{\theta}{X_0} \ll 1$:

$$\frac{d\theta}{dt} = \frac{kT}{h} CX_0 \exp\left(-\frac{W}{RT}\right) \exp\left(-\frac{S}{X_0}\theta\right) \quad . \tag{7}$$

This expression can be identified with equation (1) with "a" = $\frac{S}{X_0}$.

It is readily seen that many adsorption reactions have kinetics similar to that of Elovich without any phenomenon of variation of adsorption activation energy being involved. It is therefore absolutely necessary to take dissociation parameters and surface factors into account in order to evaluate the terms of any intermolecular interaction that may exist.

Equation (7) expresses effects of surface activity and of dissociation. It can be expanded to allow for a possible variation of adsorption activation energy.

Thus:

$$\frac{d\theta}{dt} = \frac{kT}{h} CX_0 \exp\left(-\frac{W_0}{RT}\right) \exp\left(\frac{S+\alpha}{X_0}\theta\right) \quad , \tag{8}$$

in which α represents the interaction coefficient and W_0 is the initial adsorption activation energy.

The relationship must now be established between the experimental value q obtained by a coulometric method and the electrode coverage θ. θ is defined here as the proportion of the number of sites occupied by the adsorbed substances to the total number S_0 of sites existing on the surface of the electrode. If S is the number of sites occupied after the adsorption of one hydrocarbon molecule (an adsorption which may have created other adsorbed particles by dissociation), and Z the number of electrons exchanged in the oxidation of the adsorbed molecule and its products of dissociation, the relation between q and θ is written:

$$\theta = q \frac{NS}{ZFS_0} \quad , \tag{9}$$

in which N is the Avogadro number and F the Faraday.

The quantity q is determined in such a way that it is exclusively a function of the oxidation or of the reduction of the adsorbate. Corrections should be made to eliminate any effects of double layer or of electrochemical transformation of the surface.

Combining equations (8) and (9), we obtain:

$$\text{Ln} \frac{dq}{dt} = \text{Ln}\left[\frac{kT}{h} CX_0 \frac{ZFS_0}{NS} \exp\left(-\frac{W_0}{RT}\right)\right] - q \frac{NS}{ZFS_0}\left(\frac{S+\alpha}{X_0}\right) \quad . \tag{10}$$

The comparison of the experimental curves with this equation leads to the determination of the characteristic parameters of the adsorption kinetics.

By way of example, Fig. 1 shows the curves obtained from the tests made on a smooth platinum electrode in a normal solution of sulfuric acid at 80°C.

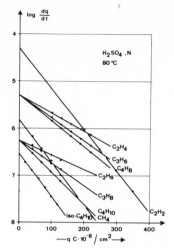

FIG. 1 *Variation of the coverage speed* $\frac{dq}{dt}[C \cdot cm^{-2} \cdot s^{-1}]$ *as a function of the coverage of the electrode in* H_2SO_4 *at 80°C.*

RESULTS

The study was made on five alkanes (CH_4, C_2H_6, C_3H_8, n-C_4H_{10}, iso-C_4H_{10}), three alkenes (C_2H_4, C_3H_6, n-C_4H_8) and on acetylene. The temperature of the trials was fixed at 80°C and 90°C. The electrolytic medium was a normal solution of sulphuric acid or phosphoric acid. The electrode was a plate of polished platinum. Before each adsorption operation, this electrode was pretreated electrochemically in order to assure a perfect reproducibility of the state of the surface of the metal. The parameters measured under these conditions are given in Tables I, II and III.

TABLE I

Characteristic constants of the adsorption kinetic of various hydrocarbons on smooth platinum

Hydrocarbon	Z	S	W_0 cal/mole	α
CH_4	8	2.1	22,500	11
C_2H_6	14	3.9	23,300	0.94
C_3H_8	20	4.7	23,200	3.80
n. C_4H_{10}	26	5.6	22,900	5.59
iso. C_4H_{10}	26	6.6	23,200	8.85
C_2H_4	12	3.8	22,300	1.34
C_3H_6	18	4.7	21,700	3.38
n. C_4H_8	24	5.7		4.22
C_2H_2	10	3.9	22,400	5.81

in H_2SO_4 at 80°C

TABLE II

Characteristic constants of the adsorption kinetic of various hydrocarbons on smooth platinum

Hydrocarbon	Z	S	W_0 cal/mole	α
CH_4	8	2.1	22,500	11
C_2H_6	14	4.2	23,300	1.66
C_3H_8	20	5.1	23,100	3.91
n. C_4H_{10}	26	6.1	22,900	9.99
C_2H_4	12	4.1	21,900	4.29
C_3H_6	18	5.1	21,200	5.87
n. C_4H_8	24	6.15		9.08
C_2H_2	10	4.2	21,300	5.14

in H_2SO_4 at 90°C

TABLE III

Characteristic constants of the adsorption kinetic of various hydrocarbons on smooth platinum

Hydrocarbon	Z	S	W_0 cal/mole	α
CH_4	8	1.9	22,400	15
C_2H_6	14	3.7	22,100	7.6
C_3H_8	20	4.6	22,000	13.9
C_2H_4	12	3.6	21,500	5.6
C_3H_6	18	4.6	21,000	12.1
C_2H_2	10	3.8	19,200	11.9

in H_3PO_4 at 80°C

DISCUSSION

A certain number of observations concerning the reaction mechanism of the fixation of hydrocarbons on the platinum can be drawn from an examination of the results.

It can be seen that in every case studied, except for that of isobutane, the number of sites occupied by the adsorption products of one molecule is equal to the number of carbon atoms in that molecule plus two. It can be assumed that the adsorption process is dissociative and provokes the rupture of C-H bonds in such a way that two atoms of hydrogen are liberated and occupy two sites. This preferential dissociation of the C-H bonds has been assumed by several authors.

For the alkanes and alkenes, as well as for acetylene, the dehydrogenation probably takes place at the chain ends. This hypothesis is
also borne out by the fact that the fixation of one molecule of isobutane leads to the occupation of seven sites. The presence in this
molecule of an extra methyl group appears to lead to the breaking of
a third C-H bond.

This occupation level must not necessarily be interpreted as being
due to the interaction of each site occupied with an adsorbed particle.
The feeble variation of W_0 as a function of the number of carbon atoms
in the molecule for hydrocarbons of the same family would suggest
rather that the number of bonds between the platinum and the dissociatio
products remains constant. These bonds are established, on the one
hand, between the platinum and the dehydrogenated terminal groups, and
on the other hand, between the platinum and the hydrogen atoms formed
by the dissociation. The other sites occupied are not bonded, but
simply covered by the central section of the chain.

A certain number of remarks can be made as to the values found
for the interaction coefficient. In effect, this coefficient increases
as a function of the length of the adsorbed molecule at a given temperature and in a given medium. It also increases sharply when, for
molecules containing the same number of carbon atoms, the number of
radicals undergoing a bond cleavage increases.

It is usually less for alkenes than for alkanes for a given number
of carbon atoms and at an equal degree of dissociation. Finally,
acetylene gives rise to a very high value of α. This is probably
related to the fact that the dehydrogenation of this molecule gives
rise to a highly reactive carbon residue which can react with the
environmental medium. It should however be noted that, even for the
highest values of α, the variations of the activation adsorption
energy W as a function of coverage remain relatively feeble. Thus, for
isobutane, the alkane having the highest interaction, the variation of
W is about 620 cal/mole as the coverage changes from 0 to 0.1.

The effect of the type of electrolyte used on the value of α is
marked and differs considerably for sulphuric and phosphoric acids.
Thus, the interaction term is related not only to the attractive or
repulsive forces between adsorbed molecules, but also to the intervention of electrolyte ions. It is probable that the effect of anions
is preponderant. This would be in accordance with the specific adsorption of H_3PO_4 ions demonstrated by other studies.

REFERENCES

[1] D. Doniat, *Thesis*, Paris (1969).

[2] M. Bonnemay, G. Bronoel, D. Doniat and E. Levart, *International
     Conference on Electrochemical Power Sources Systems for Space
     Applications*, Paris (December 1967).

[3] S. Ju. Elovich and G. Zhabrova, *Zh. Fiz. Khim.*, **13**, 1716, 1775
     (1939).

COMMENT ON PAPER OF D. DONIAT
BY J. O'M. BOCKRIS

In the way in which Dr. Doniat has carried out his experiments, the
relationship to the potential of the metal-solution interface may be

somewhat suppressed. It is by no means only for practical reasons
that a knowledge of such a potential dependence is so important. One
of the more significant aspects in understanding the mechanism of
adsorption is to know whether there is a primary dissociation of the
organic molecule to radicals and/or ions in its step from the solution
to the electrode. Is the substance itself (*i.e.*, the organic molecule
in solution) present at greatest concentrations on the electrode
surface? Or is the principal adsorband a radical?

One seems to meet both cases from the evidence of the literature.
For example, methanol on mercury is certainly mainly to be treated
as a molecule in equilibrium with the mercury surface. Conversely,
propane on platinum is largely a case of dissociation.

One of the more important factors by which to distinguish the two
kinds of mechanism (molecule in solution → molecule adsorbed; or
molecule in solution to dissociated radical + ions + e) is the dependence of adsorption upon potential. If the molecular case is present,
the coverage-potential relation is parabolic and fairly symmetrical:
if dissociation occurs, the relationship is highly dissymmetrical,
falling much more rapidly on desorption than the rise during adsorption.

The other aspect of the organic adsorption which seems to be
neglected in the discussion by Dr. Doniat, is the position of the
maximum on a coverage-potential plot.

Theoretical relations exist which connect the difference between
the potential of the adsorption maximum and the potential of zero
charge [1]. At low concentrations, such relations can be indicative,
for example, of the mechanism of adsorption.

REFERENCE

[1]   J. O'M. Bockris and D. A. J. Swinkels, *J. Electrochem. Soc.*, 111,
      743, 736 (1964).

THE STRUCTURE OF HYDROPHOBIC GAS DIFFUSION ELECTRODES

J. Giner

Tyco Laboratories, Inc., Waltham, Massachusetts

ABSTRACT

The "flooded agglomerate" model of the Teflon-bonded gas diffusion electrode is discussed. A mathematical treatment of the "flooded agglomerate" model is given; it can be used to predict the performance of the electrode as a function of measurable physical parameters.

INTRODUCTION

The Teflon-bonded electrode [1,2] represents a very efficient structure which allows extensive utilization of the catalyst and therefore very high current densities at relatively low polarization. Based on our experience with this type of electrode, we have developed a qualitative description of the mechanism based on a double porosity model [3]. We assume (Fig. 1) that the catalyst particles form porous, electronically conductive agglomerates which under working conditions are

GAS

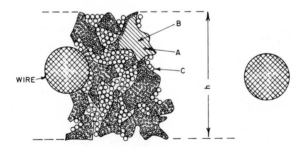

ELECTROLYTE

FIG. 1 *Schematic representation of a hydrophobic, Teflon-bonded electrode. (A) catalyst particle; (B) agglomerate; (C) Teflon particle.*

completely flooded with electrolyte. These catalyst agglomerates are held together by the Teflon binder which also creates hydrophobic gas channels. When current is drawn from the electrode, reactant gas diffuses through the hydrophobic channels, dissolves in the electrolyte contained in the agglomerates and reacts on available sites of the catalyst particles.

Some experimental measurements which confirm the validity of this qualitative model are discussed in [4]. In the following, a discussion of the mathematical treatment of a simplified model is presented. This discussion was first presented in [5].

215

MATHEMATICAL TREATMENT

One way of quantitatively treating this working mechanism is to substitute a column of flooded agglomerates perpendicular to the electrode surface by a porous cylinder of radius r_0 and length h (as shown in Fig. 2), in which catalyst particles and electrolyte are homogeneously dispersed as a continuum.

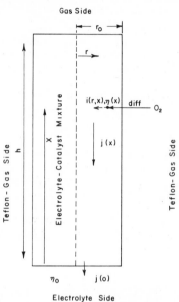

FIG. 2 *Schematic representation of flooded cylinder.*

During operation, gas arrives at the lateral surface of the cylinder and diffuses radially to its center; simultaneous reaction occurs on catalyst particles in the diffusion path. Ionic current is conducted in the axial direction of the cylinder.

For the mathematical treatment we assume that:

1) The electrode is made up of a number of porous cylinders of catalyst flooded with electrolyte. These cylinders are perpendicular to the external surface of the electrode.

2) Electrolyte and catalyst are homogeneously mixed as a continuum.

3) The intrinsic activity of the catalyst is constant throughout the cylinder.

4) Equilibration of electrolyte concentration in cylinders occurs efficiently *via* an evaporation-condensation process.

5) The local current density is directly proportional to the local concentration of reactant; *i.e.*, an expression such as Eq. (1) is pertinent.

6) The voltage in the cylinder changes only in the axial direction, and diffusion of dissolved gas occurs only in the radial direction.

7) There are no transport limitations in the gas phase.

8) There are no kinetic limitations in the process of gas dissolution.

9) There is no electronic iR-drop in the cylinders.

10) Convection inside of the cylinders is low and has negligible effect on current.

In order to extend the theory of a single cylinder to the complete electrode, we will further assume at this point that:

11) The radius of all porous cylinders has the same value. (Under these conditions the number of cylinders per cm^2 of electrode (N) is related to a measurable macroscopic factor, which we will call macroporosity (β), by the expression $\beta = 1 - N\pi r_0^2$).

Based on these assumptions, we can write Eqs. (1) to (4) as governing equations:

$$i = i_0\left[\frac{C(x,r)}{C_0} \exp[\alpha z\eta(x)/\phi] - \exp[-(1-\alpha)z\eta(x)/\phi]\right] \quad , \tag{1}$$

$$\overline{D}\left(\frac{\partial^2 C}{\partial r^2} + \frac{1}{r}\frac{\partial C}{\partial r}\right) = \frac{i\gamma}{nF} \quad . \tag{2}$$

Boundary conditions: $\frac{\partial C}{\partial r} = 0$ at $r = 0$, $C = C_0$ at $r = r_0$,

$$\frac{d\eta}{dx} = \frac{j(x)}{\pi r_0^2 \kappa} \quad , \tag{3}$$

$$-\frac{dj(x)}{dx} = 2\pi\gamma \int_0^{r_0} i r \, dr \quad . \tag{4}$$

The solutions of these equations are Eqs. (5) to (7):

$$C = C_0 \exp\left(-\frac{z\eta}{\phi}\right) + C_0 \left[1 - \exp\left(-\frac{z\eta}{\phi}\right)\right] \frac{I_0\left(q\frac{r}{r_0}\right)}{I_0(q)} \quad , \tag{5}$$

$$i = i_0\left[\exp\left(\frac{\alpha z\eta}{\phi}\right) - \exp\left(-\frac{(1-\alpha)z\eta}{\phi}\right)\right] \frac{I_0\left(q\frac{r}{r_0}\right)}{I_0(q)} \quad . \tag{6}$$

With $q = \left[\frac{\gamma i_0 r_0^2}{nF\overline{D}C_0} \exp\left(\frac{\alpha z\eta}{\phi}\right)\right]^{1/2}$

$$\frac{d^2\eta'}{dx'^2} = \frac{2nF\overline{D}C_0 h^2}{n_0\kappa r_0^2}\left[1 - \exp\left(-\frac{z\eta_0}{\phi}\eta'\right)\right]\frac{qI_1(q)}{I_0(q)} \quad . \tag{7}$$

Boundary conditions:

$\eta' = 1$ at $x' = 0$; $d\eta'/dx' = 0$ at $x' = 1$.

Equations (5) and (6) give the value of the local concentration and local current density along the radius of the cylinder, while Eq. (7) gives the variation of the local electric potential along the axis of the cylinder. A parameter of high importance to define the radial utilization of a catalyst agglomerate is the parameter q. The physical meaning of q can be understood by defining

$$I_{act} = \gamma i_0 r_0 \exp \frac{\alpha z \eta}{\phi}$$

and

$$I_{diff} = \frac{nF\overline{D}C_0}{r_0} \quad ,$$

so that

$$q = \left(\frac{I_{act}}{I_{diff}}\right)^{1/2} \quad .$$

I_{act} is the current that would be obtained under exclusive activation control from a block with a 1 cm^2 surface and r_0 thickness, made of the catalyst of bulk area γ; and I_{diff} is the diffusion limiting current density if the reactant were to be consumed at one of the surfaces of the same porous block after diffusing through it. Obviously q will be large when diffusion control is more important than activation control.

It can be seen that Eq. (7) expresses η' only as a function of x' and a series of measurable constants. This equation can be solved with a relatively simple computer program which will tell us the effect of parameters on the following:

1) The radial distribution of the local current at constant x'. This is a measure of the depth utilization of the cylinder (or agglomerate), which is controlled mainly by diffusion.

2) The current production at different values of x'. This is a measure of the depth utilization of the electrode, which is controlled mainly by ohmic drop.

3) The current density per unit area of electrode, *i.e.*, the measurable current density.

DEPTH UTILIZATION OF CYLINDER

The local current can be obtained from Eq. (6). Thus the equation relating i and r can be written as:

$$i = pI_0\left(q\,\frac{r}{r_0}\right) \quad , \tag{8}$$

where p and q, distribution parameters, are suitably defined. From q the utilization of the catalyst as a function of r for a given x can be obtained, as shown in Fig. 3.

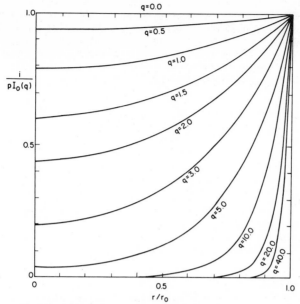

FIG. 3 *Radial utilization of agglomerate.*

DEPTH UTILIZATION OF ELECTRODE

Other information of a pertinent nature is the current production for different cross sections of the cylinder. This current is, of course, $dj(x)/dx$. The total current per cylinder across the surface x = constant, is:

$$j(x) = \frac{\pi \bar{\kappa} r_0^2 n_0}{h} \frac{dn'}{dx'} . \tag{9}$$

ELECTRODE CURRENT DENSITY

The current density per unit area of electrode surface $I(0)$ can be obtained from the total current of a cylinder $j(0)$ by:

$$I(0) = j(0) \frac{1 - \beta}{\pi r_0^2} , \tag{10}$$

or

$$I(0) = \frac{-\bar{\kappa} n_0 (1-\beta)}{h} \left(\frac{dn'}{dx'}\right)_{x'=0} \tag{11}$$

APPLICATION OF THE MODEL TO OXYGEN REDUCTION

We applied this theory to a Teflon-bonded platinum electrode operating as an oxygen electrode in 30 percent KOH at 80°C. [5] Under these conditions, the exchange current of Pt is $i_0 = 5 \times 10^{-7}$ A/cm^2. The electrode was assumed to be 0.02 cm thick and to contain 20 mg Pt/cm^2.

A computer program was used to study the effect of microporosity (θ), agglomerate radius (r_0) and exchange current (i_0) on: 1) measurable electrode current density, 2) radial distribution of current at constant x (radial utilization of agglomerate), and 3) internal ionic iR drop (transversal utilization of electrode). For these calculations an electrode polarization of 0.3 volts (*vs.* reversible potential) was assumed.

Decreasing the agglomerate radius from 10^{-3} to 10^{-4} cm has a large effect on current, but a further decrease below 1 micron has little effect on performance. This finding applies, of course, only under the diffusion conditions of this example. The current densities calculated for $i_0 = 5 \times 10^{-7}$ A/cm^2 and agglomerate sizes $r_0 \leq 10^{-4}$ cm are similar to those obtained experimentally under the same conditions.

Regarding the effect of agglomerate size, microporosity and exchange current on internal ionic iR drop, for a microporosity $\theta > 0.7$, $\eta < 20$ mV even for $i_0 = 10^{-6}$ A/cm^2 and $r_0 = 10^{-5}$ cm, *i.e.*, for conditions where the measurable current is higher than 300 mA/cm^2. This is in agreement with the experimental results obtained with $\theta > 0.8$.

Regarding the effect of microporosity, agglomerate size and exchange current on the radial utilization of the agglomerate, it was shown that for $\theta > 0.7$ and $i_0 < 10^{-6}$ A/cm^2, reasonable radial utilizations of the agglomerate are obtained even when the agglomerate size is 10^{-4} cm. The radial utilization is excellent for $r_0 = 10^{-5}$ cm.

An analysis of this mathematical treatment, using approximations, showed that for good transversal utilizations of the electrode one obtains the same Tafel parameter b when the radial utilization is good (q < 1) as those obtained from smooth electrodes while the Tafel parameter b of the porous electrode is twice that of the smooth electrode for poor radial utilization (q >> 1).

Further, an analysis of the assumptions showed that the model can be used only for electrode reactions which have low i , such as the O_2 electrode. The model has to be drastically modified when used with reactions with high i_0, *e.g.*, the hydrogen electrode.

assumed Tafel kinetics

CONCLUSIONS

1) The model allows one to predict not only the utilization of the catalyst across the thickness of the electrode (transversal utilization) but also the utilization of the catalyst along the radius of the flooded agglomerate. As a consequence, in addition to often-studied variables such as porosity (macro- and micro-), bulk area, *etc.*, the importance of agglomerate size is shown. A parameter (q) which determines the radial current distribution in an agglomerate is introduced and quantitatively defined as a function of diffusion coefficient, microporosity, solubility of reactant gas, exchange current, bulk surface area, local potential and agglomerate radius.

2) For the specific example of the oxygen reduction on Teflon-bonded Pt electrodes in 30 percent KOH at 80°C, a good radial distribution is obtained for agglomerate sizes below 1 micron and polarizations up to 300 mV. Under similar conditions the transversal utilization of the electrode is very good (very low internal iR drops).

3) For the same example, both the radial utilization of agglomerates and the transversal utilization of the electrode become poor at

higher current drains ($i > 300$ mA/cm^2), and with poor structures (low microporosity and large agglomerate size).

4) For good transversal utilization of the electrode and good radial utilization of the agglomerate, the Tafel plot is obviously the same as that obtained with a smooth electrode. If, on the other hand, the transversal utilization of the electrode is good but the radial utilization of the agglomerate is very poor, a linear relationship is predicted with a Tafel slope twice that of the smooth electrode.

5) The present model and, more specifically, the parameter of radial distribution (q) introduced here can be used to design more efficient hydrophobic gas diffusion electrodes by predicting the maximum agglomerate size tolerable. When due to a high ratio of diffusion to activation control for a certain electrode reaction, high values of q are obtained for all reasonable agglomerate sizes; the model suggests the use of porous, conductive, but catalytically inactive agglomerates which have been activated with catalyst only on their periphery.

6) The proposed model cannot be used without further modification to predict quantitatively the performance of electrodes with high exchange current, as in the case of the hydrogen electrode in acid electrolyte. This is because the parameter q becomes too large and the assumption of a continuum distribution of catalyst and electrolyte does not apply.

ACKNOWLEDGMENT

This work was supported by the National Aeronautics and Space Administration (Headquarters - Washington) under Contract No. NASW 1233.

LIST OF SYMBOLS

C, $C(r,x)$	Concentration of reactant gas at a point (r,x); (mol $\times$ cm^{-3})
C_0	Solubility of reactant gas (mol $\times$ cm^{-3})
D	Diffusion coefficient of reactant gas in liquid (cm$^2 \times$ sec^{-1})
$\bar{D}$	Effective diffusion coefficient of reactant gas in liquid; affected by microporosity and tortuosity
F	Faraday constant
h	Thickness of electrode, (cm)
i, $i(r,x)$	Local real current density (A $\times$ cm^{-2})
i_0	"Real" exchange current density (A $\times$ cm^{-2})
I_0	Bessel function of order zero
I_1	Bessel function of order one
$I(0)$	Electrode current density (A $\times$ cm^{-2})
j, $j(x)$	Current flowing through plane (x) of cylinder, (A)
$j(0)$	Total current produced by cylinder, (A)
N	Number of cylinders in one cm^2 of electrode
n	Number of electrons involved in electrode reaction
p	First distribution parameter
q	Second distribution parameter
R	Gas constant
r	Radial coordinate in cylinder (cm)
r_0	Radius of cylinder (cm)
S	Surface area of catalyst (m$^2 \times$ g^{-1})
T	Absolute temperature (°K)
w	Catalyst load in electrode (g $\times$ cm^{-2})

x	Axial coordinate in cylinder (also in electrode)(cm)
x'	Scaled axial coordinate
z	Stoichiometric number
α	Transfer coefficient
β	Macroporosity
γ	Surface to volume ratio, (cm^{-1})
$\eta, \eta(x)$	Overvoltage at plane (x), (volt)
θ	Microporosity
κ	Ionic conductivity, $(ohm^{-1}cm^{-1})$
$\bar{\kappa}$	Effective ionic conductivity, affected by microporosity and tortuosity, $(ohm^{-1}cm^{-1})$
ρ	Catalyst density $(g \times cm^{-3})$
ϕ	RT/F, (volt)

REFERENCES

[1] L. W. Niedrach and H. R. Alford, *J. Electrochem. Soc.*, 112, 117 (1965).

[2] R. G. Haldeman, W. P. Coman, S. H. Langer and W. A. Barber, "Fuel Cell Systems", *Advances in Chemistry Series*, 47, 106 (1965).

[3] J. Giner, *Proceedings of 21st Annual Power Sources Conference*, PSC Publications Committee, Red Bank, N. J. (1967) p. 10.

[4] J. Giner, J. M. Parry, S. Smith and M. Turchan, *J. Electrochem. Soc.*, 116, 1692 (1969).

[5] J. Giner and C. Hunter, *J. Electrochem. Soc.*, 116, 1124 (1969).

THIN FILM FUEL CELL ELECTRODES

W. J. Asher and J. S. Batzold

Esso Research and Engineering Company, Linden, New Jersey

ABSTRACT

The possibility of increased platinum utilization through the con-trolled deposition of platinum in the optimum position in a fuel cell electrode was demonstrated by Dr. Boris Cahan at the University of Pennsylvania in studies supported by NASA. Electrodes prepared by sputtering thin films of platinum on porous Vycor substrates were shown to avoid diffusion limitations even at high current densities. Our studies have been aimed at gaining a better understanding of this sys-tem with the eventual goal of extending the concept to the construction of practical fuel cell electrodes and total cells. The specific activ-ity of sputtered platinum was not found to be unusually high. However, even at low loadings, performance limitations were found to be control-led by physical processes, confirming the Cahan work. Catalyst activ-ity is strongly influenced by platinum sputtering parameters, apparently as a result of changes in the surface area of the catalyst layer. Work with a more practical substrate, porous nickel, showed that the pore size of the substrate was an important parameter, having a major effect on performance at constant catalyst loading. Surprisingly, electrode performance increased with increasing loading for catalyst layers up to two microns thick, showing the physical properties of the sputtered layer to be different from platinum foil. Electrode performance was also very sensitive to changing differential pressure across the elec-trode. More recently, applications of sputtered catalyst layers to fuel cell matrices has been investigated with the intent to produce thin total cells. Problems of mass and heat balance with such systems are now being evaluated.

INTRODUCTION

The research described in this paper was aimed at exploration of the application of thin-film techniques for the preparation of fuel cell electrodes and total cells [1]. It was based on the results of a fundamental study carried out by Dr. Boris Cahan [2,3]. His work was aimed at gaining a better understanding of the mechanisms opera-tive in the electrochemical dissolution of gaseous reactants in porous electrodes. He was able to obtain experimental data with a simulated single pore defined between a sputtered platinum film on an optically flat silica surface and a silica window. This allowed simultaneous optical and electrochemical data to be obtained as a function of the spacing between the window and the electrode. The space itself was partially filled with electrolyte, with the reacting gas above the electrolyte. Thus, the nature of the meniscus at the electrolyte-platinum film interface, *i.e.*, the reaction site, could be defined accurately during the experiment.

A key finding in this work was the observation that the stable configuration of the meniscus gave a <u>finite</u> contact angle of between one
and three degrees. This was true over a wide range of potential, and
the formation of thin films of electrolyte wetting the catalyst surface
above the meniscus was not observed. In effect, this meant that under
most conditions, over 90 percent of the current was produced at the
very narrow band across the electrode at the meniscus. The implication
is that, to be effective at high current density, catalyst had to be
present near a meniscus edge in a porous electrode.

An important extension of this work was then made. Cahan prepared
electrodes by sputtering platinum films on porous Vycor. This material
is obtainable in flat plates about one mm thick, and has a porosity of
25 percent, nearly all in 50 Å pores. These electrodes could be tested
as interface-maintaining structures, with electrolyte on the noncatalyzed side and reactant gas on the catalyzed side. Currents for the
oxygen reduction reaction could be measured out to 100 mA/cm^2 at reported platinum loadings of 5 to 20 micrograms/cm^2. Furthermore, if
diffusion and resistance limitations due to the <u>structure</u> of the po-
ous Vycor electrode were taken into account, the oxygen reaction itself appeared to be only activation limited even at the very low loadings.

Comparison of the results reported by Cahan with the performance of
conventional fuel cell electrodes of high loading show that the specific
activity of the platinum in the Cahan structure is not significantly enhanced. At the same polarization, the decrease in current with the
Cahan structure is essentially of the magnitude expected from the decrease in loading (Fig. 1). Thus, the high utilization must be attributed to physical rather than catalytic effects.

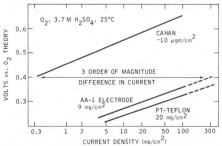

FIG. 1 *Comparison of performances: Cahan vs. conventional electrodes (O_2, 3.7M H_2SO_4, 25°C).*

The objective of our research was to confirm the Cahan findings and
to attempt to prepare more practical electrodes while maintaining the
high effectiveness of the catalyst in the Cahan structure.

CHARACTERIZATION OF SPUTTERED PLATINUM

The sputtering process [4] is a very effective technique for the
preparation of thin metal films. Material to be sputtered is mounted
as the cathode, and the substrate to be coated is placed on the anode
in a vacuum chamber. At low argon pressure (30-40 microns), a glow
discharge is produced between the electrodes by the application of high
voltage. Argon ions are accelerated to the cathode, and by momentum
transfer, target atoms are eroded. Depending on the geometry of the
system, some fraction of these atoms collide with the substrate.

Some preliminary characterization work was done on platinum films, particularly to devise a relatively simple technique for determining the loading. In this work, films were deposited on microscope slides, and both visible light transmission and electrical resistance were compared with the results of direct wet chemical analysis. The results showed that light transmission measurements correlated linearly with loading up to about 200 micrograms/cm^2. Resistance measurements, on the other hand, were useless at loadings below 100 micrograms/cm^2 and relatively insensitive at higher loadings (Fig. 2). Above 200 micrograms/cm^2, loadings could be determined quite accurately by differential weights.

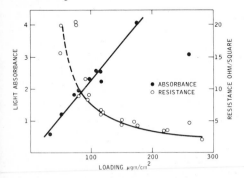

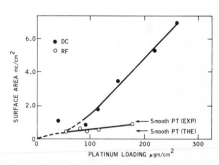

FIG. 2 *Correlation of platinum loading with electronic resistance and light absorbance. (Platinum sputtered on microscope slides)*

FIG. 3 *Comparison of surface area of RF and DC sputtered platinum as a function of loading.*

Electrochemical characterization methods were also used on the films prepared both on microscope slides and on electrodes. In order to maintain adherence of the platinum film to the glass, a thin layer of tantalum was sputtered on the glass before the platinum film. Interesting results were obtained when such films were characterized by voltage scanning and the relative surface area calculated from, for example, the size of the platinum surface oxide reduction peak. Large differences were observed between films produced using direct current power and radio frequency power. The former, at the same loading, produced films having significantly higher surface area than the latter. In fact, the radio frequency film was essentially equivalent to platinum foil in this test (Fig. 3).

No differences were apparent in the physical characteristics of the two types of film. Light transmission and electrical resistance were the same at equivalent loading. An explanation of the differences observed may lie in the introduction of some porosity by argon trapped during DC sputtering; in this case, the argon pressure is normally 30 to 40 microns compared with only 7 microns for the RF case. Confirmation of the electrochemical difference between the two types of film was obtained when DC films were found to give significantly more active electrodes, as will be seen later.

POROUS VYCOR ELECTRODES

Electrochemical evaluation of electrodes prepared by the Cahan technique included both direct measurement of activity for electrochemical reduction of oxygen, and measurement of resistive losses

introduced by the structure. Two resistive losses were identified and
measured independently: the electrolytic loss due to restricted ionic
transport through the small pores of the structure, and the electronic
loss due to the significant resistivity of the thin metal film. The
latter could be reduced by sputtering a patterned gold current collector on the catalyst film. When these losses were taken into account,
the results of Cahan — $i.e.$, activation limited performance out to high
current densities — were confirmed (Fig. 4), although not at the very
low loadings (<25 micrograms/cm^2). For the purposes of our study, however, operation at such low loadings was neither necessary nor practical.

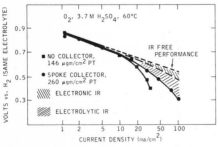 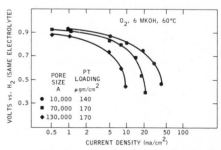

FIG. 4 *Porous glass electrode*   FIG. 5 *Porous glass electrode*
performance (O_2, 3.7M H_2SO_4, 60°C).   *performance (O_2, 6N KOH, 60°C).*

Platinum films at the same loading, prepared on the porous Vycor,
appeared to have about three times the surface area of films prepared
on microscope slides; this was due to the increased surface roughness
of the substrate. Also, films prepared by RF sputtering gave about
20 percent of the current produced by films prepared by DC sputtering;
this was due to the apparent porosity of the latter films. The performances of the DC films were sensitive to loading even when thicknesses equivalent to more than 100 atomic layers were reached, again
indicating porosity in the film. Work with the porous Vycor was not
pursued further, since practical electrodes using this material did
not seem feasible.

ELECTRODES PREPARED ON POROUS NICKEL

The use of porous nickel as an electrode substrate had a number of
advantages. Electronic resistance limitations were no longer a factor
and electrolytic resistance limitations were greatly reduced due to decreased substrate thickness and increased pore size. However, the increased pore size was also a disadvantage since the physical environment of the film was no longer equivalent to that on the smaller pore-
size Vycor substrate.

Electrode performance was found to be very dependent on the pore
size of the six to nine mil nickel substrates. Over the range in pore
size studied, the performance was found to increase as the pore size
was decreased (Fig. 5). Attempts were made to optimize the performance
by further decreasing the pore size of the substrate. These attempts
included sintering and chemical treatments of the substrate at hand;
preparation of substrates at higher temperature and pressure; and the
addition to the substrate of layers of nickel, silver or gold by sputtering and vapor deposition. However, none of these approaches resulted
in improvements in activity.

Although optimization of the <u>physical structure</u> of the electrode by these techniques failed, significant increases in performance could be obtained by <u>external adjustment</u> of the catalyst-electrolyte interface. This was accomplished by changing the pressure differential between the electrolyte and gas sides of the electrode. When the pressure on the electrolyte side was increased relative to that on the gas side, drastic performance loss resulted. Conversely, with higher pressure on the gas side, significant improvement in performance resulted (Fig. 6). The shape of the performance curves and the effect of pressure clearly show that the performance with this structure is limited by a physical process. In view of the much larger pore sizes compared with those of porous Vycor, diffusion of oxygen through relatively thick electrolyte films on the catalyst surface is the probable cause.

Again, with these substrates, performance was sensitive to the loading used. Surprisingly, increases were still seen at loadings up to 5 mg/cm^2 where the catalytic film is in the range of two microns thick (Fig. 7).

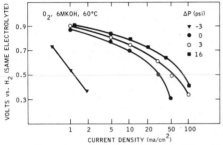
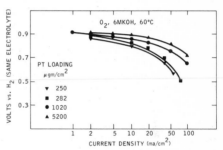

FIG. 6 *Differential Pressure Effects, Nickel Substrate (O_2, 6N KOH, 60°C).*

FIG. 7 *Performance as a Function of Loading, Nickel Substrate (O_2, 6N KOH, 60°C).*

The effect of increasing the concentration of the electrolyte and the temperature of operation were also explored. Simply increasing the concentration of KOH from 27 percent (6M) up to 50 percent caused a significant loss in performance. With the latter electrolyte, increasing the operating temperature from 60°C to 100°C gave only a marginal improvement. However, humidification of the inlet oxygen under these new operating conditions resulted in greatly increased performance (Fig. 8). This is again strong indication that conditions at the catalyst surface are critically important and very sensitive to operating conditions.

Based on photomicrographs of the surface of the porous nickel, a simplified model of the electrode was established. The surface appears to consist of concave, roughly hemispherical depressions approximately six to ten microns in diameter. The measured pore size of about one micron corresponds to the connections between these depressions and chambers of similar size in the interior (ink-bottle porosity). The thin sputtered film can be visualized as covering the surface of these depressions. During operation, these depressions can be filled either with electrolyte, or only with a thin film, or can be left partially dry, depending on conditions. Performance will, of course, be very sensitive to the surface state. For example, the major effect of differential pressure is thought to result from an expansion of the gas-electrolyte interface down into the depressions (Fig. 9), reducing the dissolved oxygen diffusion path to the catalyst film.

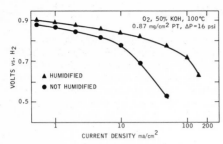

FIG. 8 *Effect of humidification (O₂, 50% KOH, 100°C).*

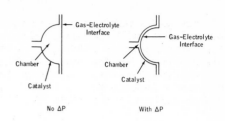

FIG. 9 *Differential pressure across pore.*

SPUTTERED FUEL CELL MATRICES

Recently, studies have been extended to an evaluation of catalyst films sputtered on conventional fuel cell matrices. Since the electron flux at the anode can cause relatively high temperatures at the substrate surface, particularly where the substrate has poor thermal conductivity, only reasonably thermally stable materials have given encouraging results. A Teflon-bonded potassium titanate matrix from TRW was evaluated with a platinum film (.83 mg/cm^2) as catalyst layer sputtered directly onto the surface. During testing, the matrix was backed with uncatalyzed porous nickel on the electrolyte side to allow high differential pressures to be used without destroying the mechanically weak structure. Using this technique, very good performance levels could be reached with this quite practical structure (Fig. 10). Similarly effective electrodes were made using an asbestos matrix, showing that the technique was quite generally applicable.

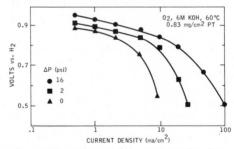

FIG. 10 *Potassium titanate matrix electrode (O₂, 6N KOH, 60°C).*

CONCLUSION

This study has shown that thin film techniques can be used to prepare effective electrodes, and can be applied to existing fuel cell matrices. The performance of such electrodes is very sensitive to the physical state at the surface, which in turn can be optimized both by structural variation and proper choice of operating conditions. The use of sputtering and other thin film techniques to prepare electrodes on both sides of a matrix to produce total cells thus appears feasible.

ACKNOWLEDGMENT

The work described in this paper was supported by the National Aeronautics and Space Administration through Contract No. NAS 8-21311. The authors wish to thank Mr. John Morgan, the Contract Monitor, for his continued cooperation, ideas, and interest; Dr. John O'M. Bockris for valuable consultations during the course of the work; and Mr. William Kobasz and Mr. John Phillips for their expert assistance in the preparation and testing of electrodes.

REFERENCES

[1]   J. S. Batzold and W. J. Asher, "Study on Thin Electrode Fuel Cells," *Contract No. NAS 8-21311, Final Report,* November 1, 1968 to January 15, 1970.

[2]   B. Cahan, "The Mechanism of Electrode Reactions on Porous Surfaces," *Dissertation,* University of Pennsylvania (1968).

[3]   J. O'M. Bockris and B. Cahan, *J. Chem. Phys.,* <u>50</u>, 1307 (1969).

[4]   R. W. Berry, P. M. Hall and M. T. Harris, "Thin Film Technology," <u>in</u> *Bell Telephone Laboratory Series D,* Van Nostrand, Princeton, N. J. (1968) Chapt. 4.

COMMENT ON PAPER OF J. BATZOLD BY J. O'M. BOCKRIS

As I understand Dr. Batzold's account of his work, he did essentially verify that the design of the fuel cell electrode given in the University of Pennsylvania thesis of Cahan did indeed produce an unusually low loading for certain power density. Thus, he also essentially verified the marked advantages (about 100 times) of the Cahan design in terms of milliwatts per milligrams of platinum.

The attainment of this low loading factor was, of course, the main aim of Cahan's work. It seems a great pity that in Dr. Batzold's follow-up of it and his attempt to develop it, the aims were changed from exploiting that exceedingly desirable feature which it offered, to investigating it under conditions (thick Pt deposits) in which it does not have any special advantage over other systems.

There <u>are</u> considerable difficulties, both in getting a sufficiently thin piece of suitably porous material on which to put down platinum, and then in dealing with the diluted electrolyte that arises between the plates as a result of the electrode reactions; both these problems are solvable only in principle at the present time. The surmounting of these difficulties realized in Cahan's work would give a solution to "the platinum problem", because the amount of platinum would then be one to ten percent of what is usually said to be needed per unit of power in a fuel cell. The cost problem would also be solved.

I do not underestimate the difficulty of overcoming the two problems stated above. However, the results of a successful solution would be remarkably attractive, and the only attempt yet made to develop the concept did verify but not develop it.

REPLY: J. S. BATZOLD

The research was oriented toward space rather than commercial
applications. Thus, efficiency was a much more important variable
than catalyst cost, and the potential at which the electrodes operated
at practical current densities was all important. Hence, while Dr.
Bockris' comments are indeed generally valid, the specific program
goal was best served by operation at substantially higher platinum
loadings than those used by Cahan. The development of electrodes with
very low platinum loading would certainly be advantageous for cost-
sensitive applications, such as large-scale terrestrial uses.

SUMMARY OF DISCUSSION AND COMMENTS: SESSION CHAIRMAN IV

E. J. Cairns

Argonne National Laboratory, Argonne, Illinois

Apart from the first paper by S. Gilman, who gave a survey on the fuel cell and battery development program at USAECOM, Fort Monmouth, the papers of Session IV fall into three groupings. 1) The hydrogen/air cell with phosphoric acid electrolyte in a matrix (O. J. Adlhart), the related thermo-catalytic hydrocarbon cracker for the production of hydrogen for fuel cells (M. A. Callahan). 2) Problems associated with the direct anodic oxidation of hydrocarbons (D. Doniat), especially the rates and extents of adsorption on platinum from sulfuric acid. 3) The electrode papers, dealing with PTFE-bonded platinum electrodes (J. Giner), and thin-film electrodes (J. S. Batzold).

The discussion of Dr. Adlhart's contribution centered around the use of the phosphoric acid matrix cell with impure hydrogen, especially with respect to the CO content, and how startup and fuel utilization were affected by the presence of CO. Dr. Adlhart commented that about 6 minutes are required for startup from room temperature to an operating temperature of 125°C, and that the fuel battery can tolerate up to three percent CO in the fuel, consuming about 70 percent of the hydrogen fed to it. The CO content of the product from Mr. Callahan's thermal cracker was a point of some discussion. In response to questions concerning the relatively high levels of CO (tens of percent) in the middle of the cycle, Mr. Callahan stated that the packing of the bed (Al_2O_3 and Ni rings) and the cycle time and temperature were presently being optimized with the intent of decreasing the CO content.

Dr. Doniat's paper concerning the adsorption of hydrocarbons on platinum in acid electrolytes raised a discussion on the various rate expressions that have been used to describe the adsorption process. Langmuir rate expressions have been used by some investigators, whereas Dr. Doniat used a Temkin expression, and reported a straight-line relationship between $\log (\frac{dq}{dt})$ and q. Another point of discussion was α, the interaction coefficient, which was found by Dr. Doniat to vary with the identity of the acid electrolyte. The appropriateness of carrying out adsorption experiments at open circuit rather than at a controlled potential was discussed by Drs. Cairns, Bockris and Doniat. Some participants were of the opinion that controlled-potential adsorption can yield more useful information.

Some rather spirited discussions were held between and following the papers on the various electrode structures. The main points discussed were the advantages and disadvantages of carrying out experiments relating to kinetics and mechanisms of electrode reactions on electrodes of complicated geometry. After a good deal of discussion, it was generally agreed that one should use the simplest geometry of electrode consistent with the objectives of the experiment. It was pointed out that sometimes complicated geometries are made necessary by the nature of the electrocatalyst (*e.g.*, supported electrocatalysts, single crystals which are hard to shape, highly strained structures, high specific area electrocatalysts prepared *in situ*); therefore, it is not always possible to prepare an electrode of simple geometry. The advantages

and some methods of locating the electrocatalyst in the immediate
vicinity of the electrolyte-reactant interface were discussed in connection with the thin-film electrode and other electrodes. In this
report C. E. Heath made a short contribution on the engineering
limitations of slurry electrodes.

SESSION V

HIGH TEMPERATURE FUEL CELLS

Chairman

S. Gilman

PROBLEMS OF HIGH TEMPERATURE ZrO$_2$ - SOLID ELECTROLYTE FUEL CELLS

H. Tannenberger

Institut Battelle, Geneve, Switzerland

ABSTRACT

An attempt is made to give a succinct description of the problems which arise in the development of the high-temperature solid-electrolyte fuel cell. References to the literature are given for those requiring more detailed information about the present state of development. The solid-electrolyte fuel cell has inherent qualities which makes it a promising energy-conversion device. In spite of this, relatively little effort has been put into developing it. The only well-established fact is that cubic stabilized zirconia has all the properties required of a solid electrolyte. Electrode development is only in its infancy. Because of the working temperature of 800°C to 1000°C, materials for all components of the cell pose a problem. For further development it would be highly desirable to know more about the mechanism of the cathodic and anodic reactions. The optimalization of the characteristics of the whole system needs a thorough engineering approach.

INTRODUCTION

High-temperature fuel cells, working at between 700°C and 1000°C, are supposed to offer the advantage of burning electrochemically conventional carbonaceous fuels with air, without posing any particular electrocatalytical problems. By comparison with molten salt fuel cells, solid-electrolyte cells are expected to be lighter for a given power output. They should support temperature cycling more easily than molten salt cells because no phase transformation occurs more between room temperature and working temperature. Taken as a whole, an entirely solid system looks attractive from an engineering point of view. In spite of these advantages, comparatively little effort has been spent on the development of ZrO$_2$ fuel cells since they made their appearance about ten years ago. Reports in the literature mainly concern single cells; a very small number of papers have dealt with battery-forming cell stacks; and only one prototype, producing 100 W, is known to have been built and tested [1-7].

THE HIGH TEMPERATURE ZrO$_2$ FUEL-CELL SYSTEM

In order to judge the present state of development and to assess the possibilities for future development, it is necessary to consider the fuel cell system as a whole. The system should be a machine which is fed either by a liquid or gaseous hydrocarbon or by coal and air, which delivers electrical energy and gives off heat, oxygen-depleted air and exhaust gases containing CO_2, H_2O, and the smallest possible amount of noxious components such as CO, other carbon compounds, and

nitrogen oxides. This machine should be equipped with a starter device
and a control system, the most important function of which is to maintain the temperature of the elements within optimal limits.

Two main types of industrial application may be considered:

1) Mobile — for instance, in heavy trucks, buses and railways.
The power range may be from some tens to some hundreds of kW. A relatively short life of some 3,000 to 10,000 hours could be tolerated.
However, the power-to-weight ratio should be high, say from 0.3 to
1 kW/kg, not only for considerations of weight but also because of the
heat content (heating-up time),

2) Stationary, as for power plants, ranging from 100 kW to tens
of megawatts. A life of at least some tens of thousands of hours is
essential. The power-to-weight ratio may be less high; however, heat-
capacity requirements make too low a power density undesirable. The
cost per unit area of element imposes another lower limit for power
density.

The basic layout of a high-temperature ZrO_2 fuel-cell system is
shown in Fig. 1. Regardless of the condition of the carbonaceous fuel
(solid, liquid or gaseous), it is necessary to treat it so as to produce a gaseous fuel which does not precipitate solid carbon at the

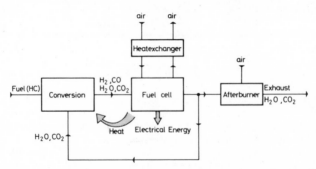

FIG. 1 *Basic layout of a high temperature ZrO_2 fuel cell system.*

working temperature of the fuel cell. In the case of hydrocarbon, this
treatment may be partial oxidation with air, or reforming with a mixture of water and carbon dioxide (*e.g.*, exhaust gas). Fig. 2 shows the
equilibrium conditions in the ternary system C-H-O at 800°C, assuming
that only C, H_2, CO, CO_2, H_2O and CH_4 will be present. Fig. 3 gives
an example of calculated gas composition during its electrochemical
oxidation in the fuel cell. In addition, the open-circuit voltage is
given as a function of the degree of combustion. Table I gives some
data pertaining to the reforming of CH_4 with H_2O and CO_2.

Quite independently of the kind of fuel, the following conclusions
are generally valid:

Even for the most stable hydrocarbons, conversion, *e.g.*, by reforming
with H_2O or CO_2, is thermodynamically necessary to avoid carbon precipitation. The amounts of H_2O or CO_2 needed for reforming are of the
same order of magnitude as the amount of fuel to be converted. It is
probable that in a practical system even higher amounts of H_2O or CO_2
would be necessary to avoid degradation of the fuel cell anodes and
fuel ducts by carbon precipitation. The reforming reaction is endothermic. Reforming enthalpy is rather high (about 25 percent of the
total energy content of the fuel involved). It is therefore necessary
to ensure heat exchange between the fuel cell and the converter.

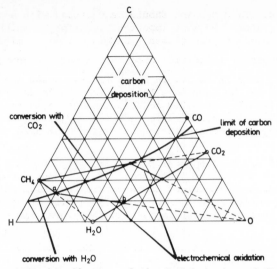

FIG. 2   *Equilibrium diagram of the C-H-O system at T = 800°C.*

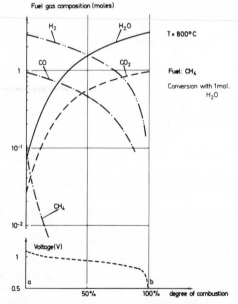

FIG. 3   *Fuel composition and open-circuit voltage for CH₄ converted with 1 mol H₂O. The abscissa (degree of combustion) corresponds to the straight line a-b in the ternary diagram of Fig. 2.*

Due to the nature of the electrolyte, which is an oxygen ion conductor, the combustion products form on the fuel side and so dilute the fuel. The open-circuit voltage remains fairly constant, even for high degrees of combustion. However, the reaction kinetics will be seriously affected by the dilution of the fuel. It does not seem worth speculating about degrees of combustion higher than 85 to 90 percent in practice, so that an afterburner is necessary to ensure clean exhaust gases. Since the exhaust temperature is high, post-combustion will probably be easy. The heat produced may help to keep the heat balance of the whole system within acceptable limits.

TABLE I

Conversion of CH₄

T(°C)	X$_{min}$ (moles)		Conversion Enthalpy (Kcal)		ΔG (Kcal)*		$\eta = \dfrac{\Delta G}{\Delta H}$ (%)**	
	650°	800°	650°	800°	650°	800°	650°	800°
$CH_4 + xH_2O \rightarrow$ CH_4, H_2, CO H_2O, CO_2	1.48	1.00	-35.3	-48.8	177.8	169.0	92.5	88
$CH_4 + xCO_2 \rightarrow$ CH_4, H_2, CO, H_2O, CO_2	4.7	1.19	-66.7	-60.5	173	168	90	87.5

Free enthalpy for the oxidation of the gas mixture after reforming.
**The efficiency η is defined as the ratio between the free enthalpy noted in the third column and the enthalpy of combustion of CH₄.*

These considerations make it clear that a very thorough engineering approach is necessary in order to ensure high overall efficiency of the fuel cell system and to define the optimal power range between idle and maximum output.

THE FUEL CELL AND SINGLE ELEMENTS

The heart of the system is the fuel cell built up from battery modules formed by a number of single elements connected in series. The performance of the fuel cell is determined by its geometric configuration (in particular the amount of element surface housed per unit volume of the fuel cell) and by single-element performance.

Temperature distribution, heat transfer, gas flow and current drain must be considered in the design of the fuel cell. As to the current drain, the solid electrolyte offers a particular advantage: it seems to be relatively easy to build batteries (tubular or flat) with very narrow elements — not more than one to two centimeters wide. This implies that the total current to be drained remains in the order of magnitude of the current densities (Fig. 4).

Rough estimation, taking the above mentioned factors into account, shows that the single elements should have the characteristics indicated in Fig. 5, in order to give the fuel cell power densities of about 0.5 kW/l at maximum output. These are rather stringent requirements.

SINGLE ELEMENT AND BATTERY PERFORMANCE

Single-element performance depends on the ohmic resistance of the electrolyte, and on anodic and cathodic polarization. These three parameters will be discussed in more detail in the following chapters.

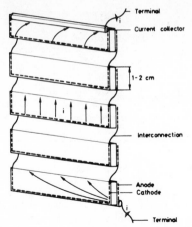

FIG. 4 *Current drain in a solid electrolyte battery (schematic).*

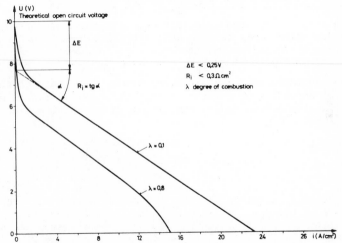

FIG. 5 *Performance requirements for single elements.*

High-temperature solid-electrolyte elements and batteries pose a particular problem with respect to life, namely the compatibility of the different materials involved. There exist interfaces between different materials which should remain unaltered during the life of the fuel cell at working temperatures of 700°C to 1000°C and after frequent cycling between ambient and working temperature. This means that the materials involved should not chemically react, that interdiffusion should be slow, and that their coefficient of thermal expansion should match. The last is determined by the use of the cubic zirconia electrolyte which has a coefficient of $\alpha = 11 \cdot 10^{-6}$ (°C)$^{-1}$. [8] It is clear that these side conditions limit the choice of materials which have the required specific (*e.g.*, catalytic) properties for the different components of the elements and battery.

As an example, the interconnection between the anode and the cathode in a battery is submitted to particularly severe conditions. It must be compatible with three different materials and must withstand oxidizing and reducing atmosphere simultaneously (Fig. 6). Replacing an interconnection made of only one material by an interconnection composed of

two materials, each better adapted to the respective conditions, introduces a new interface as a weak point.

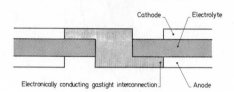

FIG. 6 *Cathode-anode interconnection.*

THE ELECTROLYTE

The only serious choice of material for the solid electrolyte [9]
is zirconia stabilized in its cubic form by doping with CaO, Y_2O_3,
Yb_2O_3 or a mixture of heavy rare earth ranging from dysprosia to lute-
tia. Scandia gives very good conductivities as doping oxide, but is
out of the question at present because of its price.

Electrolytes on a thoria or ceria basis were investigated, but
proved not to be competitive with zirconia. Ionic conductivity is
smaller, electronic conductivity occurs, and chemical stability is
lacking at certain ranges of partial pressure of oxygen occurring in
the fuel cell.

It is well established that maximum oxygen ion conductivity is obtained in the zirconia system if the concentration of the doping oxide
is at the lower limit, assuring the stability of the cubic phase. It
is possible to obtain a metastable cubic phase with less doping oxide
(*e.g.*, 6 percent Yb_2O_3). These metastable phases have higher conductivity, but at 800°C to 1000°C decompose rather quickly to the stable
state (mixture of tetragonal phase and cubic phase with higher amount
of doping oxide than the initial mixture) and their conductivity decreases [10].

Calcia-stabilized zirconia has a conductivity about ten times less
than the yttria or rare-earth-stabilized electrolytes. Moreover, at a
temperature of 800°C to 1100°C, an aging process occurs which decreases
the conductivity still further. Nevertheless, a certain interest in it
persists owing to the low price of calcia and the fact that it has considerably better sinterability than yttria or rare-earth-stabilized
zirconia.

With regard to the required characteristics of a single cell, it
can be deduced that the resistance of the electrolyte in a practical
fuel cell should not exceed about 0.1 to 0.2 Ω cm^2. The nature of the
electrolyte once given, the only parameters which may be changed freely
are the working temperature of the cell and the thickness of the electrolyte. The interdependence of these two parameters is given in Fig. 7
for the two types of electrolyte. It will be seen that for the working
temperature of 800°C to 1000°C the thickness of the electrolyte must be
between 30 to 200 μ.

Different techniques for the preparation of electrolytes of varying
thickness have been reported.

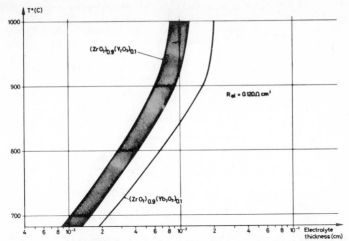

FIG. 7 *Relationship between working temperature and thickness of the electrolyte for constant resistance* R_{el}.

Conventional sintering techniques with firing temperatures of up to 2000°C are convenient to produce tubes or plates of electrolyte with a minimum thickness of 0.3 to 0.4 mm. Unsupported layers of electrolyte having thicknesses of 50 to 150 µ were prepared by flame or plasma spraying [11]. It seems that subsequent heat treatment up to 1200°C to 2000°C is necessary to assure gastightness. Electrolyte layers of 30 µ and less are not stable mechanically and should be deposited on a porous support which serves at the same time, directly or indirectly, as one of the electrodes. Sintering of a slurry, applied at the support, at temperatures lower than 1600°C have been reported to be successful [12]. Chemical vapor deposition of the electrolyte is another promising possibility [13].

It is evident that the side conditions imposed by the presence of a porous support which acts simultaneously as one of the electrodes may limit the techniques for depositing gastight electrolyte layers.

THE ELECTRODES

Little is known about the mechanism occurring at the cathode and anode [14-17]. Some basic considerations can however be formulated.

In the case of all known solid electrode materials used in anodes and cathodes, it is necessary for the electrodes to be porous and to form a multitude of electrolyte-electrode-gas interfaces because the transport of the reactants through a compact electrode by diffusion would be too slow. The lateral conductivity of the electrode should reach some minimum value to make possible a current drain of the order of amperes without too great a loss.

Silver cathodes might be formed by a continuous layer covering the whole electrolyte surface. The oxygen would then diffuse through the silver layer and react electrochemically at the silver electrolyte interface. In spite of the relatively high solubility of oxygen in silver and its high diffusion coefficient [18], the thickness of the silver electrode should be less than 1 µ, in order to avoid limiting currents of less than 1 A/cm^2. Unfortunately, silver layers of 1 µ thickness are probably not sufficiently stable at 800°C. However, a continuous layer of liquid silver some millimeters thick was successfully employed as a cathode; solubility and diffusivity of oxygen were one order of magnitude higher than with solid silver [19].

The problem of current drain is illustrated in Fig. 8. By introducing as relevant parameters the lateral resistance $R_\square$ of anode and cathode,

$$R_\square^{A,C} = \frac{\rho}{\delta}$$

where $R_\square^{A,C}$ are the lateral resistance of anode and cathode respectively,

  ρ is the resistivity of the electrode material,

  δ is the thickness of the electrode,

and the "internal resistance" R_i of an infinitesimal element, approximated by a constant, it is possible to calculate [20] the "internal resistance" of an element R_p as measured on its terminals, as a function of its geometric configuration and of a parameter α which is defined as follows:

$$\alpha = \sqrt{\frac{R_\square^A + R_\square^C}{R_i}}$$

The ratio of the mean current-density of the element and the maximum local current-density can also be expressed in terms of the same parameters.

If the concept of porous electrodes is accepted, two basically different models of electrode mechanism may be considered (Fig. 9).

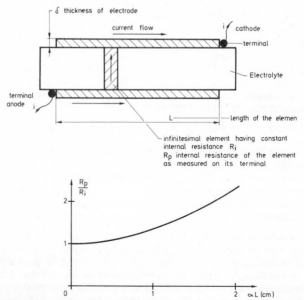

FIG. 8 *Current drain in an element.*

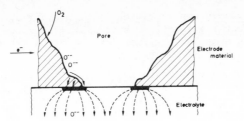

(a) *Oxygen adsorbs on the pore wall, accepts electrons, migrates to the bottom of the pore and penetrates into the electrolyte in the vicinity of the electrode-electrolyte-gas interface. The electrode material may have an influence on the overall reaction rate.*

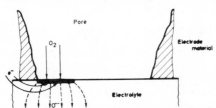

(b) *Oxygen adsorbs on the electrolyte at the bottom of the pore. Electrons flow from the electrode through the electrolyte and combine with oxygen atoms at the electrolyte surface, forming oxygen ions. The electrode material should not influence the overall reaction rate.*

FIG. 9 *Porous electrode models for cathode reaction (schematic).*

In both cases, the current originates in very small areas around the electrode-electrolyte interface; hence the real current densities are much higher than the mean or apparent current density. Moreover, near the electrode-electrolyte interface, the current lines of oxygen ions will go through only a fraction of the available cross-section of electrolyte. This current-restriction phenomenon introduces additional ohmic resistance [21]. The influence of the electrode material on the rate of the electrode reaction should appear only if the mechanism involves adsorption, dissociation on, and possibly diffusion along a short path through the electrode material. Such an influence was observed in the anodic reaction with hydrogen and carbon monoxide on platinum [3] and nickel [4] electrodes. Whereas the rate of hydrogen and carbon-monoxide oxidation varies little on nickel electrodes with identical morphology, there are drastic differences between the rates of the two reactions on platinum electrodes.

As it is very difficult to build electrodes with a controlled morphology which is quantitatively known and stable under working conditions, it is at the moment not possible to derive the relationship between overvoltage and real current densities by experiments with porous electrodes, or even to compare two different electrodes with a high degree of accuracy. A way of getting more quantitative information would be to study reactions at a geometric, well-defined contact between the electrode material and the flat electrolyte surface.

Measurement of electrode polarization of some 100 mV at current densities up to some amperes per square centimeter causes difficult problems. Ideally, the overall characteristics of a solid-electrolyte cell having a specific geometric configuration should be decomposed into: the contribution of the IR losses in the electrolyte; the IR losses for the current drain in the electrodes parallel to the electrode-

electrolyte interface; and anodic and cathodic polarization. Moreover,
it would be desirable to determine the current distribution over the
element, which may be highly uneven.

Calculating IR losses for cells with electrolyte one mm thick may
lead to considerable error. In fact, the error in the electrolyte resistance being

$$\Delta R = \Delta \rho \cdot d + \Delta d \cdot \rho$$

where R is the resistance, ρ is the resistivity, and d is the thickness,
and supposing that

$$\Delta d = 0$$

$$\frac{\Delta \rho}{\rho} \sim 0.1 \qquad \rho \sim 30 \ \Omega \ \text{cm at } 800°C$$

$$d = 0.1 \ \text{cm}$$

we obtain

$$\Delta R = 300 \ \text{m}\Omega$$

leading to an error in the calculated IR loss of 300 mV for 1 A/cm^2,
which is of the order of magnitude of the polarization to be measured.
In addition, the real temperature of the electrolyte plate is very much
higher than the temperature of the furnace at these current densities.
The exact measurement of the real temperature is again difficult and
causes further error in the determination of ρ.

Pulse methods may be employed to distinguish between ohmic loss and
polarization [22]. However, current-restriction phenomena which are
directly linked to the morphology of the electrode, cannot be distinguished from the "normal" IR losses by this method.

Separation of anodic and cathodic polarization of the cell may be
achieved by applying probes on the cell (Fig. 10). The measured potential difference between the probe and the working electrode indicates
the sum of polarization of the working electrode and the IR loss in the
part of the electrolyte between the electrode-electrolyte interface and
the equipotential surface going through the probe. It can be predicted,
and has been experimentally verified, that the geometrical configuration
of the working electrode, the counter electrode and the probe determine
the location of the sensed equipotential surface. If one tries to measure the effect of current restriction, which may be equivalent to the
resistance of a 10 to 50 µ thick electrolyte layer, it becomes clear

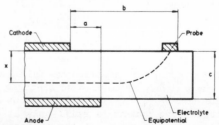

FIG. 10. *Measurement of the polarization of a single electrode by
applying probes. Parameters a, b and c influence the location x of
sensed equipotential surface.*

that the location of the equipotential should be known to a degree of accuracy practically impossible to achieve in an experiment.

The most reliable results may be obtained by measuring the overall characteristics of cells having an electrolyte thickness of about 100 μ. Polarization phenomena can be separated from IR losses by applying galvanostatic pulses. The resistance of the electrolyte can be calculated with an accuracy which may be sufficient to make at least an estimate of the effect of current restriction. Lateral resistance of the electrodes ($\frac{\rho}{\delta}$) should be measured separately and kept at sufficiently low values to assure a sufficiently even distribution of the current across the element. However, separating anodic and cathodic polarization by introducing a probe will cause considerable practical difficulty with cells in which the electrolyte is only 100 μ thick.

Anode materials were chosen from among metals such as nickel or cobalt, which remain stable even in a fuel mixture containing a high concentration of combustion products. Lateral conductivity is no problem, even for thin and very porous electrodes. However, adherence of these metals to the electrolyte may cause trouble. Coefficients of thermal expansion do not match. Hydrogen may diffuse through the nickel and be oxidized at the nickel-electrolyte interface. The water thus produced cannot diffuse back through the nickel and tends to blow off it.

The choice of cathode material [23] is more difficult because of the necessary chemical stability under oxidizing conditions. Noble metals such as platinum or palladium must be rejected because of their price. Silver causes considerable difficulties because its melting point is very near the working temperature, and because of its vapor pressure and its very high diffusivity on surfaces. This surface mobility can cause short-circuiting of elements connected in series and similar effects. Stable oxides such as cobaltites, or doped oxides such as NiO, Al$_2$O$_3$, or In$_2$O$_3$, may be considered. Apart from chemical compatibility with the electrolyte, interdiffusion and coefficient of thermal expansion, lateral conductivity causes serious problems. Most of the oxides so far proposed have too little electronic conductivity and do not meet the other requirements well. Only doped indium oxide can be considered as a satisfactory cathode material at present.

CONCLUSIONS

The solid-electrolyte high-temperature fuel cell system has inherent qualities which make it a very promising energy-conversion device. Consuming cheap and easily distributed fuels, providing a reasonable overall thermal efficiency and promising high power/weight ratios and low cost, it may be attractive for both mobile and stationary applications. To achieve maximum overall performance, the whole system must be optimalized. This optimalization makes it possible to specify the performance requirements for single elements and battery modules of the fuel cell itself.

Cubic stabilized zirconia as electrolyte fulfills all requirements. The development of cheap techniques to produce electrolyte layers between 10 and 50 μ thick in stable form (*e.g.*, supported by a porous matrix) is of primary importance. Electrode development is still only at its beginning. It is highly desirable to learn more about the mechanisms of cathodic and anodic reaction in order to obtain data for the design of optimal electrodes. In this, materials play a fundamental role.

ACKNOWLEDGMENT

I would like to thank my colleagues Dr. P. Van den Berghe and Dr. M. Voïnov for much helpful comment during the preparation of this paper.

REFERENCES

[1]   J. Weissbart and R. Ruka, *J. Electrochem. Soc.*, 109, 723 (1962).

[2]   H. Binder, A. Köhling, H. Krupp, K. Richter and G. Sandstede, *Electrochimica Acta*, 8, 781 (1963).

[3]   D. H. Archer, L. Elikan and R. L. Zahradnik, in *Hydrocarbon Fuel Cell Technology*, B. S. Baker (Ed.), Academic Press, Inc., New York (1965) p. 51.

[4]   H. Schachner and H. Tannenberger, in *Comptes Rendus, Journées Internationales d'Etudes des Piles à Combustible*, "Revue Energie Primaire," Brussels (1965) Vol. III, p. 49.

[5]   B. Rohland and H.-H. Möbius, *Naturwissenschaften*, 55, 227 (1968).

[6]   H.-H. Eysel and H. Kleinschmager, *BBC-Nachr.*, 654 (1969).

[7]   H.-H. Böhme and F.-J. Rohr, *BBC-Nachr.*, 648 (1969).

[8]   T. H. Nielsen and M. H. Leipold, *J. Amer. Ceram. Soc.*, 47, 155 (1964).

[9]   T. H. Etsel and S. N. Flengas, *Chem. Rev.*, 70, 339 (1970).

[10] H. Tannenberger, H. Schachner and P. Kovacs, *Op. Cit.*, "Revue Energie Primaire," p. 19.

[11] J. L. Bliton, H. L. Rechter and Y. Harada, *Ceramic Bulletin*, 42, 6 (1963).

[12] E. F. Sverdrup, these Proceedings, p. 255.

[13] A. O. Isenberg, *et al.*, *Sixth Biennial Fuel Cell Symposium*, ACS, 158th National Meeting, New York (September 1969).

[14] M. Kleitz, Thèse, Université de Grenoble (May 1968).

[15] J. E. Bauerle, *J. Phys. Chem. Solids*, 30, 2657 (1969).

[16] H. Yanagida, R. J. Brokk and F. A. Kröger, *J. Electrochem. Soc.*, 117, 593 (1970).

[17] R. L. Zahradnik, *J. Electrochem. Soc.*, 117, 1443 (1970).

[18] W. Eichenauer and G. Müller, *Z. Metallkunde*, 53, 321 (1962).

[19] D. W. White, B. Hills, W. E. Tragert and W. A. Rocco, U. S. Patent No. 3,432,352, March 11, 1969.

[20] R. H. Boll and R. K. Bhada, *Energy Conversion*, 8, 3 (1968).

[21] H. Tannenberger and H. Siegert, "Fuel Cell Systems - II", *Advances in Chemistry Series*, 90, 281 (1969).

[22] E. F. Sverdrup, *et al.*, in *Hydrocarbon Fuel Cell Technology*, B. S. Baker (Ed.), Academic Press, New York (1965) p. 311.

[23] C. S. Tedmon, Jr., H. S. Spacil and S. P. Mitoff, *J. Electrochem. Soc.*, 116, 1170 (1969).

SOLID OXIDE ELECTROLYTES WITH TIME DEPENDENT CONDUCTIVITY

W. Baukal

Battelle-Institut e.V., Frankfurt am Main, Germany

ABSTRACT

Solid oxide electrolytes of the stabilized zirconia type often show a gradual decrease of their ionic conductivity. The two basic mechanisms of this effect are destabilization and aging. They are described and commented on with the aid of relevant literature. Yttria-stabilized zirconia has the best long-term properties with respect to both effects.

INTRODUCTION

In high-temperature fuel cells an oxygen-ion-conducting solid electrolyte is used. Its high ionic conductivity and the possibility of using electrolyte thicknesses of from 500 down to 50 µm make the high-temperature fuel cell preferable to other fuel cell systems for the high-power and long-term application range. For calculating long-term performance data, any time dependence of the electrolyte resistivity must be taken into account.

There are two mechanisms which may cause a gradual increase of the resistivity in cubic-stabilized zirconia: destabilization and aging. In the following, a review of these two effects will be presented.

THE DESTABILIZING EFFECT

For an explanation of the destabilizing effect, some basic principles of the phase relations in the binary systems of zirconia will be outlined with the aid of the zirconia-magnesia phase diagram. This diagram is the best established among the systems ZrO_2-MgO, ZrO_2-CaO, and ZrO_2-Y_2O_3. However, MgO-stabilized ZrO_2 cannot be used as a fuel cell electrolyte, as will be seen from the following.

BINARY SYSTEM ZrO_2–MgO

Fig. 1 shows the phase diagram according to the work of Viechnicki and Stubican [1]. Other investigators [2,3,4,5] obtained the same type of diagram; only the position of some of the phase boundaries and the position of the eutectoid point differ slightly.

Pure zirconia has three modifications: monoclinic, tetragonal, and cubic [6,7,8]. The transformation monoclinic-tetragonal is very rapid [9,10,11,12] and is accompanied by a high change in volume, so that pure zirconia cannot be used as a refractory material. It has to be stabilized by the addition of foreign oxides such as MgO to give a cubic mixed crystal. Even if stabilized in this cubic modification, zirconia containing MgO cannot be used for more than a few hours at temperature levels of about 1,000°C. Prolonged heating at these

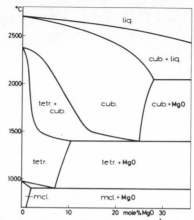

FIG. 1   *Phase diagram ZrO₂—MgO after Viechnicki—Stubican.*

temperatures will cause another phase transformation which is called
destabilization. The phase diagram (see Fig. 1) shows that the cubic
solid solution decomposes into pure MgO and a tetragonal solid solution.
In contrast to the exceptionally rapid monoclinic-tetragonal transformation of pure ZrO_2, destabilization is a sluggish reaction with a
rate similar to many other solid state reactions.

The cubic solid solution has all the electrical properties required
for an oxygen ion conductor. The crystal structure is of the fluorite
type, with vacancies in the anion sublattice; because of these vacancies
the oxygen ions are highly mobile. The ionic conductivity depends on
the temperature and on the composition. It has its maximum at or near
the lower boundary of the stability field of the cubic phase. The
reason for this composition dependence is not well understood up to
now and is still under investigation.

In the system ZrO_2-MgO the cubic solid solution is thermodynamically
unstable at the operating temperature of a high-temperature fuel cell
(800°-1,000°C). Therefore, suitable electrolyte compositions have to
be sought in other binary systems of zirconia.

BINARY SYSTEM ZrO_2—CaO

The phase relations in the system ZrO_2-CaO are not yet well established. A review of the phase boundary data reported in the literature
shows that most of the investigators considered only small portions of
the system and that their results differ widely. The existence of a
cubic modification in pure ZrO_2 has not been taken into account at all.
In Fig. 2 those measurements have been summarized [2,13-18] which
substantiate what is now generally accepted as most probable.

The extension of the cubic phase field is not exactly known; especially, it is not established whether it ends at a eutectoid point
[5,19] as in the case of zirconia-magnesia. If there is such a
eutectoid at all, its temperature seems to be very low, because zirconia
stabilized with about 15 mole percent CaO does not decompose on prolonged heating at 800° to 1,000°C.

However, a concentration as high as 15 mole percent CaO is not
the ideal composition for a fuel cell electrolyte. The maximum
conductivity is again at low CaO concentrations near the lower phase

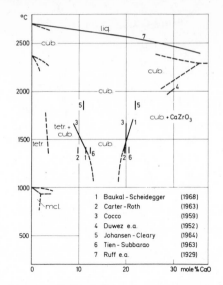

FIG. 2 *Phase diagram* ZrO_2—CaO.

boundary. From Fig. 2 it can be seen that this boundary is not independent of the composition, but rather has some slope towards the cubic-
tetragonal transformation point of pure zirconia. Therefore, it may
happen that the composition of a sample with a low CaO content falls
within the cubic field at the temperature of preparation (1,700° to
1,800°C) but not at 800° to 1,000°C. In this case the phenomenon of
destabilization will occur again.

The progress of the destabilization reaction may be observed by
measuring the conductivity of the sample; because the highly conductive
cubic phase is diluted in the reaction by a comparatively nonconducting
phase. The final conductivity corresponds to the new equilibrium.
The amount of the phases coexisting in the heterogeneous mixture may be
calculated by the lever rule, provided that the exact phase boundary
data are known. Even if the equilibrium mixing ratio is known, the
final conductivity cannot generally be calculated from a simple law
of dilution; it rather depends on how the second phase precipitates
at the individual grains of the parent phase. If the nonconductive
phase is formed primarily at the points of triple contact, the simple
dilution law will be valid as long as the conductive phase is the
continuous one. At about 50 volume percent the kind of the continuous
phase will change and a rapid conductivity decrease has to be expected.
On the other hand, if the nonconductive phase precipitates at the
boundary of the parent grains, the conductivity of the bulk will
diminish rapidly even if only a few percent of the second phase are
formed in equilibrium.

The latter type of precipitation has been shown to occur in the
case of destabilization, *i.e.*, when tetragonal zirconia is formed out
of MgO- or CaO-stabilized zirconia [1,20]. The kinetics of this
precipitation may be described by the nucleation and growth theory.
Hence, the decomposition rate will have a maximum at a certain
temperature. For the system ZrO_2-MgO this temperature is about 200°
below the eutectoid temperature [1,5].

BINARY SYSTEM $ZrO_2-Y_2O_3$

Fig. 3 shows the phase diagram of the system $ZrO_2-Y_2O_3$. The only phase relation well established is that the cubic phase is stable above approximately nine mole percent yttria. Below this concentration even the reported types of phase diagram differ. Duwez [21] found a composition-independent boundary between the cubic field and a narrow two-phase field. Lefèvre [22] reports a continuous transition from tetragonal to cubic zirconia above a critical temperature; below the critical temperature there would be a miscibility gap between two different tetragonal phases.

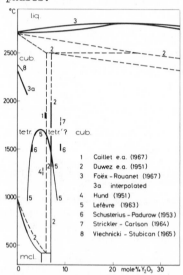

FIG. 3   *Phase diagram $ZrO_2-Y_2O_3$.*

Apart from these two investigations, the measurements summarized in Fig. 2 [25-27] relate to only small portions of the $ZrO_2-Y_2O_3$ system which, in part, are deduced from the closely related systems $HfO_2-Y_2O_3$ [23] and ZrO_2-rare earth oxides [24].

According to Lefèvre [22], the transition tetragonal-cubic is homogeneous, *i.e.*, the ratio of the crystal parameters   c   and   a   of the tetragonal phase decreases continuously and becomes unity at the transition point. This would take place at about 20 mole percent Y_2O_3 and would be independent of temperature. A vertical line in the phase diagram would represent this phenomenon. However, this theory neglects the existence of a cubic modification of pure ZrO_2. It may well be that the transition line tetragonal-cubic has a slope and tends towards the transformation point of pure ZrO_2, although the latter transition is distinctly a heterogeneous one. Further clarification of these phase relations is necessary, especially as the most important solid oxide electrolytes are found in the system $ZrO_2-Y_2O_3$.

Any time dependence of the conductivity in the low doping range may again be attributed to the fact that there is one highly conductive phase at the temperature of preparation but there are two phases at the temperature of investigation. The results of Takahashi and Suzuki [28], obtained at yttria and ytterbia concentrations of 7 mole percent and below, may be explained in this way.

THE AGING EFFECT

"Aging" of cubic stabilized zirconia occurs in the temperature range of about 700° to 1,100°C. It is a gradual decrease of the electrical, *i.e.*, ionic, conductivity in the course of several days. The following features of the aging phenomenon have been confirmed repeatedly [14,28-32].

Aging occurs within the stability field of the cubic stabilized phase.

The structure of an aged sample remains cubic; the lattice parameter does not change.

A second phase cannot be observed by X-ray or microscopic techniques.

The effect is reversible at temperature levels above 1,200°C. After this "de-aging" the conductivity assumes its initial value.

Although the conductivity decreases during aging, its nature remains purely ionic.

The activation energy of the conduction process is not altered during aging.

Aging is not due to impurities.

A number of other features have been reported, but they have not been confirmed or have even been disputed:

In CaO-stabilized ZrO_2 an order-disorder transition occurs in the cation sublattice; this was found by X-ray techniques [30]. The result could not be reproduced [14]; the accuracy of the original X-ray measurements has been doubted [29].

In aged CaO-stabilized ZrO_2, the anions and anion vacancies would form an ordered superstructure; this was observed by neutron diffraction [14]. Even if this phenomenon occurs, it cannot be rate-determining for a process which extends over several days, because the particles involved are highly mobile and would be rearranged much faster.

The aging effect seems to depend on the surrounding gas atmosphere [32].

The aging effect depends on the grain size, in a way that single crystals age very fast in comparison with polycrystalline material [14].

For long-term applications of oxide electrolytes we have to know the increase in resistivity and whether there is a stationary value.

In Fig. 4 the kinetic curves available in the literature [14,28,29, 31,32] have been combined into one diagram. There are striking differences even between measurements on samples with similar composition and similar aging temperature. The kinetic laws which were fitted to the observed curves are also different. Exponential laws were used by Takahashi [28], Cocco and Danelon [29], and Baukal [32]. Carter and Roth [14] who have studied the aging phenomenon most extensively using CaO-stabilized ZrO_2 found a square-root law.

A $\sqrt{t}$-law implies that the electrolyte in its final state has no conductivity at all. This seems not to be the case for yttria- or ytterbia-doped zirconia as can be seen in Fig. 4. After an increase of about 20 percent, the resistivity of these materials rather

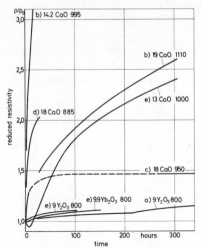

FIG. 4 *Aging effect in stabilized zirconia. The first letter at
each curve denotes the author, the first figure denotes the concentratio.
of the stabilizers CaO, Y_2O_3 and Yb_2O_3 in mole-%, the last figure
denotes the aging temperature in °C. a) Baukal, b) Carter-Roth,
c) Cocco-Danelon, d) Subbarao-Sutter, e) Takahashi-Suzuki.*

approaches an asymptotic value. In CaO-stabilized ZrO_2, on the other
hand, the resistivity assumes values of twice the initial value and
more, and does not show a tendency to level off after 300 hours. The
only exception is the measurements of Cocco and Danelon [29].

The striking differences between measurements obtained under almost
identical conditions may indicate that the aging effect is sample-
dependent. This would be the case if the grain size and the density
of the subgrain boundaries determined the kinetics of the aging process.
According to the theory of Carter and Roth [14], anions and cations
form a long-range ordered array in "domains" or "clusters" and the
progression of these domains is stopped at the grain and subgrain
boundaries.

For further clarification of the aging effect, the following two
experiments are proposed:

The kinetics of de-aging should be studied at different temperature
levels with the objective of finding an activation energy for this
process.

The influence of controlled grain size on the aging rate should be
investigated.

REFERENCES

[1] D. Viechnicki and V. S. Stubican, *J. Am. Ceram. Soc.*, **48**, 292
     (1965).

[2] P. Duwez, F. Odell and F. H. Brown, Jr., *J. Am. Ceram. Soc.*, **35**,
     107 (1952).

[3] A. Cocco and N. Schromek, *Radex-Rundschau*, 590 (1961).

[4] C. F. Grain, *J. Am. Ceram. Soc.*, **50**, 288 (1967).

[5] A. Dietzel and H. Tober, *Ber. Deut. Keram. Ges.*, 30, 47, 71 (1953).

[6] B. C. Weber, Wright-Patterson Air Force Base, ARL Report No. 64-205, (November 1964) AD 610758.

[7] D. K. Smith and C. F. Cline, *J. Am. Ceram. Soc.*, 45, 249 (1962).

[8] R. Ruh and T. J. Rockett, *J. Am. Ceram. Soc.*, 53, 360 (1970).

[9] I. Hinz and A. Dietzel, *Ber. Deut. Keram. Ges.*, 39, 489 (1962).

[10] G. M. Wolten, *J. Am. Ceram. Soc.*, 46, 418 (1963).

[11] L. L. Fehrenbacher and L. A. Jacobson, *J. Am. Ceram. Soc.*, 48, 157 (1965).

[12] C. F. Grain and R. C. Garvie, *Report of Investigations*, No. 6619, Bureau of Mines, U.S. Department of the Interior (1965).

[13] W. Baukal and R. Scheidegger, *Ber. Deut. Keram. Ges.*, 45, 610 (1968).

[14] R. E. Carter and W. L. Roth, (a) *General Electric Research Laboratory Report*, No. 63-RL-3479 M, (November 1963), (b) in *Electromotive Force Measurements in High-temperature Systems*, C. B. Alcock (Ed), The Institution of Mining and Metallurgy, London (1968) p. 125.

[15] A. Cocco, *Chim. Ind., (Milan)*, 41, 882 (1959).

[16] H. A. Johansen and J. G. Cleary, *J. Electrochem. Soc.*, 111, 100 (1964).

[17] T. Y. Tien and E. C. Subbaro, *J. Chem. Phys.*, 39, 1041 (1963).

[18] O. Ruff, F. Ebert and E. Stephan, *Z. Anorg. Allg. Chem.*, 180, 215 (1929).

[19] R. Roy, Discussion of a paper by R. Collongues, *et al.*, *Bull. Soc. Chim. France*, 1141 (1965).

[20] T. Y. Tien, *J. Am. Ceram. Soc.*, 47, 430 (1964).

[21] P. Duwez, F. H. Brown and F. Odell, *J. Electrochem. Soc.*, 98, 356 (1951).

[22] J. Lefèvre, *Ann. Chim. (France)*, 8, 117 (1963).

[23] M. Caillet, C. Déportes, G. Robert and G. Vitter, *Rev. Int. Hautes Tempér. et Réfract.*, 4, 269 (1967).

[24] M. Foëx and A. Rouanet, *Compt. Rend.*, 264 C, 947 (1967).

[25] F. Hund, *Z. Elektrochemie*, 55, 363 (1951).

[26] C. Schusterius and N. N. Padurow, *Ber. Deut. Keram. Ges.*, 30, 235 (1953).

[27] D. W. Strickler and W. G. Carlson, *J. Am. Ceram. Soc.*, 47, 122 (1964).

[28] T. Takahashi and Y. Suzuki, *Compt. Rend., Deuxièmes Journées Internationales d'Etude des Piles à Combustible*, Brussels (1967) p. 378.

[29] A. Cocco and M. Danelon, *Ann. Chim., (Italy)*, 55, 1313 (1965).

[30] T. Y. Tien and E. C. Subbarao, *J. Chem. Phys.*, 39, 1041 (1963).

[31] E. C. Subbarao and P. H. Sutter, *J. Phys. Chem. Solids*, <u>25</u>, 148 (1964).

[32] W. Baukal, *Electrochim. Acta*, <u>14</u>, 1071 (1969).

A FUEL-CELL POWER SYSTEM FOR CENTRAL-STATION POWER GENERATION USING COAL AS A FUEL

E. F. Sverdrup, C. J. Warde and A. D. Glasser

Westinghouse Electric Corporation, Pittsburgh, Pennsylvania

ABSTRACT

The Westinghouse Electric Corporation, in a joint program with the Office of Coal Research of the United States Department of the Interior, has investigated the technical and economic feasibility of a central station fuel-cell power system capable of generating electricity from coal at an overall thermal efficiency of sixty percent. The thin-film-fuel-cell battery — a solid state device using a zirconium oxide electrolyte — is the heart of this generating system. Materials and fabrication processes for all of the components of the battery have been identified. The size of the individual cell units is determined by the electrical characteristics of the electrode, electrolyte, and interconnection films. The cells are small in order to achieve power densities in excess of 5 kilowatts per cubic foot of battery volume when operated at high efficiency. The materials and fabrication costs of a thin-film battery are estimated to be less than $30 per kilowatt of fuel cell generator rating. The thin-film-battery concept makes these costs possible because of the simple processes and small materials quantities required to form a battery. Performance of a 5-cell thin-film battery is reported. The battery on open circuit generated 96% of the theoretical open circuit voltage and delivered electrical power. The research and development needed to establish the technical and economic feasibility of this new power generator is discussed.

I. FUEL CELL BATTERY CONCEPT

Fig. 1 shows a laboratory-prototype fuel cell battery. It consists of five series-connected cells fabricated as thin layers of oxides

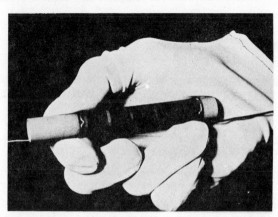

FIG. 1 *A thin-film fuel-cell battery.*

sintered on the outside of a one-half inch diameter, porous, ceramic
tube. Fuel gases, derived from coal, and consisting of carbon monoxide,
hydrogen, carbon dioxide, and water vapor, flow through the inside of
the tube. Air surrounds the tube. When air and fuel gases are fed
continuously, direct-current electrical energy can be drawn from the
terminals of the battery.

The device operates at 1000°C so that: 1) the ceramic electrolyte —
zirconia stabilized in the cubic fluorite crystal structure — will have
low oxygen ion ($O^=$) resistivity (20-50 ohm cm), 2) the heat released in
the fuel electrode reactions can be efficiently transferred and used in
the coal gasification reaction, and 3) the rate of coal gasification is
adequate for the provision of a suitable supply of fuel gas from a coal
reactor of reasonable size (0.25 - 0.4 cubic feet per kilowatt of generating capacity).

Fig. 2 shows schematically a cross section through the wall of this
battery. It consists of five components.

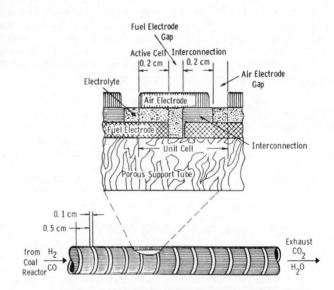

FIG. 2 *Cross section through the wall of a thin film fuel cell
battery.*

1) Porous support tube — a tube functioning as the mechanical
support for the battery — porous enough to allow counter-current diffusion of the fuel gases and their reaction products.

2) Fuel electrodes — porous films of electronically-conductive
material providing a large three-phase contact area between fuel gas,
electrolyte, and electronic conductor.

3) Electrolyte - gas-tight films of stabilized zirconia having
high ionic but negligible electronic conductivity.

4) Air electrodes - highly-permeable electronically-conductive
layers, stable in air at 1000°C, and providing a large contact area
between oxygen from the air stream, electrons, and the zirconia electrolyte.

5) Cell interconnections — thin gas-tight films of an electronic conductor which serves to connect adjacent cells in series by connecting the fuel electrode of one cell to the air electrode of the adjoining cell.

Insulation between adjacent cell units of the battery is provided by the gaps in the fuel and air electrodes, and by the relatively long paths through the thin electrolyte film. This is evident if the scale of the design is considered — the film thicknesses lie in the range of 20-80 microns (0.002 - 0.008 cm), while the gap lengths are between 0.1 and 0.2 cm.

II. OPERATING PRINCIPLE

When fuel and air are supplied, and the battery is near its operating temperature, 1000°C, oxygen from the air accepts electrons at the air electrode to form $O^=$ ions, which may enter vacancies in the crystal lattice of the electrolyte. On the fuel side of the cell, oxygen ions leave the electrolyte to react with the fuel gases, giving up electrons to the fuel electrode of the cell. In this way the air electrode operates at a positive potential with respect to the fuel electrode of a given cell. The magnitude of the voltage difference, E, between air and fuel electrodes depends on the composition of the fuel and air streams, and is given by the equation

$$E = - \frac{RT}{nF} \ln \frac{P_{O_2}(fuel)}{P_{O_2}(air)} \quad ,$$

where $P_{O_2}(fuel)$ = partial pressure of oxygen in the fuel gas stream

$P_{O_2}(air)$ = partial pressure of oxygen in air

R = gas constant, 8.314 joules $°K^{-1}$ $mole^{-1}$

T = absolute temperature, 1273°K

n = number of equivalents per mole of oxygen, 4 equiv. $mole^{-1}$

F = Faraday number, 96,500 coulombs $equivalent^{-1}$

Fig. 3 shows how the generated voltage depends upon the composition of the fuel stream. Two parameters, hydrogen-carbon ratio, H/C, and oxygen-carbon ratio, O/C are used to define the composition of the fuel-gas mixture. In a fuel gas containing no hydrogen, the values, H/C = 0, and O/C = 1, describe pure carbon monoxide, while O/C = 2 describes carbon dioxide. The voltage generated by a solid-electrolyte fuel cell operating near 1000°C falls from slightly over one volt to approximately 0.6 volts as oxygen is added and the fuel mixture changes from carbon monoxide to carbon dioxide. The addition of hydrogen to the fuel gas mixture raises generated voltages and extends the O/C ratio corresponding to complete oxidation.

A thin-film fuel-cell battery contains many series-connected cells. The flow of current adds oxygen to the fuel stream uniformly as one proceeds from the inlet of the battery to its exit. Thus, the O/C ratio linearly increases with distance along the battery and generated voltages drop accordingly.

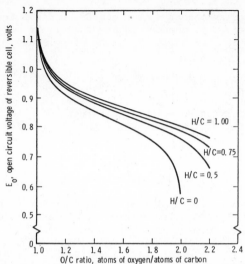

FIG. 3 *Thermodynamic predictions of the effect of H/C ratio on open circuit voltage (T = 1300°K).*

III. THE FUEL CELL POWER SYSTEM

In addition to fuel cell batteries, a complete power system must provide some means for gasifying the coal. The heat released in the fuel electrode reaction is used to supply the heat required by the endothermic gasification reactions. Fig. 4 shows such a system.

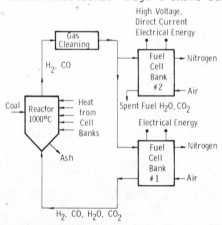

FIG. 4 *Coal-burning solid-electrolyte fuel-cell power system.*

Coal is fed to a fluidized-bed reactor where it reacts with a partially oxidized fuel stream coming from one bank of fuel cells. A fuel gas rich in hydrogen and carbon monoxide results. Excess fuel gas is completely oxidized in a second bank of cells. In practice, the fuel cell banks are immersed within the reactor so that efficient heat transfer can take place. The heat required for gasification is only slightly less than the heat released during electrochemical oxidation of the fuel, so that the overall thermal efficiency of an ideal system is nearly 100 percent. It should be noted that all of the oxygen reaching

the coal enters the system through the fuel-cell electrolyte doing
electrical work.

A practical system will operate at an overall efficiency approaching 60 percent as shown in Fig. 5. To meet a system efficiency goal

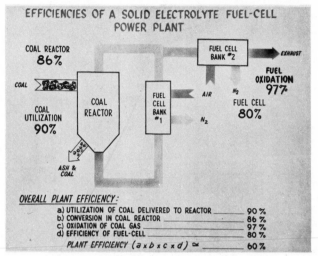

FIG. 5 *Efficiencies of a solid electrolyte fuel-cell power plant.*

of 60 percent, it is necessary to operate the fuel cell batteries at
an electrical efficiency, defined as actual electrical power output
divided by power output of a loss-less battery, of 80 percent.

$$\text{Electrical Battery Efficiency} = \frac{\overline{VI}}{\overline{EI}} = \frac{\overline{V}}{\overline{E}} \quad ,$$

where: $\overline{E}$ = energy-average generated voltage, volts

$\overline{V}$ = energy-average terminal voltage, volts

I = cell current, amperes.

Figure 6 presents the generated voltages expected in a fuel-cell
power system reacting Pittsburgh #8 Coal Char and air. The inlet and
outlet composition for the two cell banks are indicated. The energy-
average voltage from cells in bank #1 is 0.89 volts, and from cells
in bank #2 is 0.83 volts. If these cell banks are operated at 80
percent electrical efficiency, they can be loaded to average terminal
voltages of 0.69 and 0.66 volts respectively. Hence, 200 mV are available for resistive and concentration polarization losses for cells in
bank #1, while cells in bank #2 must be more lightly loaded, so that
polarization and resistive voltage losses do not exceed 170 millivolts.

The choice of how many cells to include in bank #1 in which partial
oxidation of the fuel gas occurs, and how many to include in bank #2,
rests with the engineering compromise between the higher power outputs
obtainable from bank #1 cells against the larger quantities of fuel gas
to be recycled through the reactor, the increased pump work, and reactor
gasification volume associated with a large recycle. Since the volume
required for fuel cells (~0.2 cu ft/kW) is less than the volume of coal
bed required for gasification (~0.25 to 0.4 cu ft/kW) the improvements

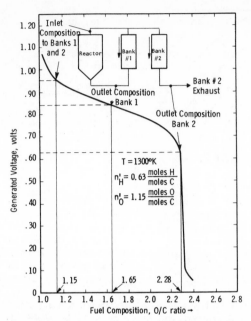

FIG. 6 *Generated voltage for 100 kW process development plant showing inlet and outlet compositions for the two banks of fuel cell batteries.*

in overall-system power density that can be achieved by using more, higher power density, bank #1 cells is not dramatic.

In the 100 kW process-plant-design, recently completed [1], 67 percent of the fuel gas leaving the reactor is recycled through bank #1, and 33 percent is oxidized in bank #2. Bank #1 cells provide 46 percent of the heat used in the gasification reactor, as well as delivering 49 percent of the electrical power.

IV. BATTERY DESIGN — EFFECT OF DIMENSIONS ON POWER PER UNIT VOLUME OBTAINABLE FROM A FUEL CELL POWER SYSTEM

To achieve an overall power system efficiency exceeding 60 percent, the fuel-cell batteries should be loaded to operate at an electrical efficiency of 80 percent. For cells using gas derived from coal, this requires that the total electrode polarization losses including the ohmic resistance losses of the cell units be limited to 170-200 millivolts. Operating-electrolyte current densities are expected to lie between 400 and 800 mA/cm^2. At these current densities the "IR"-free electrode polarizations will total about 100 millivolts leaving 70 to 100 millivolts for cell resistance losses.

The power output per unit of volume occupied by the fuel-cell batteries has been calculated as a function of the battery unit-cell dimensions and the resistivities of the electrolyte, interconnection and electrode materials. The analysis reveals that for the electrodes, the parameter of importance in the determination of the power density of the fuel-cell banks is the resistivity-thickness quotient. The resistivity-thickness product is the corresponding parameter for both electrolyte and interconnection films. Fig. 7 presents the results for a

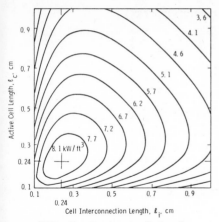

FIG. 7   *Contour plot of power density as a function of the lengths of the interconnection and the active cell in a thin-film fuel-cell battery.*

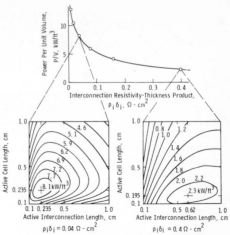

FIG. 8   *Effect of interconnection resistivity-thickness product on the power density and dimensions of a fuel-cell battery.*

particular case in which the resistivity-thickness products of the electrolyte and interconnection are taken to be 0.04 ohm cm^2 (resistivity: 20 ohm cm; film thickness: 0.002 cm), and the fuel and air electrode resistivity-thickness quotients are 0.1 and 0.2 ohms respectively. The intercell insulating gap has been assumed to be 0.1 cm. These values represent those which we consider achievable using the current materials and processing techniques. The power densities have been computed assuming that the batteries are constructed on one-half inch outside diameter support tubes and that the batteries are spaced, in the manner of shell and tube heat exchangers, with their centers at the apexes of equilateral triangles using a spacing between adjacent tubes of 1.25 times the tube diameter. The figure is a contour plot showing the effect of the length of the active cell and the length of the interconnection on the power density. Power density reaches a maximum of slightly over 8 kilowatts per cubic foot of battery volume when the cell and interconnection lengths are each 0.24 cm. As cells or interconnections are increased in length, the power density which can be achieved at 80 percent operating efficiency falls off. Because the interconnection and active cell resistivity-thickness products were assumed equal, the contour plot is symmetrical — maximum power per unit volume being achieved when the cell and interconnection lengths are made equal.

The effect of changing the interconnection resistivity-thickness product can be seen in Fig. 8, in which all of the other parameters have been held constant. As interconnection resistivity-thickness product increases, the power density that can be achieved drops off, and for maximum power density the interconnection film should be lengthened to reduce the resistance contribution of the interconnection to the overall cell resistance. The analysis emphasizes the need for a battery fabrication process that can provide for short cells with accurate registration of thin ($\approx$ 0.002 cm) component layers. These studies show that power densities greater than 5 kilowatts per cubic foot of battery volume in batteries operating at 80 percent electrical efficiency are possible.

V. MATERIALS AND FABRICATION PROCESSES

A thin-film battery consists of films of fuel electrode, inter-connection, electrolyte, and air electrode laid sequentially on a highly-porous support tube. The materials have been chosen to be economic and long-lived. They will not interact deleteriously with one another either during fabrication, or on prolonged operation at 1000°C. As the films are made in sequential processes which involve repeated heating and cooling from 1400°C, the materials have compatible thermal expansion characteristics. In the following sections, a dis-cussion is given of the materials and fabrication processes in the order of their application in the manufacturing process flow sheet shown in Fig. 9.

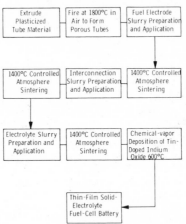

FIG. 9 *Current process sequence for the fabrication of thin-film fuel-cell batteries.*

Porous Support Tube

Calcia-stabilized zirconia is used as the support tube material to match the thermal expansion of the electrolyte film. To facilitate the rapid flow of reactants to, and products from the fuel electrode — electrolyte interface, the tube must be highly permeable. The necessary porosity is introduced by slip casting or extruding tubes using a mix-ture of two particle-size-ranges of calcia-stabilized zirconia (70 per-cent by weight of 10-44 μm diameter particles, 30 percent by weight of particles smaller than 5 μm in diameter). Bentonite, a clay comprising the oxides of silicone, aluminum, iron, calcium, magnesium, potassium, and sodium, is included at the 1-3 percent by weight level to strengthen the green tubes, which are fired, after overnight drying, for four hours in air at 1800°C.

Cross sections of tubes produced by these techniques are shown in Fig. 10. The open porosity of these tubes is approximately 30 percent. The average modulus of rupture has been determined as 8000 psi. The maximum pore diameter lies between 10 μm and 20 μm. A typical analysis of porous support tube material is shown below. The figures quoted were determined for a slip cast tube and are given in percentages by weight.

ZrO_2	CaO	MgO	SiO_2	Al_2O_3	TiO_2	Fe_2O_3
90.0	5.7	1.4	1.4	0.59	0.50	0.33

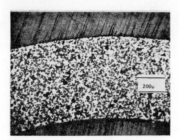

Extruded Support Tube

Slipcast Support Tube

FIG. 10 *Cross sections through a wall of porous support tubes.*

The balance of the impurities were barium, manganese, copper, and chromium oxides, all in trace quantities.

Future development work is required to improve the uniformity of porosity, and decrease the maximum and average pore diameters, so as to optimize support tube strength while retaining adequate permeability. Tight control of impurities in the tube materials is needed to avoid adversely affecting the sintering and electrical characteristics of the films making up the battery.

Fuel Electrode

The fuel electrode must also be highly permeable to gas flow to reduce concentration polarization effects. The structure must be electronically conductive — a resistivity — thickness quotient of less than 1.0 ohm is satisfactory, while a value of 0.1 would be desirable.

Porous structures consisting of a continuous metal phase and a continuous stabilized-zirconia structural skeleton to maintain porosity have shown good performance as fuel electrodes [2]. A composition containing 70 percent by weight of metal ensures good electronic conductivity. The stabilized-zirconia skeleton, by forcing the metal to deform on heating or cooling, keeps the electrode well-attached to the tube substrate, and acts to keep the electrode porous on prolonged operation at 1000°C.

The oxidation resistance of electrodes, containing nickel as the metal phase, in the fuel-gas atmosphere is shown in Fig. 11. Nickel-zirconia fuel electrodes are stable over the expected range of fuel-gas compositions and may be employed in both fuel-cell banks. Electrodes have also been prepared using cobalt as the conductive phase. The use of these electrodes is limited to cell bank #1 because of their lesser oxidation resistance.

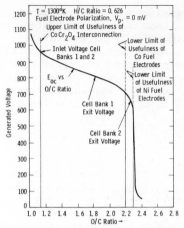

FIG. 11 *Generated voltage as a function of the oxygen-carbon ratio in the fuel-gas stream, showing the voltage levels at which cobalt and nickel oxidize and cobalt chromite reduces.*

The recommended slurry for the preparation of fuel electrode films is given in Table I.

TABLE I

Recommended Slurry for Preparation of Fuel Electrode Films

Component	Material	Quantity	Rationale
Metal Oxide Particle Size Range: 37-44 μ	NiO	200 g	To yield continuous metal phase. ~70% by weight.
Ceramic Phase: 1) 20-30 μ	$(CaO)_{0.15} (ZrO_2)_{0.85}$	46 g	Narrow size range to ensure open structure.
2) 1-5 μ	ZrO_2	12 g	To promote sintering of porous skeleton.
3) 1-5 μ	$CaZrO_2$	4 g	To promote sintering of porous skeleton.
Vehicle	n - Butyl Acetate	170 ml*	To yield stable slurry.
Mixing Method	End-over-end tumbling for 17 hours		To ensure complete dispersion.

*Disperse powders in 500 ml, allow to settle to 170 ml total volume, and decant clear supernatant liquid.

The slurry is prepared by thorough mixing of the sized metal oxide particles, calcia-stabilized zirconia particles, and fine powders of zirconia and calcium zirconate in butyl acetate. The narrow particle size ranges of the metal oxide and calcia-stabilized zirconia are selected to prevent close-packing of the metal-ceramic structure and thus introduce considerable porosity. Fine powders of calcium zirconate and zirconia amounting to 30 percent by weight of the ceramic phase, are included to promote rapid sintering of the calcia-stabilized zirconia skeleton.

A support tube, pre-wetted with butyl acetate, is dipped, as shown in Fig. 12, into a slurry containing approximately 2.5 percent by weight of solids, and withdrawn at a controlled rate. Both the drying time after wetting and the rate of withdrawal determine the coating thickness. The gaps between cells are then formed by erasing the unsintered electrode material, as illustrated by Fig. 13. The coated battery tube

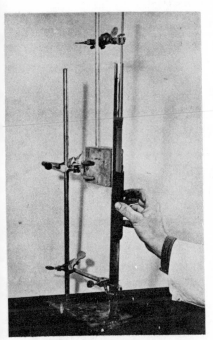

FIG. 12 *Slip casting the fuel electrode.*

FIG. 13 *Erasure of green fuel electrode from the intercell regions to define fuel electrode bands.*

is sintered at 1400°C for one hour to sinter the ceramic skeleton of the fuel electrode. Flowing air has been employed successfully as a sintering atmosphere. Reduction of the metal oxide to form the continuous metal phase is accomplished by replacing the air atmosphere by nitrogen-5%-hydrogen during the cooling step. Average heating and cooling rates of 10°C per minute are employed to avoid thermal shock to the porous tube substrate and excessive loss of metal oxide to the tube.

Typical cross sections of nickel- and cobalt-zirconia electrodes are shown in Fig. 14. The continuity of the metal and ceramic phases, and the electrode porosity are evident. Electrode tests show that these electrodes operate at "IR"-free polarization losses of 50 mV when

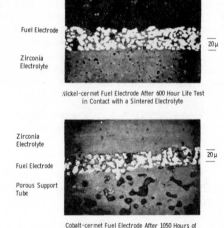

FIG. 14 *Cross section photomicrographs of spray-sintered fuel electrodes.*

loaded to current densities of 360 mA/cm^2 using a fuel gas composition characterized by an O/C ratio of 1.6 and H/C ratio of 1.0. Further improvement in electrode performance can be accomplished by careful sizing of the raw materials to ensure maximum porosity of the metal-ceramic structure. The use of gelling agents, appropriate dispersants, and improved mixing techniques, to provide stable suspensions of the multicomponent fuel electrode slurry should allow much improved uniformity and control of the electrode structure.

Interconnection

The fuel electrode is not stable in air at 1000°C, and similarly, tin-doped indium oxide, the air-electrode material, is not stable in the fuel-gas mixture. Consequently, a material is needed to serve as an interconnection between the air and fuel electrodes of adjacent cells in the battery. In both air and fuel gas at 1000°C, this material should be thermodynamically stable, and have an electronic resistivity of less than 50 ohm cm. The material should display a low cationic diffusivity to prevent loss of the metal component of the fuel electrode through the interconnection by simultaneous cationic and electronic migration induced by the oxygen partial pressure gradient.

Cobalt chromite satisfies the stability [3,4,5] and diffusivity [6,7] requirements. It reduces to cobalt metal and chromium sesquioxide only in fuel gas mixtures which are more reducing than those encountered anywhere in the fuel cell batteries (see Fig. 11). It will not react with the nickel constituent of the fuel electrode, as nickel chromite is unstable in the reducing atmosphere of bank #1 cells and throughout most of bank #2. The extent of interaction between nickel, nickel chromite, and cobalt chromite in the more-oxidizing fuel-gas atmospheres at the exit of bank #2 cells is not now known. It is estimated to be small. (This is a point requiring further study). The electronic resistivity of cobalt chromite depends upon oxygen partial pressure and upon the impurities present. Its resistivity in air at 1000°C has been reported by early investigators as 40 ohm cm [8] and as 200 ohm cm [9].

Although the sintering of cobalt chromite to near theoretical densities has been reported to be difficult [10], probably as a result of the low ionic mobility that makes this material desirable as a stable interconnection, it was found to sinter to thin, gas-tight films at 1400°C when:

1) material in the particle size range 0.1 to 0.5 μm was used.

2) well dispersed slurries were prepared using sufficient blending times and appropriate dispersants (see Table 2 below).

3) the oxygen partial pressure in the sintering atmosphere was maintained high enough to prevent reduction to cobalt and chrome oxide.

A substantial thermal expansion mismatch exists between cobalt chromite and the porous tube material. The percentage thermal contraction of the former from 1400°C, the sintering temperature, to room temperature is 1.08 percent, while the corresponding value for the tube is 1.65 percent. As the chromite film is in compression, no film-failure problems are anticipated or observed, and a stress analysis has shown that the tube, which is subject to a tensile stress, should not fail until a film-tube wall ratio of greater than 0.1 is reached. This corresponds to a film thickness of more than 100 μm, which is a factor of four or five greater than the expected value.

TABLE 2*

Recommended Slurry for Preparing Interconnection Films

Component	Material	Quantity	Rationale
Pigment 0.1 - 0.5 μ size range	$CoCr_2O_4$	640 g	Supplied by the C. J. Pfizer Co. [11].
0.001 - 0.005 μ size range**	$CoCr_2O_4$	320 g	Not used in our experimentation but indicated by packing studies of Furnas [12].
Dopant 0.001 - 0.005 μ size range**	Mn_3O_4	40 g	Lowers resistivity and promotes sintering
Vehicle	Cyclohexanone	1000 ml	Has satisfactory characteristics (evaporation index, *etc.*) for application by spraying
Dispersant	Advasperse[R]***	0.5 g	Allows preparation of stable slurry with uniform distribution
Mixing Method	End-over-end tumbling for at least 17 hours		Ref. [13]

In no case does this data imply that the choices are optimum. They represent current "state of the art" and are presented as guides for further development.

**Appreciable quantities of material in this particle size range may not be achievable by conventional pyrolysis techniques. The inclusion of material in the size range, 0.01 to 0.05 μ will also serve to increase the packing density, though not so effectively.*

****Registered Trademark of Cincinnati Milicron, Inc.*

The slurry was applied to the support tube by paint-spraying techniques. Optical masks using collimating slots formed by metal fingers were used to mask the tube and provide for the deposition of 0.2 - 0.3 cm bands spaced 0.4 cm apart. Fig. 15 shows the technique employed in spraying the interconnection slurry through the optical masks. More sophisticated spraying equipment was used in our later experimentation.*

FIG. 15 *Spraying of interconnection slurry through optical masks.*

Following application to the tubes, the cobalt chromite films were sintered. Fig. 16 presents the safe temperature-atmosphere profile, which prevents oxidation of the fuel electrode or reduction of cobalt chromite. Shown also are two experimental profiles employed in the sintering of interconnection films. Cross sections of the chromite films are shown in Fig. 17. The metallic phase in the upper photograph is cobalt, a product of the reduction of cobalt chromite. The lower photograph shows the metal-free interconnection film, well-bonded to the underlying fuel electrode obtained by sintering within the "safe" profile.

Electrical testing of thin-films of $CoCr_2O_4$ showed that the effective "resistivity" of the films exposed between air and the fuel atmosphere was in excess of 200 ohm cm. Thus, cobalt chromite, though possessing appreciable electronic conductivity, is not sufficiently conductive. Preliminary experiments [14] indicate that the replacement of the chromium cations in the B sites of the spinel lattice with manganese cations lowers the resistivity to acceptable levels. Further development work on the interconnection is needed to optimize doping, determine stability, and perfect application techniques to permit fabrication of minimum-thickness, gas-tight chromite layers.

Electrolyte

Zirconia, stabilized with calcia, yttria, scandia, or mixtures of rare earth oxides has high oxygen ion conductivity and negligible

Spraying Systems Co., #110005 1/4 J#1 pneumatic atomizing system using their #2050 fluid nozzle equipped with clean-out and shut-off needle.

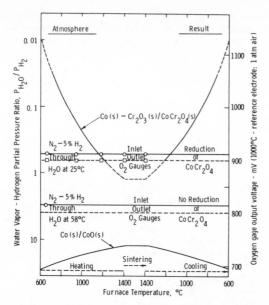

FIG. 16 *Experimental temperature-atmosphere profiles employed in*
the sintering of cobalt chromite interconnection films inside and out-
side the limits, between which cobalt and cobalt chromite may coexist
at high temperatures.

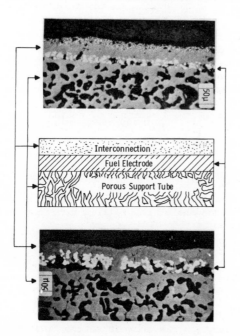

FIG. 17 *Cross sections of the batteries which underwent the*
temperature-atmosphere profiles shown in Fig. 16 Note appearance of
cobalt in cobalt chromite film sintered above the Co-Cr_2O_3/$CoCr_2O_4$
temperature-atmosphere profile (upper photo).

electronic conductivity at 1000°C. The metal ions in the cubic-fluorite crystal lattice are locked tightly in place — cation mobility is very low [15] so that the material is an ideal electrolyte at fuel cell battery operating temperatures near 1000°C. Although calcia stabilized zirconia has the highest ionic resistivity, 20-50 ohm cm (depending on purity), — when used in the thin-film battery, resistivity-thickness products of 0.04 ohm cm^2 can be approached in films of sufficient thickness to be gas tight. Yttria or scandia can also be used as stabilizing agents. Their use would lower the ionic resistivity by factors of two and five respectively. The use of these more expensive agents does not add significantly to the materials costs, due to the small quantity of material employed in the fabrication of thin electrolyte films.

A number of methods of producing gas-tight films of the electrolyte in the thin-film battery configuration were investigated [16]. These included sputtering [17,18], thermal decomposition of alkoxide vapors [19], chemical vapor deposition from chloride vapors [20], and sintering [21]. The sintering approach appears most satisfactory primarily because masking the battery to deposit the 0.4 cm length bands of zirconia separated by 0.2 cm gaps can be carried out at room temperature. (In contrast, masking for the chemical deposition process must withstand a 1100°C deposition temperature.)

The use of mixtures of calcium zirconate and pure monoclinic zirconia, which react during sintering to form the cubic phase zirconia, allows the electrolyte films to be sintered at 1400°C. [21]

Sintering was probably aided by the presence of impurity phases melting below 1400°C, introduced from the relatively impure support tubes. Microprobe analyses indicate that these phases consist of oxides of silicon, aluminum, magnesium, calcium, and iron. One source of these impurity phases is the mineral bentonite introduced in the porous support tube to give green strength to the slip cast tube.

Table III presents a recommended slurry composition suitable for preparing electrolyte films by "paint-spray and sinter" techniques.

The electrolyte slurry was applied to the battery tube using the same spraying and masking techniques used for applying the interconnection bands (Fig. 15).

Sintering of the electrolyte requires heating to 1400°C in an atmosphere which is sufficiently oxidizing to prevent reduction of the cobalt chromite. Sintering can apparently be carried out in air if:

1) Reduction of nickel or cobalt oxide in the fuel electrode is postponed to the end of the battery fabrication process, and

2) Impurity levels in the support tube are reduced appreciably so that reaction between the nickel or cobalt oxides in the fuel electrode and liquid phase introduced from impurities in the support tube does not occur.

A typical cross-section of a calcia-stabilized zirconia film is shown in Fig. 18. The film adheres to the fuel electrode and withstands thermal cycling. The resistivity of these films in the thin-film battery configuration has approached the bulk resistivity. Improved application techniques are needed to achieve gas-tight, pore-free films less than 20 μ in thickness. Particle sizing should be employed to assist in achieving dense films. If required liquid phase sintering aids to promote electrolyte densification should be identified and incorporated in the electrolyte slurry.

TABLE III*

Recommended Slurry for Preparing Electrolyte Films

Component	Material	Quantity	Rationale
Pigment 1-5 μ size range	ZrO_2	490 g	
0.01-0.05 μ size range	ZrO_2	250 g**	React to form $(ZrO_2)_{0.85}$ $(CaO)_{0.15}$ on sintering at 1400°C
1-5 μ size range	$CaZrO_3$	170 g	
0.01-0.05 μ size range	$CaZrO_3$	90 g**	
Vehicle	Cyclohexanone	1000 ml	Satisfactory evaporation index for paint spraying
Dispersant	Advasperse^{R***}	7 g	Allows preparation of stable slurry with uniform distribution of solids
Mixing Method	End-over-end tumbling for not less than 17 hours		Ref. [12]

Recommendations are not necessarily optimum, but represent 1970 state-of-the-art.

**Not used in our experimentation but suggested by work on the packing of clays by Furnas [12].*

****Registered trademark - Cincinatti Milicron, Inc.*

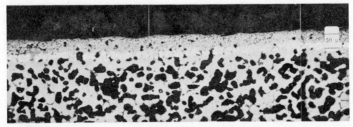

FIG. 18 *Reactively-sintered stabilized-zirconia electrolyte over fuel electrode.*

Air Electrode

A suitable air electrode material will be stable in air at 1000°C, and will display good electronic conductivity. A resistivity-thickness ratio of less than 1 ohm is desirable. Noble metals are not stable, volatilizing through gaseous sub-oxides.

Tin-doped indium oxide has been found to be a satisfactory air electrode material [22]. This electrode has been applied to batteries by chemical vapor deposition at 600-750°C, the fuel electrodes being protected by flowing a hydrogen-water vapor mixture through the inside of battery during deposition. The recommended spray solution and masking technique are given in Table IV.

TABLE IV

*Recommended Spray Solution for Chemical Vapor Deposition
of Tin-Doped Indium Oxide Films*

Component	Material	Quantity	Rationale
Indium species	$InCl_3$ (aqueous) (concentration: 1.05 g/ℓ)	240 ml	To form In_2O_3
Dopant	$SnCl_4$ (anhydrous)	2.4 ml	To reduce resistivity of In_2O_3 films
Vehicle	HCl (aqueous) (37% by weight)	60 ml	
Mixing Method	With magnetic stirrer in beaker, until solution becomes clear		To ensure thorough mixing
Masking Technique	Air-spray fine-particle ZrO_2 dispersed in n-butyl acetate in air electrode gaps		To define air electrode bands
Carrier Gas	Air		To provide oxygen for reaction with $InCl_3$ and $SnCl_4$ to form $(In_2O_3)_{0.98}$ $(SnO_2)_{0.02}$

Thin-film air electrodes of this type have been shown to display stable high-performance characteristics in life tests well in excess of 1000 hours [22].

Use of liquid phase sintering aids allows fine powders of tin-doped indium oxide to sinter at temperatures in the range 1400°-1500°C. This points the way toward fabrication of air electrode films in a manner that would greatly reduce the complexity of the application technique, and permit the easy introduction of controlled porosity into the electrode structure. If this electrode can be developed, the manufacturing sequence for making thin-film batteries becomes a series of simple "spray-sinter" processes. Experimentally it has been possible to sinter the electrolyte and interconnection simultaneously obtaining leak tight joints between these. Using simultaneous-sintering to form these films would result in a process that requires only three sintering operations. The simplified process is outlined in Fig. 19.

VI. STATUS OF THE DEVELOPMENT

Early work culminating in a 100-watt, coal-burning, fuel-cell power supply made clear that a fuel-cell power system for large-scale power generation must use:

1) materials and electrode structures stable for years operating at temperatures near 1000°C,

2) battery fabrication techniques that would avoid the making, handling, and joining of many small individual cell units in order to make reliable batteries.

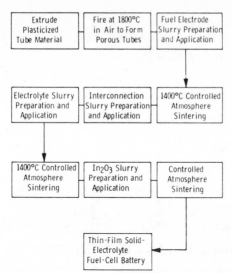

FIG. 19 *Proposed process flow chart for the manufacture of thin-film solid-electrolyte batteries.*

The thin-film battery was invented to meet these needs. Demonstrations that indium oxide cathodes and nickel-zirconia anode structures could be made that had at least 1000 hour stability and overvoltage losses approaching 50 millivolts at current densities greater than 350 mA/cm^2 followed. A stable electronically-conductive material for the interconnection was identified and methods of doping it to achieve the required conductivity were developed. A method for forming gas-tight, thin-film electrolyte and interconnection films by compatible processes that would allow masking and deposition of these films and the formation of leak-tight joints between them was developed. Five-cell batteries were constructed and tested (Fig. 20 and 21). These batteries demonstrated the feasibility of the thin-film battery concept.

Problems pointed out by the battery processing studies and electrical tests included the need for:

1) control of film thicknesses to avoid overstressing the support tube on thermal cycling,

2) improved structure and impurity control in the support tube,

3) improved methods for introducing porosity in the air electrode,

4) control of film overlap in the electrolyte-interconnection junction regions to limit parasitic currents.

VII. CONCLUSIONS

The thin-film solid electrolyte battery provides the large number of series connected fuel cells required in a commercial power system. The battery can be made by a series of spray-sinter processes in which appropriate slurries of oxides are applied in bands on a porous support tube. Controlled atmosphere high temperature (1400°C) sintering is used to form the battery. Five-cell, laboratory prototype batteries have been made and tested.

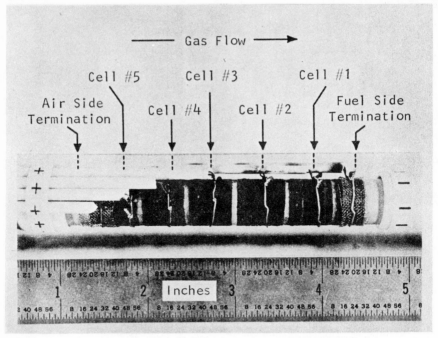

FIG. 20 *A thin-film fuel-cell battery, instrumented for testing.*

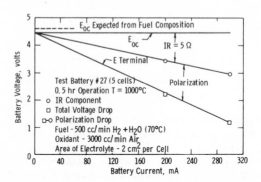

FIG. 21 *Volt-ampere characteristics of test battery #27.*

The materials currently used in the battery cost about $20 per
kilowatt of generator rating. This cost has been estimated to halve
in large scale production [16]. Processing costs to manufacture the
batteries have been estimated and amount to approximately $20 per
kilowatt of fuel cell battery. The cost of an entire fuel cell power
plant — fuel cell batteries, coal reactor, gas cleaning, and recircu-
lation equipment, and power conditioning — could be as low as $130 per
kilowatt of generator rating. A fuel-cell power generating system
that is competitive in price with conventional generating systems but
which offers a 60 percent conversion efficiency appears possible.

A large amount of development work remains to be done with the
battery components and with the battery fabrication process before a
commercial fuel-cell power plant can be realized. The principal areas

requiring effort are:

1) Interconnection Conductivity and Stability

To achieve power-densities exceeding 5 kilowatts per cubic foot of fuel cell battery volume, the cell interconnection material must have a resistivity less than 50 ohm cm. Undoped cobalt chromite displays a resistivity of 20 ohm cm in air and 2000 ohm cm in fuel gas atmospheres at 1000°C. Doping with manganese at the 2 m/o level has been shown to result in 1000°C resistivities of 6 ohm centimeters in air and 50 ohm centimeters in fuel atmospheres. Resistivity in the fuel atmosphere was stable over a testing period of 26 days. Resistivity in air appeared to be stable. The stability and rates of interaction between doped cobalt chromite and other battery components needs further detailed study.

2) Development of a Spray Sintered Indium-Oxide Air Electrode

Indium oxide air electrodes deposited by chemical vapor deposition were used in early work to form a stable air-electrode material. The process used to introduce porosity into these electrodes is not satisfactory for use in thin-film batteries. Based on the discovery of sintering aids which allow sintering of indium oxide at temperatures below 1400°C, it should be possible to fabricate porous air-electrode structures on thin film batteries by sintering. Well dispersed slurries containing coarse indium oxide particles with ultra-fine sintering aid particles would be used. Slurries in which cyclohexanone was used as the vehicle and containing 0.5 wt % of a commercial dispersant have been used to prepare slurries which were suitable for application by paint-spraying techniques. The correlation between process details and electrode performance needs to be established.

3) Fuel Electrode Performance over the Complete Range of Fuel
    Compositions

The effect of electrode structure — metal — oxide — and void distribution on the polarization, resistance, and stability of the fuel electrode over the complete range of fuel gas compositions and operating temperatures needs to be established. Measurements on early porous platinum fuel electrodes indicated that hydrogen was preferentially oxidized to water vapor on these electrodes. The water in turn reacted with carbon monoxide in the fuel gas mixtures to generate hydrogen and carbon dioxide (the water-gas shift reaction). Using platinum electrodes it was necessary to include a chrome-oxide water gas shift catalyst to allow the water gas shift reaction kinetics to supply hydrogen at the rates required to support fuel cell operation. The nickel and cobalt fuel electrodes now employed may not require additional catalysts. The detailed studies needed to establish this still need to be carried out.

Further research and development is needed to demonstrate the performance levels and life required in a commercial fuel-cell power plant. The pertinent science and materials technology available to apply to these problems is extensive. A new solid state electric power generator offering the possibility of realizing the promise of the high-efficiency of "direct-conversion" appears to be possible.

ACKNOWLEDGMENTS

    This work was largely financially supported by the Office of Coal
Research of the United States Department of the Interior. The authors
would particularly like to acknowledge the assistance of Mr. Neal P.
Cochran, Chief of Utilization and Mr. Paul Towson of the Division of
Utilization for many contributions to the work. Dr. David Archer and
later Dr. J. T. Brown supervised this work and were instrumental in
seeing that the effort was always focused toward producing a practical
fuel-cell power system. The authors would also like to acknowledge
the many contributions of the staff of Project Fuel Cell.

REFERENCES

[1]   D. L. Keairns, D. A. Archer and L. Elikan, "Symposium on Coal
       Combustion in Present and Future Power Cycles," Amer. Chem. Soc.,
       Div. of Fuel Chem., *Abstracts of Papers*, 14(2), 97 (1970).

[2]   A. O. Isenberg, *Extended Abstracts, Fall Meeting*, The Electro-
       chemical Society, October 4-8, 1970, p. 43.

[3]   J. D. Tretyakow and H. Schmalzried, *Ber. Bunsenges. Physik. Chem.*,
       69, 396 (1965).

[4]   V. A. Levitskii, T. N. Rezukhina and V. G. Dneprova, *Soviet
       Electrochemistry*, 1, 833 (1965).

[5]   L. M. Lenev and I. A. Novokhatskii, *Russ. J. Phys. Chem.*, 40,
       1097 (1966).

[6]   A. Morkel and A. Schmalzried, *Z. Physik. Chem. N.F.*, 32, 76
       (1962).

[7]   J. S. Armizo, Final Report - Task 2, NASA CR-72537 (May 1969).

[8]   H. Schmalzried, *Ber. Bunsenges. Physik. Chem.*, 67, 93 (1963).

[9]   T. E. Bradburn and G. R. Rigby, *Trans. Brit. Ceram. Soc.*, 52,
       417 (1952).

[10]   R. Sun, *J. Chem. Phys.*, 28, 290 (1958).

[11]   Chas. Pfizer and Co., Inc., Easton, Pa. 18042. Sample supplied
       through the courtesy of Mr. K. Hancock, Minerals, Pigments, and
       Metals Division.

[12]   C. C. Furnas, *Ind. Eng. Chem.*, 23, 1052 (1931).

[13]   G. A. Loomis, *J. Amer. Ceram. Soc.*, 21, 393 (1938).

[14]   C. C. Sun, E. W. Hawk and E. F. Sverdrup, paper to be presented
       at the Electrochem. Soc. Meeting, Cleveland, October 1971.

[15]   W. H. Rhodes and R. E. Carter, "Ionic Self-Diffusion in Calcia-
       Stabilized Zirconia," Abstract in *Bull. Amer. Ceram. Soc.*,
       (April 1962).

[16]   Final Report, Project Fuel Cell, Research and Development Report
       No. 57 (1970), Office of Coal Research, Department of the Interior,
       Washington, D. C. 20240.

[17]   N. J. Maskalick, Ref. 16, p. 270.

[18]   A. D. Glasser, Ref. 16, p. 268.

[19] A. D. Glasser and E. F. Sverdrup, Ref. 16, p. 270.

[20] A. O. Isenberg, Ref. 16, p. 270.

[21] N. J. Maskalick and C. C. Sun, *Extended Abstracts, Fall Meeting,* Electrochemical Society, October 4-8, 1970, pp. 44-5.

[22] E. F. Sverdrup, D. H. Archer and A. D. Glasser, "Fuel Cell Systems-II," *Advances in Chemistry Series No. 90,* pp. 301 (1969).

SUMMARY OF DISCUSSION AND COMMENTS: SESSION CHAIRMAN V

S. Gilman

U. S. Army Electronics Command, Fort Monmouth, New Jersey

Discussions under this topic were confined mainly to the area of solid electrolyte fuel cells. Activity in the area of high temperature solid electrolyte fuel cells dates back to the beginning of this century. Haber and his coworkers experimented with conducting glasses as early as 1905. Nernst and coworkers discovered the high conductivity of 85 percent ZrO_2 — 15 percent Y_2O_3 and utilized this conducting oxide mixture in the "Nernst Glower" light source before 1900. By 1914, Ruff and coworkers discovered that the mechanical integrity (as well as conductivity) of zirconia was improved by additions of Y_2O_3, CaO and MgO. Baur and Preis reported fuel cell experiments using those formulations by 1937 with patents on a number of other solid electrolyte fuel cells appearing over the intervening years.

More recent work on fuel cells utilizing zirconia electrolytes was reported in the early 1960's, notably by research groups at Battelle Institut Frankfurt, the General Electric Company and Westinghouse. Prototype devices generally operated at temperatures near 1000°C, utilizing hydrocarbons or steam-reformed hydrocarbons as fuel. Attention has been focused largely upon improvement of electrolyte conductivity (and lowering of operating temperature), electrolyte stability, contacts and system engineering.

The first speaker, Dr. Tannenberger of Battelle-Geneva, addressed himself to an evaluation of materials problems encountered in zirconia fuel cell development. He made the observation that progress in this area during the last decade has been slow and he proposed future areas of investigation.

Dr. Baukal of Battelle-Frankfurt has addressed himself to one specific material problem encountered in the development of zirconia electrolyte fuel cells — time dependent increase in electrolyte resistivity. The mechanism of this phenomenon was explored for yttria, ytterbia and calcia-stabilized zirconia.

The last speaker, Mr. Sverdrup of Westinghouse Research Laboratories discussed the overall engineering effort devoted at his establishment to the development of a 100 kilowatt zirconia fuel cell utilizing coal as fuel. An operating efficiency of 60 percent appears possible judging from the engineering analysis.

In addition to the solid electrolyte fuel cells the state of development of molten carbonate cells was discussed. The outlook for both systems was considered to be good, but much work regarding the stability of materials remains to be done.

SESSION VI

IMPLANTABLE FUEL CELLS

Chairman

S. K. Wolfson

DESIGN CONSIDERATIONS FOR AN IMPLANTABLE FUEL CELL TO POWER AN ARTIFICIAL HEART

J. Giner and G. Holleck

Tyco Laboratories, Inc., Waltham, Massachusetts

ABSTRACT

An implantable fuel cell operating on glucose and oxygen extracted from the blood is discussed. The individual cell shows five components in intimate contact with each other, and in the following sequence: 1) Glucose-selective permeable membrane, 2) porous (hydrophilic) glucose electrode, 3) electrolyte matrix, 4) porous (hydrophobic) oxygen electrode, and 5) oxygen-selective permeable membrane.

The major advantages of this cell configuration are 1) the blood does not come in contact with the electrodes and therefore both blood coagulation and electrode poisoning are minimized, 2) it appears possible to treat the surfaces exposed to blood with ultrathin, blood-compatible layers with little effect on the reaction rates, 3) the configuration allows the packing in the available volume of many very thin cells separated by thin blood channels in such a way that the cells can be operated at low current densities, 4) the needed performance (current density and voltage) seems feasible without major technical breakthrough.

The magnitude of the individual process rates occurring at each cell component is discussed based on diffusion calculation and actual experimental results. This discussion is performed for two working modes of the fuel cell: one with immobile electrolyte and the other with flow of plasma through the cell.

In order to obtain an output of 5 watts with a volume not larger than 500 cc, it is proposed to use 200 individual cells of 5 cm × 10 cm surface and 250 μ thickness (including the blood channels). A hybrid configuration, with a rechargeable battery (such as Ni Cad battery) for peak power, is proposed for better optimization of fuel cell size.

INTRODUCTION

An implantable fuel cell operating on glucose and oxygen extracted from blood is a very attractive power source to drive an artificial heart. In the following, the principal design considerations for such a fuel cell are discussed giving the magnitude of the individual process rates occurring at each cell component. The discussion is based on diffusion calculations and actual experimental results and is performed for two working models of the fuel cell: one with immobile electrolyte and the other with flow of plasma through the cell.

POWER REQUIREMENTS

The approximate power requirements of a human heart are:

Mean power of human heart < 1 W (at rest).
Mean power of human heart $\simeq$ 3.8 W (moderate work).
Instantaneous peak power (systole) 10 W (moderate work).

To meet these power requirements which have to be increased by the inefficiencies of a practical pump, we propose a hybrid system consisting of a 5-W fuel cell and a Ni-Cd battery. The Ni-Cd battery delivers the peak power for high power demand and is recharged by the fuel cell during periods of low power demand. Such a system allows a better optimization of the fuel cell.

CHARACTERISTICS OF BLOOD

Several important parameters for the operation of an implanted biological fuel cell are fixed by the physiology of the human body; they have to be the starting point of any design considerations. The main characteristics of blood which are relevant to an implantable fuel cell are:

Total blood volume (recirculated once per min): 5 to 6 liters.
Plasma: 55 vol %
Cellular components: 45%; largest white cell: 20 μ
Viscosity: 5-6 × viscosity of water
Electrical conductivity (25°C): 0.012 $ohm^{-1}cm^{-1}$

Anions:

Bicarbonate	24-30 meq/liter
Chloride	100-110 meq/liter
Phosphate	1.6-2.7 meq/liter

pH=7.4

c_{O_2} (100 mm Hg) $\sim$ 8.8 × 10^{-3} M/ℓ (in whole blood)

c_{O_2} (100 mm Hg) $\sim$ 1.35 × 10^{-4} M/ℓ (in plasma)

c_{O_2} (art.) $-$ c_{O_2} (ven) $\sim$ 3 × 10^{-3} M/ℓ

Free glucose $\sim$ 4.5 × 10^{-3} M/ℓ

Proteins $\sim$ 7% (by weight, in plasma)

ELECTRODE REACTIONS

The reduction of oxygen occurs according to the overall reaction:

$$O_2 + 2\ H_2O + 4e^- \rightarrow 4\ OH^-$$

The oxidation of glucose is more complicated. For our further discussion, we will consider two extreme cases: 1) the complete oxidation according to:

$$C_6O_6H_{12} + 6\ H_2O \rightarrow 6\ CO_2 + 24\ H^+ + 24e^-$$

and 2) the two-electron oxidation to gluconic acid:

$$C_6O_6H_{12} \rightarrow C_6O_6H_{10} + 2\ H^+ + 2e^- \quad .$$

BUFFER REACTIONS

The potential of both electrodes depends also on the pH value of the surrounding electrolyte and since, upon current flow, OH^- and H^+ ions are produced at the oxygen and glucose electrodes, respectively, we have also to consider the buffer reactions taking place in the electrolyte. They can be summarized as follows:

$$CO_2 \text{ (gas)} \rightleftharpoons CO_2 \text{ (dissolved)} \qquad\qquad H_2CO_3 \rightleftharpoons H^+ + HCO_3^-$$

$$CO_2 \text{ (dis)} + H_2O \underset{k_1'}{\overset{k_1}{\rightleftharpoons}} H_2CO_3 \qquad\qquad HCO_3^- \rightleftharpoons H^+ + CO_3^{2-}$$

$$CO_2 \text{ (dis)} + OH^- \underset{k_2'}{\overset{k_2}{\rightleftharpoons}} HCO_3^- \qquad\qquad H^+ + OH^- \rightleftharpoons H_2O$$

The ionic dissociation reactions shown in the last three equations are very fast and they will be in equilibrium at all times. The formation of H_2CO_3 and HCO_3^- from dissolved CO_2, however, is quite slow. This reaction was studied by Roughton and Booth [1]. They give the following values for the rate constants:

$$k_1 = 0.026 \text{ sec}^{-1}$$

$$k_1' = 5.82 \cdot 10^4 \text{ } \ell/\text{mol sec}$$

$$k_2 = 2.4 \cdot 10^4 \text{ } \ell/\text{mol sec}$$

$$k_2' = 5.44 \cdot 10^{-4} \text{ sec}^{-1}$$

CONDITIONS FOR AN IMPLANTABLE FUEL CELL

The data given above show that the conditions for the operation of an implantable biological fuel cell differ significantly from those of conventional fuel cells. Many are less favorable. However, not all of them are.

Unfavorable
Low reactant concentrations.
Neutral electrolyte, with low buffer capacity.
Possibility of blood clotting.
Presence of high molecular weight compounds.
Possible deleterious effects of products to body.

Favorable
Constant temperature — excellent heat transport.
Excellent water and products removal.
Constant reactant supply.
Good mechanical environment — low-pressure differentials.
Material costs relatively unimportant.

THE PROPOSED FUEL CELL

A cross section of the basic element of the proposed fuel cell is shown in Fig. 1. The individual cell shows five components in intimate contact with each other, and in the following sequence: 1) glucose-selective permeable membrane, 2) porous (hydrophilic) glucose electrode, 3) electrolyte matrix, 4) porous (hydrophobic) oxygen electrode, and 5) oxygen-selective permeable membrane. The glucose electrode can be

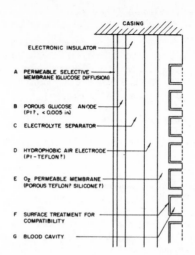

FIG. 1 *Magnified cross sectional view of building blocks perpendicular to blood flow.*

made of pure Pt or of a noble metal alloy such as Pt-Ru, while the oxygen electrode can be a Teflon-bonded structure with Pt or Au alloy as catalyst. A fuel cell array is obtained by stacking individual cells in such a way that a channel for blood flow is formed between the glucose membrane of a cell and the O_2-membrane of the neighboring cell. Under these conditions, the electrolyte in the electrolyte matrix is in equilibrium with the blood through the glucose-permeable membrane, which is ionically conductive.

The major advantages of this cell configuration are: 1) the blood does not come in contact with the electrodes, and therefore both blood coagulation and electrode poisoning are minimized, 2) it appears possible to treat the surfaces exposed to blood with ultrathin, blood-compatible layers with little effect on the reaction rates, 3) the configuration allows the packing in the available space in a way that the cells can be operated at low current densities, and 4) the needed performance (current density and voltage) seems feasible without major technical breakthrough.

KINETIC PROCESSES IN ELECTROCHEMICAL RECOMBINATION OF GLUCOSE AND OXYGEN

The overall reaction taking place in the fuel cell can be subdivided into a number of successive kinetic processes:

O_2 in Blood	Glucose in Blood
Boundary Layer in Blood	Boundary Layer in Blood
O_2-Permeable Membrane	Glucose Permeable Membrane
O_2-Transport in Electrode	Glucose Transport in Electrode
Electrode Reaction	Electrode Reaction
Transport of Products in Electrode (OH^-, CO_3H^-)	Transport of Products in Electrode (H^+, CO_2)

Separator

(Transport of products and ohmic iR drop)

In the following, we will discuss the magnitude of the process rates occurring at each cell component, based on diffusion calculations and actual experimental results.

RATES OF INDIVIDUAL PROCESSES AND DIMENSIONS OF COMPONENTS

BLOOD CHANNEL (OXYGEN SIDE)

The size of the blood channel is limited on the low side by the size of the largest white cells which are approximately 20 μ. Besides this, the actual channel depth is determined by practical considerations such as flow rate, pressure drop, and overall size of the fuel cell. A reasonable value would be approximately 100 μ.

The boundary layer is a consequence of the velocity profile of the moving blood in a channel and of the diffusion normal to the channel surface. Its effect is practically independent of the material of the channel walls, but is a function of viscosity and flow velocity. Turbulent flow greatly reduces the boundary-layer thickness and its transport characteristics.

We have observed under conditions of laminar flow a boundary layer giving rise to limiting currents for deoxygenation of blood of approximately 0.8 mA/cm^2 at a ΔP of 100 torr. Any promotion of mixing in the blood, *e.g.*, by flow normal to electrode surface, pulsating or otherwise, or by a special shape of the blood channels could reduce this boundary layer considerably.

O$_2$-MEMBRANE (HYDROPHOBIC)

The limiting currents resulting from oxygen diffusion through membranes of dimethyl silicone and Teflon at 37°C are:

25 μ dimethyl silicone at ΔP = 100 torr:

i_L = 4.5 mA/cm^2 (measured)

3 μ Teflon at ΔP = 100 torr:

i_L = 0.8 mA/cm^2 (measured)

O$_2$-ELECTRODE (HYDROPHOBIC STRUCTURE)

The transport of O$_2$ in the hydrophobic channels is very fast and thus of no particular concern. For a standard Teflon-bonded Pt black electrode (10 mg Pt/cm^2) which has a roughness factor of approximately 1,000, we measured a current density of about 1.5 mA at an activation overvoltage of 200 mV. In addition, we observed concentration polarization due to the production of OH$^-$ ions. These ions react with the electrolyte according to the equations given before, and the products are removed by diffusion. Since an equivalent amount of acidic species is generated at the glucose electrode, the observed concentration polarization is not only a function of the current density, but also of the separation of the two electrodes. For the case of electrodes of 20 μ thickness with 10 μ electrolyte between them, one can calculate a concentration polarization η_C = 100 mV at i = 1 mA/cm^2.

ELECTROLYTE MATRIX (SEPARATOR)

The main considerations concerning the separator can be summarized as follows:

12 μ porous structure (micron size pores)

— iR drop: 0.2 mV at 1 mA/cm^2

— (O_2 interference at glucose electrode i_L = 2 × 10^{-5} A/cm^2)

— No glucose interference at O_2-electrode with Au-based O_2-electrode

— Concentration polarization at O_2-electrode increases somewhat with separator thickness.

BLOOD CHANNEL (GLUCOSE SIDE)

The effect of the boundary layer on glucose extraction is considerably less than on that of oxygen, limiting current densities much above 1 mA/cm^2 can be obtained.

GLUCOSE MEMBRANE

The permeability of a typical cellulose based dialysis membrane, 12 μ cuprophane, was determined and would result in the following limiting currents:

n = 2 i_L = 0.2 mA/cm^2

n = 24 i_L = 2.4 mA/cm^2

(O_2-interference: i_L = 0.02 mA/cm^2) .

GLUCOSE ELECTRODE

The transport limited current for a 10 μ electrode would be:

n = 2 i_L = 6.6 mA/cm^2

n = 24 i_L = 79 mA/cm^2

In a complete cell, there is no pH polarization at the glucose electrode. The magnitude of the activation polarization during glucose oxidation on platinum (10 mg/cm^2 roughness factor approximately 1,000) is reflected in the measured performance:

$i_{(400\ mV)}$ = 0.3 mA/cm^2 (steady state value)

DIMENSIONS OF CELL ARRAY AND CELL COMPONENTS

Analysis of the individual steps shows that the prime factors which limit cell performance are electrode kinetic effects at the anode and mass transfer effects at the cathode which are due to the blood boundary layer.

In view of this fact, we feel that the following is a reasonable design goal:

Component thickness, μ

    O_2-membrane   12

    O_2-electrode   50

    Separator (electrolyte matrix)   22

    Glucose-electrode   40

    Glucose-membrane   20

    Blood channel   <u>100</u>

               Cell unit   244

Cell array:

    200 (10 cm × 5 cm) cell units operating at 1 mA/cm^2 and 0.5 V = 5 W

    Volume $\sim$ 5 cm × 5 cm × 10 cm (500 cc)

Exchange surface:

    1 m^2 oxygen exchange

    1 m^2 glucose exchange

ALTERNATIVE DESIGN INVOLVING PLASMA FLOW THROUGH THE CELL

An alternative approach to the fuel cell with immobile electrolyte would be a cell with plasma flow through the electrode assembly. The individual cell would consist of four components in the following sequence: 1) hydrogel layer (instead of the hydrophobic oxygen membrane), 2) porous, partially hydrophobic oxygen electrode, 3) electrolyte matrix, and 4) porous, hydrophilic glucose electrode.

The hydrogel layer would act as a filter and prevent the cellular components and the proteins of the blood from entering the electrode structure. Its thickness would have to be sufficiently thin (approximately 100 Å) in order to permit oxygen extractions at an acceptable rate. The oxygen transport in the hydrophobic part of the oxygen electrode would again be relatively fast. Here, the oxygen electrode catalyst would have to be unaffected by the presence of glucose (e.g., an Au-based electrocatalyst).

In this model, the glucose would be carried by the plasma flowing through the cell. If we assume a 24-electron oxidation at the glucose electrode, we would need a minimum total plasma flow of 50 cm^3/min; for the case of n = 2, the minimum flow would be 600 cm^3/min.

As for the kinetics of glucose oxidation, the same arguments mentioned earlier apply. This design would combine an oxygen electrode with hydrophobic channels for gas diffusion with the improved product removal of a flowthrough system. It would be preferentially located in series with the circulatory system. This would have the following advantages: 1) the blood channels would be relatively wide, thus making them less susceptible to clogging; 2) the whole pressure drop of approximately 70 to 80 torr can be used to drive the plasma through the fuel cell; 3) the oxygen extraction would take place only from highly oxygen-saturated blood; 4) the arrangement would require only the hydrogel-covered oxygen electrode to be in contact with the blood, since the outflowing plasma could be vented into the thoracic cavity.

The practical verification of a fuel cell based on this design is, however, much more complicated, especially since the state of the art concerning hydrogels is not yet sufficiently advanced.

Based on the design considerations discussed above, we are at present involved in the development of a prototype fuel cell with a design goal of 5 W power output. Considering the dimensions required for the individual components, it is obvious that new techniques for their manufacture have to be developed.

This research is supported by the U.S. National Institute of Health under Contract No. NIH 69-32.

REFERENCE

[1] F. J. W. Roughton and V. H. Booth, *Biochemical Journal*, **32**, 2049 (1939).

COMMENT ON PAPER BY J. GINER
FROM J. O'M. BOCKRIS

The excellent progress made in this research is well recognized — particularly Dr. Giner's suggestion of the electrochemical gas motor.

It seems that a severe problem is the clotting of blood on the electrode; however, a solution may come from the work of Sawyer and Srinivasan [1].

These workers showed that clotting was manageable if the potential of any parts in contact with blood was more negative than -0.6 V on the N.H.S.

It might be difficult completely to encapsulate everything in plastic to prevent clotting. Cathodic protection could be utilized in connection with the electrode. Periodic pulsing might be applied to remove the surface films from clotting.

REFERENCE

[1] P. N. Sawyer and S. Srinivasan, *J. Colloid. and Interface Sciences*, **32**, 456 (1970).

A BIOAUTOFUEL CELL FOR PACEMAKER POWER[*]

S. J. Yao, Maryanne Michuda, F. Markley and S. K. Wolfson, Jr.

Michael Reese Hospital and Medical Center, Chicago, Illinois

ABSTRACT

The goal of this work is to develop a miniaturized pacemaker unit attached directly to the heart wall. It would derive power from fuel cells functioning with endogenous carbohydrates and oxygen and should have an unlimited life span. Experiments over the past years in our laboratory have indicated that there are many naturally occurring fuels in body fluids and tissues. These fuels were studied in in vitro systems utilizing oxygen under simulated physiological conditions. Permselective membranes were utilized to separate anode from cathode so that both fuel and oxygen could be extracted from the same body fluid compartment. Typical cells contained 5 cm^2 platinized platinum electrodes with an anion-exchange-membrane internal electrolyte. These cells produced a constant power output in the 80 to 90 microwatt range at approximately 0.50 volts. These power levels were sustained for periods of weeks or until gross bacterial and fungal contamination of the external electrolyte brought on their demise in the nonsterile experiment.

INTRODUCTION

Present implantable cardiac pacemakers, while theoretically capable of 3 to 5 years' service, usually fail after 6 to 18 months requiring surgical intervention. These failures are usually related to problems with battery power or electrode leads. An alternative and attractive power source would be an implantable bioautofuel cell. A small, unitized pacemaker, completely implanted on or close to the heart and deriving power from fuel cells functioning with endogenous carbohydrates and oxygen, would eliminate the need for battery and leads replacement. In theory, such a unit could become a permanent pacemaker implant.

The possibility of direct bioelectrochemical energy conversion was suggested by experiments of Del Duca [1]. Work toward the bioautofuel cell approach to pacemaker power source has been reported by Wolfson [2], Warner [3] and Drake [4]. The work presented here describes the principle, design and laboratory studies of an implantable bioautofuel cell for lower power source medical devices.

[*]Work performed at the Beatrice F. and Melville N. Rothschild Surgical Research Laboratories, Michael Reese Hospital and Medical Center, Chicago, Illinois and Argonne National Laboratory, Argonne, Illinois. Supported in part by U.S.P.H.S. Grants GM 16654, HE 12555 and FR 5476.

PRINCIPLE AND DESIGN

The implantable cell will derive power from direct energy conversion of endogenous carbohydrates — mainly glucose — and oxygen. The
cell is to be immersed in a single-electrolyte medium: extracellular
fluid or blood. Such a device is ideal since the fuel cell consumes a
negligible amount of carbohydrates and oxygen and can, in principle,
provide a life-time power source for the pacemaker. However, previous
studies have indicated certain difficulties in the design and operation
of this type of fuel cell. The presence of both reducing agent (*i.e.*,
the carbohydrates) and oxygen in the same medium would easily lead to
electrode poisoning and also increasing polarization via mixed-potential
kinetics [5]. The poor buffering quality and lower open-circuit voltage afforded by the dilute, near-neutral medium (pH 7.4) limits the
performance of fuel cell electrodes. The physiological concentration
of dissolved oxygen is only one fiftieth of the glucose concentration.
The viscous liquid medium reduces the rate of mass transfer, especially
the diffusion of oxygen to the reaction site at the cathode. Both
concentration and diffusion factors affect the efficiency of the
cathode. The ecology of the fuel cell in the body has also to be
taken into account. Implanted materials should be compatible with the
physiological environment. Fuel cell products should not be toxic.

The theoretical reactions in a glucose fuel cell can be written
below,

at the cathode $\quad O_2 + 2H_2O + 4\ e^- \quad ---------- \quad 4OH^-$ $\qquad\qquad$ (1)

at the anode $\quad CH_2O + H_2O - 4\ e^- \quad ---------- \quad 4H^+ + CO_2$ $\qquad$ (2)

overall reaction $\quad CH_2O + O_2 \qquad\qquad ---------- \quad H_2O + CO_2$ $\qquad$ (3)

where CH_2O represents 1/6 of a glucose molecule.

The above equations reveal that OH^- ions are produced at the
cathode, H^+ ions are produced at the anode, and that water and carbon
dioxide (or gluconic acid) are products of the glucose fuel cell
reaction.

Consideration of the above has led us to develop a single-electrolyte fuel cell utilizing permselective membranes [6]. These membranes
separate anode from cathode so that both fuel and oxygen can be extracted from the same body fluid. The cell is constructed by sandwiching a
hydrophilic anode and a hydrophobic cathode with an anion-exchange membrane and with two covering membranes. The anode-covering membrane is
permeable to glucose or lower-molecular-weight carbohydrates and oxygen,
while the cathode-covering membrane is permeable only to gases, *i.e.*,
oxygen, carbon dioxide and water vapor. Sources and characteristics
of these cell parts are:
  a. Anodes:
      1. Allis Chalmers Co.   40 mg Pt/cm^2
      2. Energy Research Corp.   40 mg Pt/cm^2
      3. General Electric Co.   Pt-20, Ru
  b. Cathodes:
      1. Allis Chalmers Co.   40 mg Pt/cm^2
      2. Energy Research Corp.   Ag
  c. Anion exchange membrane: American Machine and Foundry Co.,
      Type A310.

d. Anode covering membrane: Nephrophane; cellophane with elon-
    gated pores or with elliptic pores, ∿ 50 Å pores.
e. Cathode covering membrane: GE: MEM-213, 0.5 mil thick or the
    outer surface of the cathode is painted with 0.5 mil thickness
    of Dow-Corning Silastic medical adhesive.

Both membranes prevent the diffusion of macromolecules. The anion-
exchange membrane was chosen as a means of operating the cathode dry
in order to provide better oxygen diffusion to the cathode. The OH^-
ions produced at the cathode Eq. (1) transfer across the anion-exchange
membrane to the anode where they meet with the H^+ ions, a product of
the anodic reaction Eq. (2), to form liquid water. The outward diffusion of these newly formed water molecules presumably would serve
the purpose of cleaning up the catalytic surface of the anode. Aside
from transferring the excess OH^- ions which would otherwise cause
severe cathodic polarizations [7,8], the anion-exchange membrane also
acts as a pumping device to prevent water formation in the cathode.
The recycled water for these reactions is available at the surface of
the ion-exchange membrane since it operates in a water-saturated state
and water vapor is present in the cathode compartment. However, net
water production occurs at the anode surface.

The arrangement of cell components is illustrated in Fig. 1. The
edges of these components are etched and then bound together with
epoxy. Except for the exposed covering membranes, the whole cell is
covered with medical grade silicone rubber. Experiments with the use
of a stainless steel (SS 316) clamping arrangement is proceeding.
Sterilization of cells is performed by means of either autoclaving or
Gamma-ray irradiation. The sterilized cell is then immersed in a
beaker containing 250 ml of modified Krebs-Ringer bicarbonate buffer
(see Table I), with added carbohydrates fuel and with O_2 and CO_2 concentration maintained in the physiological range (pH 7.4, Po_2 = 80-90
torr, Pco_2 = 35-40 torr). Gas composed of 83% N_2, 12% O_2 and 5% CO_2
is introduced into the beaker via a glass diffusor at a rate sufficient
for greater than 99% equilibration to maintain the desired Po_2, Pco_2
and pH. Cell performance is evaluated by measuring open-circuit (rest-
ing) potential and voltage/current density curves and by life tests.
Open-circuit and passive-load measurements are made with a Keithley
153 microvoltammeter (200 megohm input resistance). Galvanostatic
measurements are made by plotting anode half-cell potentials vs. fixed

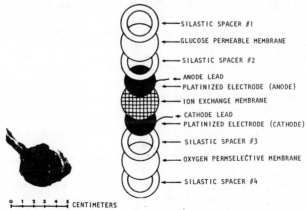

SILASTIC SPACER #1
GLUCOSE PERMEABLE MEMBRANE
SILASTIC SPACER #2
ANODE LEAD
PLATINIZED ELECTRODE (ANODE)
ION EXCHANGE MEMBRANE
CATHODE LEAD
PLATINIZED ELECTRODE (CATHODE)
SILASTIC SPACER #3
OXYGEN PERMSELECTIVE MEMBRANE
SILASTIC SPACER #4

0 1 2 3 4 5 CENTIMETERS

FIG. 1 *An implantable cell. The photograph at lower left is of
an actual working cell.*

current loads generated with a Harrison programmable power supply
(Hewlett-Packard No. 6209B). A Wenking No. 68TAl potentiostat is used
to maintain the anode at initial open-circuit potential between galvanostatic runs. A period of 30 minutes is allowed for equilibration
followed by a 30-minute recovery period. pH and P_{CO_2} are estimated by
the IL Duomatic blood gas meter and P_{O_2} by IL 125 polarographic oxygen
meter. Glucose is estimated by the glucose oxidase method (Worthington
Biochemical: Glucostat).

TABLE I

External Electrolyte Solution (pH 7.4)

Electrolyte	Molarity
NaCl	0.122
KCl	0.005
KH_2PO_4	0.001
$MgSO_4 \cdot 7H_2O$	0.001
$NaHCO_3$	0.025
	0.154
Glucose	0.005
	0.159

RESULTS

Initial voltage rising of a typical cell at 25°C is shown by curve
(a) in Fig. 2. The final open-circuit potential is 0.75-0.80 V, which
is close to that of separate half-cell (salt-bridge) systems reported
previously [2]. At body temperature (37°C), the initial voltage rising
of this cell was faster than that of the same cell at 25°C. The final
open-circuit potential, at 37°C, is 0.80-0.85 V, about 0.05 V higher
than that of the same cell at 25°C (see Fig. 2). Long-term life testing of immersible cells was difficult at the present stage, since it
was observed that glucose solution from the anode eventually leaked
along the edge of the anion-exchange membrane to the cathode compartment and caused flooding and poisoning of the cathode. This leakage
is a bonding problem and may be solved 1) by substituting more easily
bonded ion-exchange membranes and permselective membranes, 2) by
substituting better bonding agents, or 3) by means of a suitable clamping device. In any case, the ultimate life of this cell was usually
limited by growth of bacteria and/or fungi behind the permselective
membrane and directly on the electrodes. This can be prevented by
initial sterilization of the cell. The covering membranes will then
act as a microbiological barrier to maintain sterility. Ultimately,
implantation of a sterile cell into the body will provide a permanently
sterile environment. Fig. 3, shows results of a life test on a prototype immersible cell. The cell potential leveled at about 0.50 V for
20 hr. at a current density of 34 $\mu A/cm^2$, and gave an output power of
17 $\mu W/cm^2$. It should be mentioned that at 25°C the total output of
the cell, which had about 5-cm^2 geometrical area, was 85 μW at this
current density.

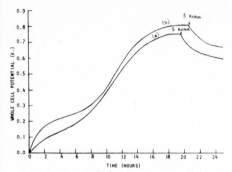

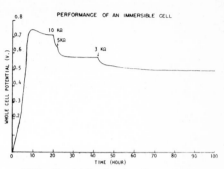

FIG. 2 *Initial voltage rising of whole cells. (a) immersible cell, 25°C. (b) immersible cell, 37°C.*

FIG. 3 *Life test of a prototype cell. Final output was 17 µW/cm² at 34 µA/cm² at 0.50 V.*

Fig. 4 shows the potential/current density (Tafel) plot for the anode of a prototype cell. Maximum power output for a typical cell was evaluated by discharging the cell at various passive loads and observing the leveling-off voltage of the whole cell. Results are shown in Fig. 5. The plots of Figs. 4 and 5 seem to indicate that the optimal current density for the prototype cell is around 75 µA/cm² at 0.56 V. This represents 210 µW total power; considerably more than enough for cardiac pacemaker or similar applications. Since the anode potential is maintained better than that of the cathode, it may be that insufficient supply of oxygen to the cathode is the major limitation of performance at higher current densities. We foresee operation at lower current density levels which may improve stability. The effect of varying temperature (25°-37°C) was insignificant. The presence of 10 percent plasma did reduce cell output, but the effect was less than 50 percent and there was evidence of bacterial growth in the plasma. The necessary inclusion of antifoaming agents to experimental cells containing plasma may have an undesirable effect.

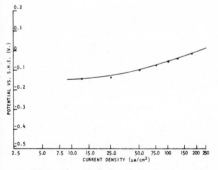

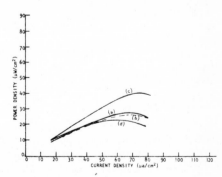

FIG. 4 *Tafel plot for anodic oxidation of glucose.*

FIG. 5 *Power density of immersible cells. (a) GE: Pt-20, Ru/Pt, 37°C (b) GE: Pt-20, Ru/Pt, 25°C (c) ERC: Pt/AA40, 37°C (d) ERC: Pt/AA40, 37°C, 10% Plasma. Curves (a) and (b) as well as (c) and (d) were runs on the same physical cells, respectively. Moreover, it required over eight hours each to record curves (c) and (d). The plasma exhibited definite signs of microbiological growth and deterioration in that period. Conditions were not sterile. Maximum cell power at peak of curve (c) is 210 µW at 74 µA/cm².*

DISCUSSION

GENERAL CONSIDERATIONS

The prototype cell reported above appears to be a first working model of an implantable pacemaker fuel cell. It meets the major requirements and is limited only by problems which seem solvable:

1) The cell is a self-contained unit which lends itself to implantability without need of entering the blood-circulating system.

2) It functions spontaneously when in contact with a medium containing fuel and oxidant of the type and of the concentration to be encountered in the human extracellular space or natural fluid spaces, such as pleura, pericardium and peritoneum.

3) Externally exposed materials used are compatible with body tissue.

4) Power output is adequate for pacemaker and similar applications.

5) The products of reactions as previously determined by chromatographic analysis are apparently restricted to normal metabolites and thus are nontoxic [9]. This is in agreement with the work of others [5,7,10].

6) Fuel and oxidant consumption are trivial when compared with normal metabolic flux.

Although the cell is projected for implantation outside the blood-circulatory system, endocardial pacemaker placement is a possibility and, moreover, many components of plasma, including protein, are often present in extracellular fluids. Thus, it is significant that the addition of 10 percent plasma to the external electrolyte did not obliterate cell function.

Commercial fuel cells usually employ three phase systems (liquid, solid and gas). The essential elements of the cell require the presence of electrolyte and catalyst; the former is commonly liquid, and the latter solid. Fuel and oxidant are introduced as a gas because of the favorable mass-transport situation in gas and because of the ease of handling reactants in this state. In our cell, the carbohydrate fuel, which is nonvolatile, requires a "drowned" hydrophylic anode. However, the cathode (inherently the weaker half-cell) can be operated dry, since oxygen, although dissolved in water in the body fluid, can be converted to gas by a permselective membrane.

ADHESIVES AND BONDING

A major problem is the means of sealing the immersible cell elements without a clamping arrangement. The center of the bond is the ion-exchange membrane, which is generally a chemically substituted polyolefin or polyfluorocarbon. These substances are noted for their inertness and poor adhesive qualities. The use of heat and/or chemical treatments threatens to destroy the ion-exchange quality. Outer elements of the bond are the permselective membranes. The oxygen membranes used thus far are silicones and copolymers such as silicone-polycarbonate, while anode membranes have been micropore carbohydrate films, such as cellulose. These substances are relatively easily bondable. Experimentation with a number of epoxy and other adhesives is proceeding.

PENETRATION OF FUEL ACROSS THE ION-EXCHANGE MEMBRANE

The bioautofuel cell in the test system and in the body is in contact with both fuel and oxidant at physiologic concentrations. Oxygen is present at both the cathode and anode, while fuel is present only at the anode. Since the physiologic molar concentration of fuel (*e.g.*, glucose) is 50 times that of oxygen, the presence of oxygen can be tolerated at the anode. This is fortunate since a membrane permeable to glucose, water and oxygen can thus be employed. On the other hand, a membrane permeable to glucose and water but impermeable to oxygen would be difficult to obtain. Transfer of fuel across the ion-exchange membrane is possible and could be a serious problem. These membranes are permeable to water and several have passively transported glucose in a static test. In this test, buffer containing glucose is separated by the ion-exchange membrane from buffer not containing glucose. Glucose did appear across the membrane after several days. However, arrangement of the cell is such that glucose must pass through the anode before reaching the membrane, and presumably could be consumed by an operating cell to lower markedly the concentration in contact with the membrane. The function of the cell for an extended period, reported above, may indicate that glucose is so consumed before reaching the membrane. On the other hand, the use of ion-exchange membranes impermeable to glucose or other carbohydrate fuel might further increase cell output. Studies are underway in which ^{14}C labeled glucose is being used to trace movement of fuel in a functioning cell.

ELECTRODES AND COVERING MEMBRANES

Electrode materials were generally obtained from commercial sources. Those with heavier catalyst loadings performed the best. This is not so great a disadvantage as in the case of commercial fuel cells, since ultimate cost per pacemaker unit will be high in any case and catalyst costs will not be a major part of this cost. Improvement is sought in anode wettability and cathode nonwettability. An oxygen catalyst that could minimize the contribution of the peroxide mechanism would be desirable [11]. Catalyst selectivity is also an area of current investigation. The advent of catalysts which are mutually and completely selective for fuel and oxidant would make it possible to substitute relatively open-pore-covering membranes on both anode and cathode, thus increasing fuel and oxidant flux. In this happenstance, the covering membrane would function only to restrict inflow of macromolecule poisons.

OTHER FUELS

We have found that a number of carbohydrates can function as well or better than glucose in the fuel cell system [12]. These include glucosamine, mannose, sorbitol, gluconic acid, glucuronic acid, fructose and ethanol (see Table II). Fig. 6 shows galvanostatic polarization on 5-mM glucose, gluconic acid and glucosamine solutions. The values quoted were taken after electrodes had remained at open circuit for 24 hr. in the solutions, and represent steady state data (less than 2 mV change in 30 min.). Repetition of the voltage/current density curves in the Fig., after as long as 24 hr. under constant load, did not indicate that poisoning was a serious problem. These curves were reproducible for that time and for varying periods beyond 24 hr. In the eventual application of this work, it is not necessary

to restrict other oxidizable fuels from the anode. Actually, the
presence of multiple fuels (with noninterfering reactions) would
enhance the ultimate objective.

TABLE II

Anodic Data: Blood Carbohydrates and Related Compounds

Fuel 5 mM in pH 7.4 Buffer	OCV (vs. SHE)	Anode Voltage (vs. SHE)	
		At 50 μA/cm²	At 100 μA/cm²
Ethanol	-0.18	-0.13	-0.10
Fructose	-0.22	-0.14	-0.10
Gluconic Acid	-0.11	-0.05	-0.02
Glucose	-0.22	-0.07	-0.04
Glucosamine	-0.42	-0.41	-0.40
Glucuronic Acid	-0.22	-0.07	-0.04
Mannose	-0.30	-0.24	-0.16
Sorbitol	-0.25	-0.20	-0.15

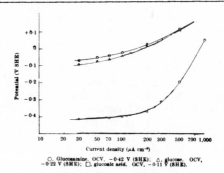

O, Glucosamine, OCV, -0·42 V (SHE); △, glucose, OCV, -0·22 V (SHE); □, gluconic acid. OCV, -0·11 V (SHE).

FIG. 6 *Tafel plots for anodic oxidation of glucose, gluconic acid
and glucosamine.*

Though many of the carbohydrates give initial OCP near the reversible hydrogen potential (-0.44 V SHE), after electrode pretreatment by
anodic and cathodic pulsing, only glucosamine gives a potential which
shows little time dependence, with low polarizations at practical
current densities. This is perhaps connected with the special adsorptive character or actual oxidation of the -NH₂ group.

Although glucosamine and other amino sugars are widely distributed
throughout the body tissues, they are present in plasma only in
combined form as mucopolysaccharides and glycoproteins. The amino
sugars might ultimately prove to be promising fuels for cardiac-assist
devices if a suitable method of making them available at an implanted

fuel cell anode were found. Other work in this laboratory has been
directed toward development of an enzymatic converter for production
of monomer from endogenous polysaccaride [13].

ACKNOWLEDGMENTS

The authors acknowledge the continuing interest of Mr. Anthony
Geisel and his valuable suggestions and technical assistance. We are
also grateful for helpful discussions and advice from Drs. A. John
Appleby and Daniel Y. C. Ng of the Institute of Gas Technology, Chicago,
Illinois. We wish to thank Dr. A. B. Laconti of General Electric Co.,
West Lynn, Mass., Drs. B. S. Baker and M. Klein of Energy Research
Corp., Bethel, Conn., and Mr. L. C. Hymes of Allis-Chalmers Co.,
Greendale, Wis., for supplying us with electrodes and other materials.

REFERENCES

[1] M. G. Del Duca, "Direct and Indirect Bioelectrochemical Energy
     Conversion Systems," in *Develop. Industr. Microbiol.*, Plenum Press,
     New York (1963), p. 81.

[2] S. K. Wolfson, Jr., S. L. Gofberg, P. Prusiner and L. Nanis,
     "The Bioautofuel Cell: A Device for Pacemaker Power from Direct
     Energy Conversion Consuming Autogenous Fuel", *Trans. Amer. Soc.
     Artif. Int. Organs*, 14, 198 (1968).

[3] H. Warner and B. W. Robinson, "A Glucose Fuel Cell", *Digest of
     the 7th International Conference on Medical and Biological
     Engineering*, Stockholm (1967), p. 530.

[4] R. F. Drake, B. K. Kusserow, S. Messinger and S. Matsuda, "A
     Tissue Implantable Fuel Cell Power Supply", *Trans. Amer. Soc.
     Artif. Int. Organs*, 16, 199 (1970).

[5] M. L. B. Rao and R. F. Drake, "Studies of Electrooxidation of
     Dextrose in Neutral Media", *J. Electrochem. Soc.*, 116, 334 (1969).

[6] S. K. Wolfson, Jr., S. J. Yao, A. Geisel and H. R. Cash, Jr.,
     "A Single Electrolyte Fuel Cell Utilizing Permselective Membranes,"
     Trans. Amer. Soc. Artif. Int. Organs, 16, 193 (1970).

[7] R. F. Drake, "Implantable Fuel Cell for an Artificial Heart",
     Proc. Artif. Heart Program Conf., U.S. Dept. Health, Education and
     Welfare (1969), p. 869.

[8] M. Beltzer and J. S. Batzold, "Limitations of Blood Plasma as a
     Fuel Cell Electrolyte", *ibid*, p. 354.

[9] S. J. Yao, Unpublished Data.

[10] J. Giner and P. Malachesky, "Anodic Oxidation of Glucose", *Proc.
     Artif. Heart Program Conf.*, U. S. Dept. Health, Education and
     Welfare (1969), p. 839.

[11] J. R. Goldstein and A. C. C. Tseung, "A Joint Pseudo-Splitting/
     Peroxide Mechanism for Oxygen Reduction at Fuel Cell Cathodes",
     Nature, 222, 869 (1969).

[12] S. J. Yao, *et al.*, "Anodic Oxidation of Carbohydrates and Their
     Derivatives in Neutral Saline Solution", *Nature*, 224, 921 (1969).

[13] L. A. Dohan, S. J. Yao and S. K. Wolfson, Jr., "An Enzymatic
     Converter for the Implanted Fuel Cell", To be Published (1971).

SUMMARY OF DISCUSSION AND COMMENTS: SESSION CHAIRMAN VI

S. K. Wolfson, Jr.

Michael Reese Hospital and Medical Center, Chicago, Illinois

The papers in this session reflect an attempt to provide power sources which differ in output by 4-5 orders of magnitude. They are both intended for implantation in the body, one to power an artificial heart and the other to provide a tiny electrical pulse for stimulating or timing the contraction of the natural heart. It would seem that if one could give serious thought to the former, then the latter must be easily accomplished. The fallacy of this superficial view becomes apparent when one considers the specific points covered by the essayists. The proposed artificial heart fuel cell can be of considerable size and weight, could be located remote from the heart and has direct access to the blood stream for convective supply of fuel and oxidant. The proposed pacemaker cell is to be integrated with the pacemaker circuit itself and the combined mass and volume need be diminitive enough for direct implantation on the cardiac surface with the implied limitation of electrode surface and total reliance on diffusion for mass transfer. Both fuel and oxidant in the latter case must be derived from the same extracellular fluid space and then only indirectly from the blood. This interposes at least two additional membrane interfaces and one additional liquid diffused phase. Thus, even at the microwatt level, mass transport engineering problems are of considerable importance. Consideration of the body's reaction and effects upon the system are problems not usually encountered in fuel cell chemistry or engineering. Living tissue has the power to mechanically rearrange implanted systems. It can construct thick, walling-off membranes which can become an effective barrier against diffusion of reactants and products. Within the blood stream, clotting, presence of macromolecules and flow patterns are brought into play.

Discussion from the audience included questions and comments on blood clotting, performance of oxygen electrodes, covering membranes and animal implantation. The importance of electric charge upon blood clotting was mentioned. The selection of surfaces with an appropriate negative charge or with such a charge artificially maintained would resist thrombus formation. This was not within the scope of the reported work but investigative efforts in the area was acknowledged. In the case of the pacemaker fuel cell, it was stressed that the device would be located extravascularly (epicardial surface) and thus clotting was not a problem. However, the possibility of pseudomembrane formation was a realistic consideration. It was pointed out that the proper choice of materials was important in this area. The supply of adequate oxygen for minimal cathode performance was discussed. One suggestion was that of an air breathing cathode which could enhance performance both from the increased diffusion of O_2 in the gas phase and from the increased P_{O_2} gradient (150 torr in air *vs.* 100 torr in blood). A percutaneous pore was acknowledged as a disadvantage in the application of this concept. The problem of cathode poisoning and mixed reactions due to presence of fuel at the cathode was raised. Each essayist stressed the particular manner in which his system handled this. Dr. Giner pointed out that the gold alloy cathode would be selective for

O$_2$ reduction and thus insensitive to the presence of carbohydrate.
Dr. Yao emphasized that fuel could only reach his cathode by passing
completely through the anode (where its concentration would presumably
be markedly reduced in proportion to current flow) and then through
the polyolefin or polyfluorocarbon ion exchange membrane which consti-
tuted a significant, but not complete, diffusion barrier.   Direct
access of fuel to the cathode was precluded by the presence of the
hydrophobic (silastic or copolymer) membrane.

The session was adjourned on a note of cautious optimism and the
recognition that considerable effort remained before realization of
the ultimate goal of either essayist: a trouble free, long-lived fuel
cell implant powering the complete artificial heart or pacemaker.

SESSION VII

DIGRESSION ON PRIMARY AND SECONDARY BATTERIES

Chairman

T. R. Beck

RECHARGEABLE Al/Cl$_2$ BATTERY WITH MOLTEN AlCl$_4^-$ ELECTROLYTE

G. L. Holleck, J. Giner and B. Burrows

Tyco Laboratories, Inc., Waltham, Massachusetts

ABSTRACT

A molten salt system based on Al- and Cl$_2$ carbon electrodes, with an AlCl$_3$ alkali chloride eutectic as electrolyte, offers promise as a rechargeable, high energy density battery which can operate at a relatively low temperature.

Electrode kinetic studies showed that the electrode reactions at the Al anode were rapid and that the observed passivation phenomena were due to the formation at the electrode surface of a solid salt layer resulting from concentration changes on anodic or cathodic current flow.

It was established that carbon electrodes were intrinsically active for chlorine reduction in AlCl$_3$-alkali chloride melts. By means of a rotating vitreous carbon disk electrode, the kinetic parameters were determined.

INTRODUCTION

A high energy density battery based on an aluminum and a chlorine electrode with a molten AlCl$_3$-alkali chloride electrolyte has been proposed and made the subject of a preliminary study [1]. The specific advantages of this battery are:

1) It shows a high theoretical energy density of 650 Whr/lb.

2) Aluminum chloride-alkali chloride melts form low melting eutectics [2-6]. Thus, problems associated with the high working temperatures of the present molten salt systems, such as Li-Cl$_2$ batteries, are minimized, while the advantages of high energy density and relatively efficient electrode processes common to molten salt systems are retained. The operating temperature of this system would be in the range of 120°-150°C, but it can be started at 90°C when working with an AlCl$_3$-KCl-NaCl eutectic, or even at 61°C by adding lithium chloride [7].

3) Aluminum, being a solid electrode, requires no special retaining structure such as is required, for instance, for liquid lithium electrodes.

4) The reactants are inexpensive, readily available elements.

SUMMARY OF OUR WORK ON THE SYSTEM

As a consequence of our preliminary study of this system under a NASA contract [1], we have been able to demonstrate the feasibility of the proposed system and to define the major potential problem areas. A summary of our findings is given in the following:

1)   FeCl$_3$ is the major impurity in all commercially available AlCl$_3$ that may affect the performance of the Al electrode. A standard procedure for AlCl$_3$ purification was developed which involves the treatment of a melt of high AlCl$_3$ concentration with *in situ* electrochemically generated, finely divided Al. After this step, AlCl$_3$ is sublimed and used to prepare AlCl$_3$-KCl-NaCl melts, which stay clear (nearly colorless) and show no appreciable background current.

2)   The aluminum electrode is highly reversible. Measurements of the activation overvoltage using galvanostatic pulses indicated that over 200 mV/cm^2 anodic or cathodic current density can be obtained at only 10 mV of activation overvoltage. This is reflected in an apparent exchange current of 260 mA/cm^2 in an AlCl$_3$-KCl-NaCl (57.5-12.5-30 mol %) melt at 130°C [8].

3)   Upon anodic polarization of the Al electrodes in quiescent molten salt electrolyte, the electrode passivates. Typical current potential curves at different temperatures are shown in Figs. 1 and 2. We have been able to show that this passivation is caused by the formation of a salt layer (probably AlCl$_3$) due to concentration changes at the electrode during current flow. In spite of these passivation phenomena, sizable steady state currents can be obtained for the anodic dissolution of Al after partial passivation of the electrode. The magnitude of the steady state currents depends strongly on operating temperature and on melt composition [8].

The increase of the limiting current with rising temperature (compare Figs. 1 and 2) is the result of: 1) a lower concentration gradient at a given current density, and 2) the ability of the melt to accommodate larger concentration changes before precipitation of a solid phase occurs. For example, with a melt AlCl$_3$-KCl-NaCl 55.5-12.5-30 mol % at 157°C steady state, anodic currents of 100 mA/cm^2 have been obtained. Since there is a correlation between the anodic limiting currents in quiescent melts and the AlCl$_3$ content of such melts, even considerably higher current densities can be expected for a melt containing approximately 50 mol % AlCl$_3$.

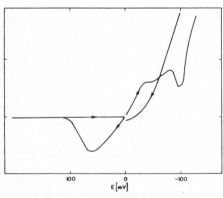

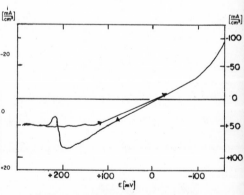

FIG. 1 *Triangular cathodic potential scan (0 mV to -150 mV back to 0 mV) and anodic potential scan (0 mV to 200 mV back to 0 mV) at Al electrode in melt II at 105°C and 20 mV/min (no i-R correction).*

FIG. 2 *Triangular anodic potential scan (0 mV to 300 mV back to 0 mV) and linear cathodic potential scan (0 mV to -150 mV) at Al electrode in melt II at 157°C and 20 mV/min (no i-R correction).*

The linear increase of the anodic limiting currents obtained at a rotating aluminum disk electrode with rising rotation rate (Fig. 3) demonstrates clearly the extent of transport limitation in the melt.

4) A similar passivation occurs during charge of the Al cathode due to the formation of a layer of solid $AlCl_4^-$ salt, (compare Fig. 1). An increase in temperature has the same beneficial effect on the cathodic currents as discussed above for the anodic currents.

5) In addition, dendrite formation was observed at the Al electrode during charge. The occurrence of dendrites was a function of concentration gradients at the electrode. Thus, below a certain current density, no dendrite formation was detected. The critical current density at which dendrites appear varies with temperature and melt composition and was not investigated in this study.

6) Carbon electrodes are intrinsically active for the reduction of chlorine. The open circuit potential of an Al/Cl_2 cell is readily established and was measured to be 2.1 V at 130°C and 2.06 V at 157°C [9]. A typical current-potential curve for the reduction of chlorine at a rotating vitreous carbon disk electrode is shown in Fig. 4. Fig. 5 contains the current-potential data in the form of a Tafel plot. From it the exchange currents for chlorine reduction, which are important parameters to allow the design of efficient porous electrodes, were determined as 1×10^{-4} A/cm² at 130°C and 2×10^{-4} A/cm² at 157°C (as compared with only $i_o = 5 \times 10^{-7}$ A/cm² for O_2 electrode in KOH at 80°C) [10,11].

7) From experiments with the rotating disk electrode, we obtained a transport coefficient (C x D factor) for the chlorine electrode, which is nearly temperature independent and has a value of approximately 5×10^{-11} mol/cm sec. This is about 50 times larger than the same parameter for the well-studied oxygen electrode in KOH at 80°C [10,11].

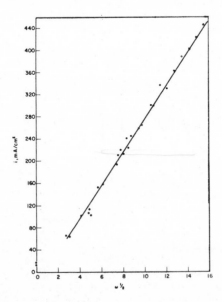

FIG. 3 *Anodic limiting current at rotating Al disk electrode in AlCl₃-KCl-NaCl (57.5-12.5-30 mol %) at 125°C.*

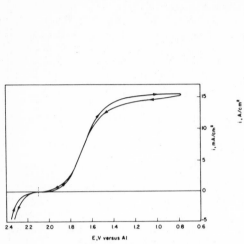

FIG. 4   *Reduction of Cl$_2$ at ro-*
tating carbon disk electrode in
AlCl$_3$-KCl-NaCl (57.5-12.5-30 mol %)
at 130°C, 20.8 rps, and 200 mV/min.

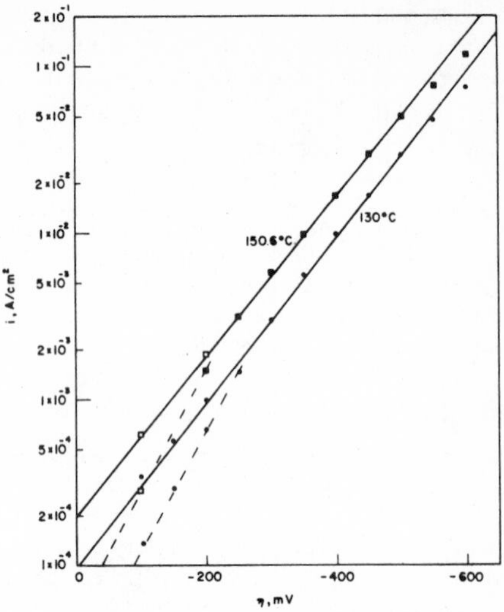

FIG. 5   *Diffusion corrected*
Tafel plots for chlorine reduc-
tion (open symbols at low over-
voltage = returning sweep).

8) An extensive study on the compatibility of different carbon
electrodes led to the results summarized in Table I. Graphite is
attacked by the melt especially in the presence of chlorine gas. However, carbon samples free of graphite proved to be compatible with the
melt, and we found samples which are suitable for use as porous electrodes for chlorine reduction.

9) Using a rotating carbon disk electrode, we were further able
to determine the kinetic parameters for the reduction of Cl$_2$. It was
found to occur according to the paths Cl$_2$ + e$^-$ $\rightleftharpoons$ Cl$^-$ + Cl$_{ads}$ and
Cl$_{ads}$ + e$^-$ $\rightleftharpoons$ Cl$^-$, with the first step being most likely rate determining [9].

10) The specific conductivity as a function of temperature for an
AlCl$_3$-KCl-NaCl (57.5-12.5-30%) melt was measured. The values increase
from 0.167 ohm^{-1} cm^{-1} at 128°C to 0.279 ohm^{-1} cm^{-1} at 183°C. The temperature dependence can be represented very well by:

$$K_{sp} = 11.4 \pm 0.9 \exp \frac{-3360 \pm 68}{RT} \quad ohm^{-1} \; cm^{-1}$$

11) The feasibility of the proposed aluminum-chlorine battery was
demonstrated using a cell with a porous carbon electrode, as shown in
Fig. 6. The electrode had a cylindrical shape (4.5 cm high and 1.9 cm
in diameter) and was machined out of Speer Pure Carbon no. 37. For
the aluminum electrode, a concentrical cylinder (4.3 cm diameter) of
Al sheet was used. The cell was operated at 150°C with a chlorine
pressure of 10 psi. The current voltage curve (Fig. 7) data obtained
with this cell showed that an Al/Cl$_2$ battery can be built which can be
discharged at a current density of 20 mA/cm^2 with a voltage of over
1.8 V, using a noncatalyzed carbon electrode.

TABLE I

Stability of Carbon-Graphite Materials in AlCl$_3$-NaCl-KCl Eutectic in the Absence and Presence of Chlorine

Material Identification	Source	Conditions		Remarks on Cl$_2$ Effect
		AlCl$_3$ Eutectic at 120°C	Eutectic Cl$_2$ Passed Over Melt	
No. 37 porous carbon	Speer carbon	No attack	No attack	In contact over 48 hr
P-6038-C carbon-graphite	Pure carbon	No attack	Failed	Failed in less than 15 min
PO2 dense carbon	Pure carbon	No attack	No attack	
103G dense carbon	Speer carbon	No attack	Failed	Failed in less than 10 min
104 dense graphite	Speer carbon	No attack	Failed	Same
108 dense carbon-graphite	Speer carbon	No attack	Failed	Same
37G porous graphite	Speer carbon	No attack	Failed	Same
Graphite "A"	Carborundum	Failed	Failed	Rapid failure
L-50 porous carbon-graphite	Pure carbon	No attack	Failed	Failure in 10 to 15 min
L-56 porous carbon-graphite	Pure carbon	No attack	Failed	Same
PO3 porous carbon-graphite	Pure carbon	No attack	Failed	Same
P3W porous carbon-graphite	Pure carbon	No attack	Failed	Same
CS grade dense graphite	National carbon	No attack	Failed	Failed in less than 30 min
Vitreous carbon	Atomergic Chemetals	No attack	No attack	No attack after more than 72 hr contact with AlCl$_3$
Pyrolytic graphite	Ultra carbon	No attack	No attack	Pyrolytic graphite coating (a piece of plain graphite) appears to have resistance to attack

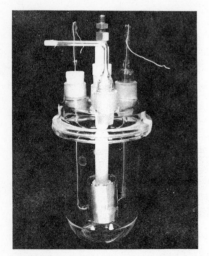

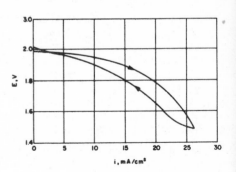

FIG. 7 *Stationary current-voltage behavior of experimental Al/Cl$_2$ cell (porous Speer carbon no. 37 electrode; Cl$_2$ pressure = 10 psig; AlCl$_3$-KCl-NaCl = 66-14-20 mol %).*

FIG. 6 *Cell arrangement for feasibility test of Al/Cl$_2$ battery.*

ACKNOWLEDGEMENTS

The authors wish to thank Mr. M. Turchan for valuable contributions. This research was supported by the electronics Research Center, U.S. National Aeronautics and Space Administration under Contract No. NAS 12-688.

REFERENCES

[1]   J. Giner and G. L. Holleck, *Final Report on Contract #NAS 12-688,* Tyco Laboratories, Inc., Waltham, Mass. (1970).

[2]   W. Fischer and A. Simon, *Z. Anorg. Allg. Chem.,* **306,** 1 (1960).

[3]   R. Midorikawa, *J. Electrochem. Soc., Japan,* **23,** 127 (1955).

[4]   H. Grothe, *Z. Elektrochem.,* **54,** 216 (1950).

[5]   R. Midorikawa, *J. Electrochem. Soc., Japan,* **23,** 352 (1955).

[6]   T. Chao, *Diss. Abstr.,* **12,** 459 (1952).

[7]   W. E. Trout, Jr., and W. J. Triner, Jr., *University of Richmond Report No. CCC-1024 TR 139* (October 3, 1955).

[8]   G. L. Holleck and J. Giner, paper submitted to *J. Electrochem. Soc.*

[9]   G. L. Holleck, paper submitted to *J. Electrochem. Soc.*

[10] J. Giner and C. Hunter, *J. Electrochem. Soc.,* **116,** 1124 (1969).

[11] J. Giner, J. M. Parry, and L. Sweete, in "Fuel Cell Systems II, Proceedings of the 5th Biennial Symposium on Fuel Cells", *Advances in Chemistry,* Series No. 90, p. 102 (1969).

THE CHARACTERISTICS OF METAL-AIR SYSTEMS

K. D. Beccu

Institut Battelle, Geneve, Switzerland

ABSTRACT

In recent years metal-air systems have received large research and development expenditures because of their attractive energy densities, their promising performance and their ability to be electrically, chemically or mechanically recharged. The objective of the research and development is a high energy density power source for military or civilian applications and in particular for use in an electric vehicle. Although a large variety of systems has been investigated and numerous technologies have been set up, this objective has not yet been attained. The major problems still to be overcome are not only of a fundamental nature, such as the passivation phenomena of certain metals during discharge, but also of a more technological aspect, such as the optimal recharge process of the anode, the heat evolution during high power drain and the corrosion reaction at high temperature. In this paper emphasis is put on the still-remaining problems of the most advanced system: zinc/air operating in alkaline solution. The shortcomings and the advantages of the other systems — Cd-, Fe-, Ga-, Al-, Mg-, Li-, Ca-, Pb- and Na-air — are briefly outlined. The cathode of the cell, the air-oxygen electrode, has been considered only when incompatibility with the electrolyte or the anodic discharge product have been reported.

INTRODUCTION

Metal/air systems can be considered as special types of fuel cell consisting of a high energy density metal as fuel on the anode side and a gas electrode operating on air/oxygen as the cathode. The striking feature of these systems is the possibility of recharging the anode electrochemically, chemically, or mechanically. In the case of electrochemical recharging with a satisfactory cycle-life, energy can be provided at approximately the attractively low cost of electrical power plants. Although considerable R & D effort has been devoted to various systems of this kind, a number of serious obstacles still have to be overcome, particularly to reach the objective of an electrically rechargeable metal/air cell. However, there is also a considerable market for primary and mechanically rechargeable metal/air batteries because of their outstanding performance and their high energy density.

This paper tries to give a review of the advantages and shortcomings of available systems and to identify some of the main problems of the metal anodes. The characteristics of the air electrodes will not be dealt with in this paper since they have been thoroughly discussed elsewhere in these proceedings.

Table I gives a survey of the characteristics of metal/air systems that have been reported in the literature. The first row contains

TABLE I

Characteristics of Metal/Air Systems

Metal-Anode	Electrolyte	Energy Density Theoret. Ah/Kg Anode	Energy Density Practical Wh/Kg	Max. Power Density (W/Kg)	Cell Potential (V) O.C.V.	Cell Potential (V) Discharge (c/10)	Main Problems
Zn	KOH	825	100-350	65/90	1.65	1.2	Passivation at low temp., corrosion at high temp., dendritic Zn deposition.
Cd	KOH	480	100	80	1.22	0.8	Fadeout, Cd-penetration and reduced performance of air electrode.
Fe	KOH	965	100	40	1.27	0.8	Low discharge efficiency; low H_2-overvoltage resulting in high self-discharge.
Ga	KOH	1170	(50)	-	1.68	0.8	High cost.
Al	KOH, NaCl	2990	500-900	150	2.71	1.1	Strong corrosion, excessive gas and heat evolution.
Mg	NaCl, $MgBr_2$ Li-, K-, Mg $(ClO_4)_2$	2210	500-1000	-	3.09	1.3	Delayed action, excessive heat evolution, corrosion at high temp.
Ca	CH_3OH/H_2O + $(NH_4)_2SO_4$ + LiCl	1340	110-170	-	2.6	2.0	Strong corrosion.
Li	NDA + $PhMe_3NPF_6$	3860	-	-	-	-	Moist air necessary for air electrode incompatible with Li anode
Pb	H_2SO_4	260	40	-	1.08	0.6	Self-discharge, low efficiency
Na/$NaHg_x$	NaOH	1170	300	100	1.9	1.4	Two-step process, works at 135°C, poisonous Hg.

systems operating in a 7N KOH solution; the second, those operating in neutral, semiorganic or acid solutions; and the third, a system operating at high temperatures.

In order to insure that the weight and performance of the air electrode have no influence on the theoretical energy density (Wh/kg), only the coulombic capacity (Ah/kg) of the metal anodes are indicated; these decrease in the following sequence:

Li > Al > Mg > Ca > Ga > Na > Fe > Zn > Cd > Pb.

When the energy density (Wh/kg) with the open-circuit voltage (O.C.V.) is considered, this theoretical sequence changes slightly as follows:

Li > Al > Mg > Ca > Na > Ga > Zn > Fe > Cd > Pb.

However, the practical energy density values differ completely from the theoretical values since:

The discharge voltage is in several cases much lower than the
    O.C.V.
The theoretical capacity can be utilized only to a small extent
    which varies with the anode and the discharge conditions.
The performance and weight of the air/oxygen electrode contributes
    to the overall energy density as a function of the anode
    capacity.

The maximum power density depends essentially on the performance of the air electrode and hence on the selected catalyst, the electrolyte, and, of course, the cell design. Therefore, W/kg-values have only been indicated where calculations have been reported based on a laboratory prototype. For primary systems, Al remains highest; while for secondary systems, Zn/air — with a still limited cycle life — shows the highest power density.

Furthermore, a number of factors must be taken into consideration, such as whether the system is primary or secondary, the availability of metal, the consumption of electrolyte, the operating temperature, the compatibility of the reaction products of the metal anode with the air cathode, and so on.

The main problems of each system will now be discussed separately.

ZINC/AIR

In the past, zinc/air has been the metal/air system receiving the largest research and development expenditure. In this case, a wide variety of primary and secondary concepts has been developed [1-7] and in small series, already marketed. This system should be therefore reviewed in more detail.

The outstanding problems of the zinc/air systems are:

Passivation at rather low current densities and low temperatures
    during anodic polarization.
The formation of a nonadherent powdery or dendritic zinc deposit
    during recharge.
The dimensional change of the electrode during cycling.
Corrosion and consequently self-discharge, particularly at high
    temperature.

Several of the cited problems are caused by the comparatively high solubility of the oxidized Zn species in KOH. In addition, Zn tends to supersaturate the electrolyte during oxidation to twice the equilibrium value. The high solubility leads directly to the problems of shape change, dendritic growth, separator penetration, and the precipitation of a discharge product in the air cathode.

The phenomenon of passivation can be quantitatively described by the anodic overvoltage/current curve which is shown in Fig. 1. As the current density increases, the electrode becomes strongly polarized, and at the limiting current density (l c d), the potential changes drastically to the passive state. The l c d as well as the overvoltage for the zincate reaction depend on the electrolyte concentration, the temperature, and the surface morphology (Hg), as shown in Figs. 2 and 3 by Dirkse, *et al*. [8].

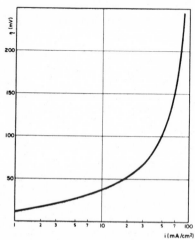

FIG. 1   *Anodic overvoltage/current behaviour of Zn in 7 N KOH saturated with ZnO [8].*

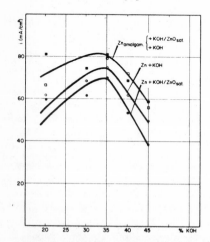

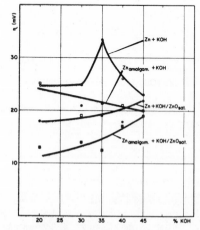

FIG. 2   *Limiting anodic current density for Zn electrodes (22°C) in solutions of various KOH concentrations [8].*

FIG. 3   *Overvoltage for zincate formation at Zn electrodes in solutions of various KOH concentrations [8].*

In general, the l c d reaches its maximum at about 35 percent KOH and decreases markedly as the concentration of KOH increases and somewhat less as it decreases. Amalgamated zinc has a higher l c d than pure zinc except at temperatures near 0°C. Decrease of temperature decreases the l c d markedly, and more so in the case of amalgamated Zn than in that of pure Zn (see Fig. 4).

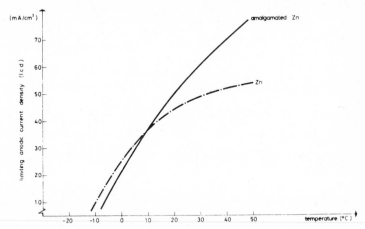

FIG. 4 *Limiting current density as a function of temperature and surface morphology of the zinc electrode.*

Irregularities are observed [8] for the overvoltage of the zincate reaction in KOH (see Fig. 3). Particularly in the absence of zincate ions, the overvoltages are rather uncertain since it is possible that no equilibrium potential for the zinc/zincate reaction is obtained.

The mechanism of anodic zinc behaviour shows the rate-determining step at which the passivation of the electrode must be accounted for and a way determined for overcoming it.

$$Zn_{kink} \longrightarrow Zn_{ad} \tag{1}$$

$$Zn_{ad} + 2OH^- \longrightarrow Zn(OH)_{2ad}^{--} \tag{2}$$

$$Zn(OH)_{2ad}^{--} \longrightarrow Zn(OH)_{2ad} + 2e^- \tag{3}$$

$$Zn(OH)_{2ad} \longrightarrow Zn(OH)_{2\ diss} \tag{4}$$

$$Zn(OH)_{2\ diss} + 2OH^- \longrightarrow Zn(OH)_4^{--} \tag{5}$$

$$Zn(OH)_4^{--} \longrightarrow ZnO + 2OH^- + H_2O \tag{6}$$

In accordance with the fundamental work of Farr and Hampson [9,10], it is generally agreed that the anodic zinc process in an alkaline solution consists at least of six steps, namely:

Step 1. The oxidation of Zn starts with the movement of Zn atoms from kink sites (lattice dissolution points, for example screw dislocations) to become mobile surface atoms with a low coordination number ($Zn_{kink} \longrightarrow Zn_{ad}$).

Steps 2 The formation of a charge-transfer product $Zn(OH)_{2ad}$.
and 3.

Steps 4   The dissolution of this product in the electrolyte.
and 5.

Step 6.   The precipitation of ZnO when the solution is saturated with
           zincate.

The formation of a passivating layer on the electrode takes place
either by precipitation of a dissolved product from the electrolyte or
because of the inability of the electrolyte to dissolve the charge-
transfer product as fast as it is formed. From the high exchange-
current density (i_0 = 200 mA/cm^2), it is obvious that Step 3 is very
fast and cannot be the rate-determining step. That Steps 1 and 2 are
likewise very fast may be concluded from the similar l c d behaviour
of pure zinc and amalgamated Zn on which no kink sites and no Zn(OH)$_{2ad}$
should exist. Step 5 is also a very fast reaction and Step 6 cannot
take part in the reaction because of the high oversaturation. Thus
Step 4 seems to be the rate-determining step.

Hampson, Farr and Dirkse [8,9,10] conclude that the important
factor must be the rate of dissolution of the charge-transfer product
in the electrolyte, in spite of its high solubility. This also depends on the zincate concentration in the boundary layer. The diffusion of zincate away from the electrode thus controls the l c d and
this means that any process that enhances the diffusion of the charge-
transfer products into the bulk electrolyte can minimize the passivation; for example, electrolyte circulation, fluidized bed electrodes,
and thin, highly porous zinc electrodes.

According to this hypothesis, the passivation at different current
densities is directly related to the zinc utilization rate. In order
to optimize the energy output, the anode must be thin and of high
porosity. As shown in Fig. 5, at low current densities, the zinc
utilization is quite independent of the electrode thickness. At high
current densities, however, the difference becomes significant.

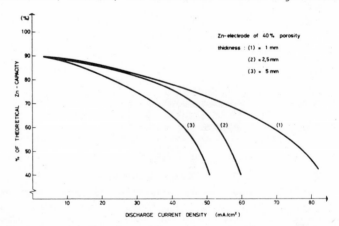

FIG. 5 *Utilization of theoretical Zn capacity as function of discharge current density and electrode thickness.*

Considerable work has been done on the anodic behaviour of Zn
electrodes [8,11,12,13,14] and more recently also on the cathodic
behaviour. The latter is closely connected with the rechargeability
of Zn electrodes. In accordance with much published work [15-21] it
can be said that:

The hydrogen overvoltage on zinc is significantly high: at 40 mA/cm^2 (1 N KOH) the overpotential (*vs*. Hg/HgO) is about 700 mV.

Tafel lines are obtained between 500 µA and 40 mA/cm^2.

The deposition of Zn occurs at rather low efficiency (20-50%) owing to simultaneous hydrogen evolution in spite of the high overvoltage.

The deposited Zn layer is very mossy, dendritic, and not very adherent, as it contains ZnO or Zn(OH)$_2$ inclusions.

It is evident frqm previous work that dendritic deposits are found when the rate of the electrode reaction is controlled by the mass transport of the electro-active species to the surface of the electrode, *i.e.*, when the concentration of the electroactive species on the electrode surface is lower than the bulk concentration.

The growth of Zn dendrites depends on the cathodic overvoltage applied to the electrode, so that: 1) at 20 mV fine filaments (0.2-1 µ) begin to form, causing spongy deposits; 2) at 100 mV dendrites of hexagonal crystals 50-200 µ in diameter are formed.

Various techniques have been proposed and developed to make Zn deposition more adherent, and to avoid dendrites. Among these we may list:

Minimizing the quantity of electrolyte in order to decrease the total zincate concentration.

Applying special membranes which retard the diffusion of zincate [22].

Circulating the electrolyte [23,24] which modifies the deposit morphology as proposed by General Dynamics (Fig. 6).

Charging with controlled current [25] which influences the concentration gradient between bulk and electrode surface.

Adding foreign ions which change the deposit morphology [26].

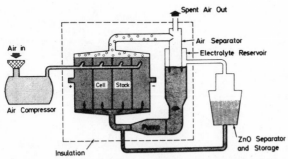

FIG. 6 *Zinc-air system with circulating electrolyte and separate storage of the discharge product ZnO (General Dynamics).*

Attempts have been made to overcome the penetration of the separator by the dendritic Zn growth by development of new microporous ceramic separators which are, moreover, resistant to oxidation and capable of withstanding 50 deep discharge cycles without failure. The conventional film-type separators such as cellophane give no satisfactory results in batteries with Zn electrodes since they have a relatively high resistance which accentuates the shape change.

A serious problem with rechargeable Zn electrodes is the dimensional change during cycling; this is caused by the sinking of the
dense zincate solutions on the bottom of the cell, as a result of
convection and gravity. Zn is then preferentially deposited at the
bottom of the electrode, thus forming zinc dendrites which penetrate
the separator and shorten the cell.

In order to overcome the problems of shape change, various techniques have been proposed:

Addition of binders (teflon) to Zn electrode.

Minimization of amalgamation.

Oversizing the electrode geometry in order to reduce edge effects.

Eisenberg, *et al.* [27] have investigated the influence of gravity
on the anodic polarization of Zn in 30% KOH by comparing horizontally
and vertically placed electrodes. For high current densities (0.25-
0.6 A/cm^2), they observed (in agreement with Hampson *et al.* [28])
a relationship of the formula

$$(i - i_c)\sqrt{\tau} = K$$

where: i_c = l c d

τ = transition time (sec.)

K = constant of passivation depending on KOH
concentration

independently from the electrode position. Below 250 mA/cm^2 the results at the two positions were quite different (see Fig. 7), as the
vertical position favors the convection phenomena affecting the linear
diffusion of the zincate. The extrapolation at $\tau^{1/2}$ = 0 allows the
l c d value to be defined below which no passivation is observed.

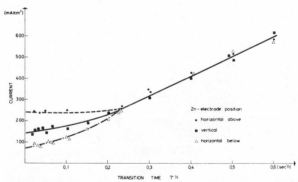

FIG. 7 *Influence of electrode position and current density on
the transition time of the passivation of Zn electrodes [17,27,28].*

Different authors [29-33] have studied the corrosion of zinc
electrodes in alkaline solutions and have found rather discrepant
results. From the work of Dirkse and Timmer [32], which is in agreement with some of our unpublished work (see Fig. 8), it can be concluded that:

Increasing temperature reduces slightly the corrosion rate.
Increasing KOH concentration decreases strongly the corrosion rate.
Amalgamation reduces the zinc corrosion.

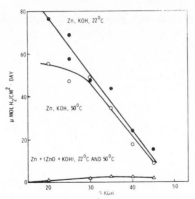

FIG. 8 *Corrosion of Zn (10 cm^2) in KOH solutions as a function of temperature, KOH concentration and ZnO addition.*

Saturation with ZnO results in a markedly lower corrosion rate both at amalgamated and nonamalgamated electrodes as well as at lower and higher temperatures.

While the influences of KOH concentration, amalgamation, and ZnO saturation of the electrolyte on the corrosion rate can be explained easily, the effect of temperature on the corrosion rate has been un-accounted for. Dirkse suggests the following hypothesis: the cor-rosion product formed at higher temperature is more dense than that formed at lower temperature, thus bringing about a more complete cover-ing of the zinc surface. In fact, below 38°C, Zn(OH)$_2$ is the stable form which is more soluble than the ZnO formed above 38°C [34]; thus, the corrosion rate is increased at low temperature.

A further point of interest is the interaction between the dis-charge products of the zinc anode and the air electrode. ZnO pene-trates and precipitates in and on the air electrode and changes its wetting characteristics and pore structure. This happens particularly at the end of charge, when the electrolyte in the O$_2$ electrode compart-ment is diluted by generation of water which reduces the zincate solu-bility. It happens, however, more often at the end of the discharge, when the electrolyte is very concentrated and large amounts of zincate are available. Rather long stand-by periods are more dangerous than rapid cycling, since the zincate supersaturation tends to diminish in these periods.

The presence of H$_2$O$_2$ generated at the air electrode has dramatic effects on the passivation of low surface-area zinc anodes. In a zinc/carbon-air cell, containing 30 ml KOH for example, the concentra-tion of H$_2$O$_2$ after 5 hours discharge was measured as 0.1 molar. The solid products on the discharged Zn anode are visibly different from those generated in cells with a low H$_2$O$_2$ content. However, this problem may be overcome by using air electrodes which have a high catalytic activity in the peroxide decomposition.

After all, it seems that Zn/air cells still present a number of serious problems connected with their electric recharge. When deep discharge is required, cycle life for a large-scale commercial ap-plication and high current density is still unsatisfactory; the only solution for overcoming these problems seems to be to apply dispersed or fluidized bed Zn electrodes such as proposed and developed by various companies [35,36,37]. Fig. 9 shows the battery assembly of a

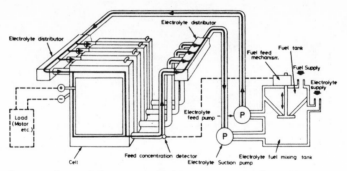

FIG. 9   *Zn-slurry/air system of the Sony Company [35,37].*

Zn-powder system which puts the finely dispersed Zn powder into circulation when discharge is required. Inside each cell, collectors are
arranged in such a way that the Zn particles are projected and discharged onto them. Although the discharge performances are rather
interesting, the auxiliary equipment necessary to circulate the slurry
is very complicated and heavy, reducing the overall energy density of
this system with respect to the weight by a factor of four compared
with the standard Zn/air cells. However, the advantage is the easy
mechanical recharge of the battery, which is accomplished merely by
pumping fresh Zn-slurry into the fuel tank.

CADMIUM/AIR

The problems related to secondary Zn electrodes have led several
companies to turn from Zn to Cd, thus benefiting from the long experience with Cd electrodes in rechargeable Ni-Cd accumulators, and
the technical advantages involved. Although the theoretical energy
density of Cd is much lower than that of Zn, stability and cycle life
are better, the potential behaviour during discharge is excellent, and
the self-discharge is very low.

Cd electrodes in nickel batteries are generally overdimensioned
with respect to the nickel oxide electrode; since, when discharged to
100 percent of their available capacity, they show a loss of capacity,
known as fade-out. Since a capacity ratio cannot be established in a
metal/air cell, a coulometric control has to prevent the complete discharge of the Cd electrode.

The capacity loss is generally attributed to a loss in real surface
area by the growth of Cd crystals and subsequent blockage of the electrode pores [38]. The problem can be partly solved by adding α-Fe_2O_3
that works as extender. Fig. 10 [39] shows the influence of various
sponge compositions on the cycling behaviour of a Cd electrode. The
loss in capacity can often be prevented by charging the electrode at
a higher current density. Increasing the current density from 9 mA/cm^2
to 23 mA/cm^2, for example, increases the capacity by 8-10 percent.

Another reason for capacity loss is due to the carbon dioxide
introduced both by the air electrodes and by the direct contact of the
electrolyte with air. The capacity of the anode falls by 30 percent
in 35 cycles in a carbonate-saturated electrolyte, but only by 3
percent in pure KOH [39]. Thus, air must be CO_2-scrapped before supplying the air electrodes. It has been demonstrated that the carbonate

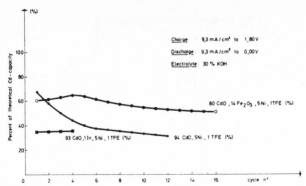

FIG. 10 *Influence of the composition of the Cd electrode on its cycling behaviour [39].*

accelerates the recrystallization process of the Cd electrode by increasing the solubility of Cd in the $CO_3^=$-saturated electrolyte; this is due to the formation of a highly soluble carbonate species $[Cd(CO_3)_3]^{-4}$ [41]. The recrystallization results in a loss of porosity and the subsequent loss of capacity of the Cd electrode [40].

Another problem is the Cd penetration into the separator. This happens particularly on overcharged electrodes and on those of the sponge type. Impregnated electrodes based on a sintered nickel body behave better. Since the build-up of Cd crystals is influenced by the solubility of Cd in the electrolyte, a high carbonate content favors penetration into the separator. Hence, special separators must be applied and overcharge reduced to a minimum.

Moreover, it has been observed [42] that the dissolved Cd species have a detrimental influence on the behaviour of the air electrode. This represents another serious drawback of the system, a drawback which, together with the already-mentioned shortcomings and the relatively high cost of cadmium, limits it to very special applications.

IRON/AIR

Although the theoretical energy density and the cost of the anode material of iron/air cells are quite attractive, considerable technical difficulties are encountered in this system, notably:

The high self-discharge.

The low discharge efficiency.

The extreme purity of iron required.

Fig. 11 shows the voltage/time characteristics of an alkaline iron electrode during discharge and charge [43]. Two voltage levels are observed:

discharge: 0.88 V *vs.* Hg/HgO (theoretical: 1.03 V)

     0.63 V " (theoretical: 0.72 V)

charge: 0.93 V " (theoretical: 0.72 V)

     1.08 V " (theoretical: 1.03 V).

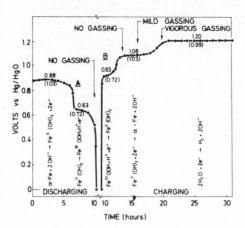

FIG. 11 *Voltage/time characteristics of a sintered iron electrode
under anodic and cathodic polarization in alkaline electrolyte [43].*

The corresponding electrode reactions are indicated in Fig. 11, as
well as the potentials where gas evolution starts. In the discharge
reaction only the first voltage level is utilized in practice.

The major hindrance to extensive application of iron-active material in alkaline accumulators and in metal/air cells has been its high
self-discharge. This discharge is largely the result of the couple
action between the impurities with low hydrogen over-voltage and the
active iron. In addition to this, iron has a more cathodic potential
than required for hydrogen evolution and also a very low H_2 overvoltage.

A typical self-discharge behaviour of an iron electrode at different temperatures is shown in Fig. 12, which demonstrates that at
room temperature, 20 percent of the initial capacity is lost in a
stand-by period of about 14 days after charge.

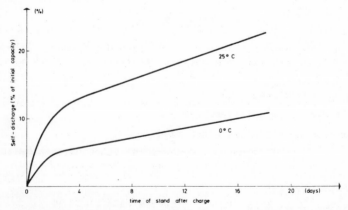

FIG. 12 *Self discharge of alkaline iron electrode at different
temperatures.*

Various additives are reported to decrease self-discharge:
Burshtein [44] proposes the partial passivation of the electrode by a
treatment with benzene or other volatile organic liquids. Other

beneficial treatments seem to occur with sodium phosphate. Generally
it is beneficial to employ materials with high H_2 overvoltage forming
no resistive oxide layers or dissolving actions in the electrolyte.

Discharge is also relatively inefficient, at best approaching 40
percent of theoretical capacity based on a two-electron reaction.
LiOH increases the efficiency to about 50 percent by a still rather
unknown mechanism. Cycling of iron electrodes results first in an
increase of capacity, as a result of a formation process, and then in
a decrease, owing to irreversible electrochemical reactions. It is
reported in the literature [45,46] that the active mass utilization
can be stabilized for a certain number of cycles by addition of graphite, Mo - , W - and S - additives.

The elevated loss of water [47] in iron/air cells presents a
serious problem for constructing a maintenance-free, vented battery,
because of the self-discharge reaction and because the extensive overcharge is approximately 50-75 percent.

Moreover, the process of producing active iron material [48] is
very costly, since it requires a low-carbon, high-purity iron powder
containing small amounts of FeS. The final powder product is eight
times as expensive as the raw iron material.

Much work has been done to replace the Edison pocket electrodes
with new sintered structures [43,47] that would be capable of increasing the performance of the iron electrode. Energy densities of 50-60
Wh/lb and a cycle life of about 200 cycles at 30 percent depth of
discharge have been obtained.

GALLIUM/AIR

The possibility of using Ga in primary or secondary cells has been
discussed by Jahn and Plust [49]. This idea is interesting for the
following reasons:

Ga in equilibrium with $H_2GaO_3^-$ has the very negative potential of
   -1220 mV (25°C).

The reaction $Ga + 4\ OH^- \rightleftharpoons H_2GaO_3^- + H_2O + 3e^-$ proceeds
   rapidly with a rather small charge-transfer overpotential.

Gallium has a high H_2- overvoltage.

The current efficiency at 100 mA/cm^2 (80°C) is near 100 percent.

The overpotential changes insignificantly with a Ga content of the
   electrolyte up to 300g Ga/l.

The corrosion rate is quite low (8 mA/cm^2).

The polarization at 0.3 A/cm^2 is still acceptable: 400 mV.

However, the cost of Gallium would prohibit the commercial application of this system.

ALUMINUM/AIR

Al anodes would be of significant interest for application in
metal/air batteries from the point of view both of the high energy
density and of the relatively low cost of this metal. Aluminum is

superior to zinc and magnesium with regard to the equivalent weight,
the electrode potential, and its electrochemical activity. However,
the utilization of Al in alkaline electrolytes brings about only a
small current efficiency; this is because of the high corrosion rate
insofar as it is not retarded by appropriate inhibitors, alloying, or
adjustment of working conditions [50,51,52].

In concentrated alkaline solutions, the electrochemical reaction
proceeds in the following manner:

$$Al + 4OH^- \longrightarrow H_2AlO_3^- + H_2O + 3e^-$$

with $E_0 = -2.35\,V$. However, in most electrolytes, the O.C.V. is only
of the order of $-1.5\,V\ vs.$ SCE.

The self-discharge owing to the high corrosion rate of Al is the
most limiting factor of this electrode. Chromate addition to the
electrolyte reduces the corrosion rate markedly, as it forms a protective layer which can still be penetrated by anions. The highest
penetration of the oxide film is shown by ions of small size, high
solubility and high diffusivity. This means, in terms of corrosion,
that the rate is much higher in the presence of Cl^--ions than with
SO_4^{2-}-ions. Moreover, the fact that the potential is the most anodic
in aluminum chloride proves also that the chloride ions have a great
ability to penetrate the oxide layer. The behaviour of OH ions on the
oxide layer of Al must be strictly distinguished from the above
effects, since the very anodic potentials in alkaline solutions are
due to the dissolution of the oxide, which itself results in an excessive corrosion, as well as in heat and gas evolution.

It should be noted that the formation of a protective layer often
results in a delayed action, depending however on the cation of the
chromate used [54]; the delay formed with ammoniumchromate differs
entirely from that with potassiumchromate. Other inhibitors used are
potassiumhypochlorite [55] and sodiumstannate.

Additives to the electrolyte to prevent the precipitation of insoluble anodic products and to increase the uniform dissolution of the
anode are:

Citric and tartaric acid [56] and

Pyridine, acridine, quinoline [56].

While the inhibitors indicated above function by an anodic mechanism, cathodic inhibitors, such as mercury, operate by changing the
hydrogen overpotential and the anodic polarization of the active metal
dissolution. Since amalgamation completely removes the oxide film on
Al, the corrosion rate is increased rather than reduced, this is therefore undesirable. Another possibility of minimizing the corrosion
reaction of Al is to avoid the presence of impurities with low hydrogen
overvoltage in the electrode [53], or to alloy Al with other metals
[57].

In particular, the Al-Zn alloys have been studied intensively and
and it has been found that their potential in neutral chloride solutions approaches that of zinc. Akimov and Clark [58] have examined
the relationship between the electrode potential of a solid solution
and its composition, and have concluded that no clear relationship
exists between the potential of a component and the effect on the
pocential in a composition.

It should be noted that the corrosion rate of Al electrodes is reduced as the electrodes are discharged at high current densities. This is a result of the shift to a more cathodic potential. Moreover, the discharge efficiency increases in the same manner. Thus Faust reports discharge current efficiencies approaching 100 percent when Al electrodes are discharged in 3 and 10 N KOH at 300 mA/cm^2. Only very limited discharge current densities can be obtained in neutral or acid electrolytes.

MAGNESIUM/AIR

Similar to aluminum anodes, magnesium electrodes can only operate in aqueous electrolytes in primary cells. Since considerable success has been achieved with Mg anodes in dry cells [59-61], attempts have also been made to use magnesium in metal/air systems.

General Electric has set up laboratory prototypes that operate with NaCl, CaCl$_2$, and MgCl$_2$ electrolyte from local water resources [62]. The resistivity of a 7 percent NaCl solution is of the order of 10 Ω cm, and thus about five times as high as KOH-36 percent, which limits the performance of the cells. The electrolyte volume is practically established by the volume of the wet reaction product. This generally requires an excess of water and brings down the specific capacity to about 50 Wh/lb. Current densities are of the order of 10-20 mA/cm^2 and can hardly exceed 30 mA/cm^2.

The main problems with Mg anodes are the delayed action of Mg, and the corrosion and consequent excessive heat and gas evolution. The delayed action is a transient anodic polarization phenomenon. With an increase of the anodic current, the potential moves sharply in the passive direction, but then recovers rapidly to its active steady state value. The magnitude of the transient polarization is related to the applied and the corrosion current. The shape of the polarization curve can be explained by continuous damage-and-repair processes of the protective layer.

A number of alloys have been developed for use as Mg anodes. They fall into two general classes. In the first, pitting action is avoided by adding metal components which mask or remove the heavy metal impurities, but which do not impair the ability of the electrode to form protective films in saline electrolytes. In the other, the alloy prevents the formation of the film, or at least makes the film prone to disruption, thereby reducing the delayed action. The most useful alloys in Mg/air systems are ternaries with aluminum and zinc, namely: AZ 10, AZ 21, AZ 31B, AZ 61. The length of the delay also depends on the nature of the electrolyte. Perchlorate solutions generally behave more favorably than chlorides or bromides.

Anode efficiency is likewise affected by the composition of the Mg anode, owing to the property of the Mg alloy of being more or less stable to the corrosion reaction. The following reactions occur simultaneously on Mg:

(a) $Mg + 2\ OH^- \longrightarrow Mg(OH)_2 + 2\ e^-$:   current production

(b) $Mg + 2\ H_2O \longrightarrow Mg(OH)_2 + H_2$:   corrosion reaction.

The effect of pH on the corrosion rate of Mg in solutions of
chlorides, NaOH, HCl, and distilled water has been studied [63]. It
has been found that below pH 3, there is a sharp increase in the corrosion rate; whereas from pH 3 to 11, the corrosion curve slopes
smoothly. At pH 11, the stability of the protective film increases
strongly preventing any corrosion above pH 11.5. The OH^--ion concentration controls the precipitation of $Mg(OH)_2$, which occurs very close
to the anode at high pH values, but at a sufficient distance in an
intermediate pH range. Only in the latter case can the anode function
normally. In this range certain anions can become active and penetrate
the passive film as already outlined in the case of aluminum.

The relative rates of the current production and the corrosion
reaction determine the efficiency of the Mg anode in producing useful
energy. This may be between 20 percent and 80 percent.

Fig. 13 shows the electrode efficiency of various Mg alloys in 2N
$Mg(ClO_4)_2$, at two current densities and in a temperature range between
20° and 70°C. As the temperature increases, efficiency is markedly
reduced owing to the increased corrosion rate, but the efficiency is
significantly improved by the AZ alloys.

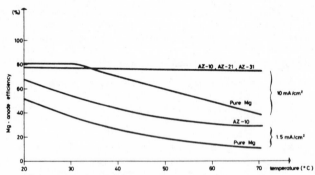

FIG. 13 *Discharge efficiency of various Mg alloys at different
temperatures and current densities [60].*

Fig. 14 illustrates the influence of the current density on the
efficiency of the anode. At low current densities the potential of
the electrode is rather negative, hence the corrosion reaction is more
pronounced and decreases the current efficiency. At higher current
densities, the potential becomes more positive, favoring the current
production reaction.

It is important to overcome the excessive heat evolution which
results mainly from the corrosion reaction of Mg to avoid premature
battery failure. The evolved heat Q is the sum of the irreversibility
of the Mg anode and of the heat of the corrosion reaction ΔH expressed
in the following equation:

$$Q = \Delta V \cdot i \cdot t \; 860 + \Delta H$$

ΔV = difference between theoretical and working potential of Mg
     electrode in volts

i = current drain in A

t = time in hours.

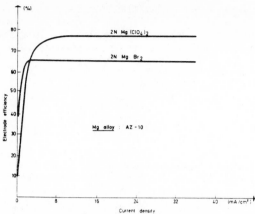

FIG. 14   *Influence of current density on the efficiency of AZ-10*
Mg electrodes in 2 N Mg(ClO$_4$)$_2$ and 2 N MgBr$_2$ [60].

Fig. 15 shows the heat evolution (kcal/h) as a function of the
current and anode efficiency for an irreversible discharge voltage of
1.1 volt [60].   This figure is valid for any electrolyte or alloy with
this voltage.   Since normally the electrode potential is 1 V or more
below the theoretical potential, it will be valid for most Mg cells.
As an example, it can be demonstrated that in a thermally insulated
battery designed for a ten-hour discharge, the electrolyte is heated
to boiling point in three hours, a circumstance which rapidly causes
the battery to fail.

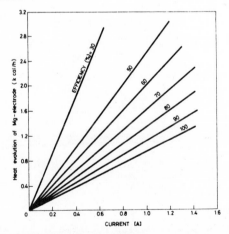

FIG. 15   *Heat evolution of Mg electrodes for a discharge voltage*
of 1.1 V as a function of the current drain and the anode efficiency
[60].

The performance of the air electrodes in Mg(ClO$_4$)$_2$ electrolyte
presents serious problems, since MgO tends to precipitate in the pores
of the cathode, thus limiting the performance of the air electrode.
It appears that chloride electrolytes are able to eliminate this
problem to a large extent, but can affect the performance and the life
span of the air electrode.

OTHER METAL/AIR COUPLES

All the other cells indicated in Table I are in quite an early state of research and pose a number of serious problems as regards:

Stability in an aqueous or semi-aqueous electrolyte.

Performance in a low-conducting electrolyte.

Low energy density (as in the case of the lead/oxygen couple).

Research has been reported on calcium/air cells operating in a mixture of methanol and water, with $(NH_4)_2SO_4$ + LiCl as solute [64]. Ca is theoretically superior to Zn as regards both voltage and energy density, but its high electronegativity has limited its use to non-aqueous electrolytes. Since the air cathode only works in an aqueous medium, an electrolyte is required which is compatible with both electrodes. A mixture of two parts methanol plus one part H_2O gave good activity and acceptable corrosion rates. Fig. 16 shows the cell voltage as a function of current density (O_2-electrode Pt). At about 70 mA/cm^2, the curve shows an inflection which is due largely to the polarization of the air electrode. At 1.55 mA/cm^2, the discharge curve indicates an anode utilization of about 25 percent of the theoretical Ca capacity, as shown in Fig. 17. The corrosion problem, which essentially limits this system in aqueous electrolytes, probably can be partly overcome by development of appropriate Ca alloys.

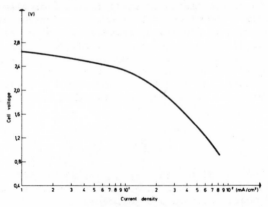

FIG. 16 *Cell voltage of Ca/air cell as a function of current density. Electrolyte: CH_3OH/H_2O = 2:1*
10% $(NH_4)_2SO_4$
5% LiCl

Another exotic metal/air system reported in the literature [65] works with Li in nitrosodimethylamine plus phenyltrimethylammonium-hexafluorophosphate, and with moist air, necessary in order to obtain acceptable air electrode performance. The most serious problem with this system is the incompatibility of the Li anode with even the small quantities of water. Moreover, the performance of the air electrode does not exceed current densities of 10 mA/cm^2.

The Pb/air system [66] is mainly a laboratory curiosity. On account of its low energy density, it is of technical interest only when low cost and durably stable air electrodes can be set up that work at sufficiently high current density and that do not affect the hydrogen overvoltage of the lead anode.

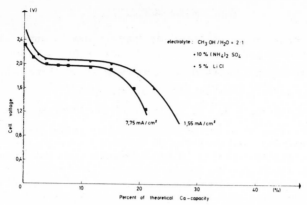

FIG. 17   *Discharge behaviour of Ca/air cells at different current densities.*

One special metal/air system, developed by Yeager [67], which operates at high temperature (135°C) is the sodium amalgam/air cell. Since Na would react immediately with an aqueous electrolyte, it first has to be transformed into an amalgam which is then oxidized in alkaline electrolyte to NaOH. The current-limiting process is the diffusion of sodium ions through the amalgam layer, while concentration and charge-transfer overpotentials are small in vertical electrode arrangement even at current densities of 1 A/cm². The current efficiency of the amalgam electrode is nearly 100 percent, and owing to the high overpotential for hydrogen on the amalgam, the corrosion reaction is very small. This two-step process poses considerable technical problems which may only be overcome when the two cells are completely separated. The short life of air electrodes at 135°C would then be improved and the system could compete with other low-temperature metal/air cells.

OUTLOOK

The metal-anode problems are only a part of the difficulties to be overcome. The complete cell with the air electrode poses a number of additional problems such as the following:

1)   The air electrode floats in stand-by conditions.

2)   The use of a third electrode for internal recharge must be considered for secondary cells.

3)   The cathode catalyst (Pt, Ag), present in trace quantities in the electrolyte, affects the hydrogen overvoltage of the anode, which consequently reduces its charge efficiency and increases the self-discharge.

4)   The geometrical dimensions for the air-to-metal electrodes must be appropriately selected since these determine the cell design and internal resistance of the cell.

This brief survey of metal/air systems should end with the conclusion that, although a large number of systems and technical approaches have been tried in this field, only primary or mechanically rechargeable systems are so far available. Electrochemical recharging still presents tremendous problems.

REFERENCES

[1] W. Vielstich, *Fuel Cells*, Wiley Interscience, London, (1970) p. 323.

[2] A. Charkey and R. DiPasquale, *Proceedings, 22nd Annual Power Sources Conference*, PSC Publ. Comm., Red Bank, N. J. (1968) p. 117.

[3] J. D. Voorhies, *Proceedings, ibid.*, p. 124.

[4] E. G. Katsoulis and B. Randall, *ibid.*, p. 120.

[5] R. A. Powers, R. J. Bennett, W. G. Darland and R. J. Brodd, in *Power Sources 2 (1968)* by D. H. Collins (Ed.), Pergamon Press, Oxford, (1970) p. 461.

[6] S. M. Chodosh, *et al.*, *Journal of Spacecraft and Rockets*, 4, No. 5 (1967).

[7] H. R. Knapp, *ECOM Report-3182*, Oct. 1969.

[8] T. P. Dirkse, D. DeWit, and R. Shoemaker, *J. Electrochem. Soc.*, 115, 442 (1968).

[9] J. P. G. Farr, and N. A. Hampson, *Trans. Far. Soc.*, 62, 3493 (1966).

[10] J. P. G. Farr, and N. A. Hampson, *J. Electroanalytical Chem.*, 13, 433 (1967).

[11] M. N. Hull, J. E. Ellison, and J. E. Toni, *J. Electrochem. Soc.*, 117, 192 (1970).

[12] E. A. Ivanov, *et al.*, *Elektrokhimiya*, 5, 695 (1969).

[13] P. L. Howard and J. R. Huff, in *Power Sources 2 (1968)* by D. H. Collins (Ed.), Pergamon Press, Oxford (1970) p. 407.

[14] M. A. V. Devanathan, and S. Lakshmanan, *Electrochimica Acta*, 13, 667 (1968).

[15] J. E. Oxley, C. W. Fleischmann, and H. G. Oswin, *Proceedings, 20th Annual Power Sources Conference*, PSC Publ. Comm., Red Bank, N. J. (1966) p. 123.

[16] J. W. Diggle, and B. Lovrecek, *J. Electroanal. Chem.*, 24, 119 (1970).

[17] F. Jolas, *Electrochimica Acta*, 13, 2207 (1968).

[18] J. W. Diggle and A. Damjanovic, *J. Electrochem. Soc.*, 117, 65 (1970).

[19] J. W. Diggle, A. R. Despic and J. O'M. Bockris, *J. Electrochem. Soc.*, 116, 1503 (1969).

[20] A. R. Despic, J. W. Diggle and J. O'M. Bockris, *J. Electrochem. Soc.*, 115, 507 (1968).

[21] J. T. Clark and N. A. Hampson, *J. Electroanal. Chem.*, 26, 307 (1970).

[22] G. A. Dalin and Z. O. J. Stachurski, in *Power Sources 1966*, by D. H. Collins (Ed.), Pergamon Press, Oxford, (1967) p. 21.

[23] R. D. Naybur, *J. Electrochem. Soc.*, 116, 520 (1969).

[24] "Technology Newsletter," *Chemical Week*, May 6, p. 99 (1970).

[25] S. Arouete, K. F. Blurton and H. G. Oswin, *J. Electrochem. Soc.*, 116, 166 (1969).

[26] F. Mansfeld and S. Gilman, *J. Electrochem. Soc.*, <u>117</u>, 588 (1970).

[27] M. Eisenberg, H. F. Baumann and D. M. Brettner, *J. Electrochem. Soc.*, <u>108</u>, 909 (1961).

[28] N. A. Hampson, M. J. Tarbox and J. T. Lilley, *Electrochem. Technology*, <u>2</u>, 309 (1964).

[29] S. A. Awad and K. H. M. Kamel, *J. Electroanal. Chem.*, <u>29</u>, 217 (1970).

[30] T. P. Dirkse and F. DeHaan, *J. Electrochem. Soc.*, <u>105</u>, 311 (1958).

[31] R. N. Snyder and J. J. Lander, *Electrochem. Technology*, <u>3</u>, 161 (1965).

[32] T. P. Dirkse and R. Timmer, *J. Electrochem. Soc.*, <u>116</u>, 162 (1969).

[33] P. Rüetschi, *J. Electrochem. Soc.*, <u>114</u>, 301 (1967).

[34] P. T. Gilbert, *J. Electrochem. Soc.*, <u>99</u>, 16 (1952).

[35] *Chemical Eng.* May 1970, p. 74.

[36] J. R. Backhurst, J. M. Coulson, F. Goodridge and R. E. Plimley, *J. Electrochem. Soc.*, <u>116</u>, 1600 (1969).

[37] French Patent 1.593.288.

[38] Y. Okinaka, Electrochemical Society Meeting, Chicago, Oct. 1967.

[39] O. C. Wagner, *Proceedings, 22nd Annual Power Sources Conference*, PSC Publ. Comm., Red Bank, N. J. (1968) p. 72.

[40] E. Lifshin and J. Weininger, *Electrochem. Technology*, <u>5</u>, 5 (1967).

[41] P. G. Lake and J. M. Goodings, *Can. J. Chem.*, <u>7</u>, 1089 (1958).

[42] O. C. Wagner in *Metal-Air Battery Symposium*, Interagency Advanced Power Group, Washington, D.C.

[43] E. R. Bowerman, *Proceedings, 22nd Annual Power Sources Conference*, PSC Publ. Comm., Red Bank, N. J. (1968) p. 70.

[44] N. A. Shumorskaia and R. K. Burshtein, *Conference Moscow 1956*, p. 768-72.

[45] V. P. Galushko, *et al.*, *Zhur. Priklad Khim.*, <u>32</u>, 1591 (1959).

[46] N. N. Voronin, *et al.*, *Ukrain Khim. Zhur.*, <u>20</u>, 182 (1954).

[47] A. Fleischer, *Techn. Report AF APL-TR-68-6*, March 1968, p. 27.

[48] G. W. Vinal, *Storage Batteries*, J. Wiley & Sons, New York (1965).

[49] D. Jahn and H. G. Plust, *Nature*, (London), <u>199</u>, 806 (1963).

[50] S. Zaromb, *J. Electrochem. Soc.*, <u>110</u>, 253 (1963).

[51] R. A. Foust, *Abstract 49*, Electrochemical Society Meeting, Boston (1962).

[52] G. R. Drengler, *et al.*, British Patent 875,977.

[53] U. R. Evans, *Metallic Corrosion, Passivity and Protection*, Longmans, Green & Co., New York (1946) p. 23.

[54] J. J. Stokes, Electrochemical Society Meeting, Pittsburgh, 1955.

[55] A. V. Kuzmina and A. N. Demidova, *Zh. Pril. Khim.*, <u>38</u>, 1038 (1965).

[56] P. L. Joseph, and B. A. Shenoi, *J. Electrochem. Soc.*, Japan **32**, 73 (1964).

[57] R. B. Mears and C. D. Brown, *Corrosion*, **1**, 113 (1945).

[58] G. W. Akimov and G. B. Clark, *Trans. Faraday Soc.*, **43**, 685 (1947).

[59] E. B. Cupp, *Proceedings, 19th Annual Power Sources Conference*, PSC Publ. Comm., Red Bank, N. J. (1965) p.92.

[60] G. S. Lozier and R. J. Ryan, *Proceedings, 16th Annual Power Sources Conference*, PSC Publ. Comm., Red Bank, N. J. (1962) p. 134.

[61] J. L. Robinson, *Proceedings, 17th Annual Power Sources Conference*, PSC Publ. Comm., Red Bank, N. J. (1963) p. 142.

[62] C. E. Kent, and W. N. Carson, *Proceedings, 20th Annual Power Sources Conference*, PSC Publ. Comm., Red Bank, N. J. (1966) p. 76.

[63] G. V. Akimov and J. L. Rozenfeld, *Chemical Abstracts*, **39**, 0424 (1945).

[64] A. Charkey and G. A. Dalin, *Proceedings, 20th Annual Power Conference*, PSC Publ. Comm., Red Bank, N. J. (1966) p. 82.

[65] J. Toni, *in Metal-Air Batteries Symposium*, Interagency Advanced Power Group, Washington D.C.

[66] International Lead Zinc Research Organization, Research Project LE 156, (1970).

[67] E. Yeager, "The Sodium Amalgam/Oxygen Continuous Feed Cell," *in Fuel Cells*, by W. Mitchell (Ed.), Academic Press, New York (1963).

ORGANIC CATHODES WITH AIR REGENERATION

H. Alt, H. Binder, A. Köhling and G. Sandstede

Battelle-Institut e.V., Frankfurt am Main, Germany

ABSTRACT

Reversible redox systems are needed in order to obtain cathodes which can be regenerated or recharged. Quinones/hydroquinones are known to act as completely reversible redox couples; however, most of the quinones are slightly soluble and partly unstable in electrolyte solutions. In order to test their stability electrochemically, we mixed the quinones with carbon. The open-circuit potentials of the solid quinone/hydroquinone systems measured in 2N H_2SO_4 are very close to the values of the redox potentials measured in alcoholic solutions. Diphenoquinones, which have a potential of about 950 mV (vs. NHE), are not sufficiently stable. Only tetrachloro-p-benzoquinone (chloranil) and tetramethyl-p-benzoquinone (duroquinone) have been found to be sufficiently insoluble and completely stable; Their redox potentials are 668 mV and 478 mV, respectively. In the case of galvanostatic discharge, the potential is almost constant. With chloranil the polarization is only 30 mV at a current density of 60 mA/cm^2. There is a voltage plateau at a utilization of up to 75 percent. Even at a current drain of 600 mA/cm^2, 50 percent of the active material is available for discharge at potentials exceeding 200 mV. It should be noted that this current density corresponds to the unusual discharge condition of 10 C. In an electrolyte consisting of concentrated aqueous ammonium chloride solution, the reduction takes place in two one-electron steps, separated by approximately 70 mV. This does not happen in zinc chloride solution. The ampere-hour capacity (Ah/kg) of the quinones is in the same range as that of the inorganic depolarizers, but the hydroquinones can be reoxidized with oxygen, using hydrophobic electrodes. Thus the quinone electrode is regenerated by air and its capacity is practically unlimited. The reoxidation with oxygen is catalyzed by active carbon, whereas for the reoxidation with hydrogen peroxide, other conductive supports can likewise be used. Electric recharging of quinones is not only possible in aqueous electrolytes but also in organic electrolytes made from aprotic solvents as we observed in experiments using acetonitrile and propylen-carbonate. Particular quinones have the advantage of being completely insoluble in the organic solvents and are therefore superior to the heavy metal salts as cathode materials in high energy density, secondary batteries with organic electrolytes.

INTRODUCTION

Apart from the latest results obtained with phthalocyanines and porphyrins, only platinum metals and — to a certain extent — carbon are known as electrocatalysts for the oxygen electrode in acid electrolyte. Current densities achieved at room temperature are not high, however, we therefore investigated organic materials in order to find

333

out about the existence of a redox system which might be chemically
regenerable with oxygen (air) or hydrogen peroxide, and which could
thus be used for insoluble cathodes of high current density. In the
case of the regeneration with oxygen, the redox potential has to be
below the oxygen potential.

If electric recharging in a secondary battery is considered, the
capacity of the substance is an important property. All the organic
electrode materials investigated so far cannot be recharged, since the
reaction is irreversible. Yet there is one class of compounds well
known as reversible redox systems in the dissolved state: the quinones.
They have, however, not yet been investigated as rechargeable electrode materials.

COMPARISON OF QUINONES WITH OTHER SUBSTANCES

Fig. 1 shows the theoretical capacities of some inorganic and
organic compounds [1-5]. The simplest aromatic nitro compound listed
is nitrobenzene. When we compare the values for nitrobenzene with
those for manganese dioxide and lead dioxide, we find that this
organic depolarizer is distinctly superior to the two dioxides. The
theoretical capacity indicated can however not be utilized; this is
because the energy density of the whole cell is rather low as a result
of the low open-circuit potentials of nitro compounds measured in
practice.

	Potential ε_H [V]	Capacity [Ah/kg]
PbO_2	1.68	224
MnO_2	0.74 (pH = 4.3)	308
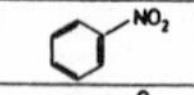 (structure: benzene with NO_2)	0.8 (not measurable)	1 300
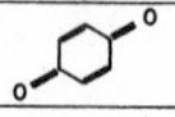 (structure: p-benzoquinone)	0.70 / 0.45 (pH = 4.3)	496
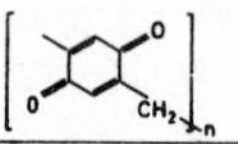 (structure: polymeric quinone with CH_2)$_n$	0.70	446
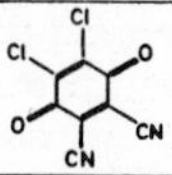 (structure: dichloro dicyano quinone, Cl, CN)	1.27	236
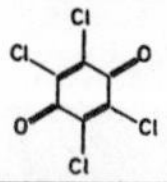 (structure: tetrachloroquinone, Cl)	0.67	218
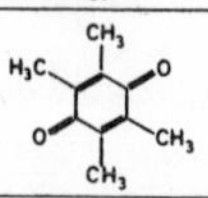 (structure: tetramethylquinone, H_3C, CH_3)	0.48	326
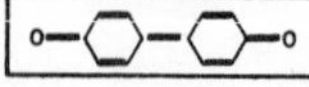 (structure: O=⟨⟩-⟨⟩=O)	0.95	291

FIG. 1   *Theoretical potentials and capacities of inorganic and
organic substances.*

The theoretical capacities of quinones per unit weight are in the
same range as those of the inorganic materials, but most of the quinones are not stable enough, at least in aqueous electrolytes. In
the reduced state, *p*-benzoquinone is slightly soluble; it can be made
insoluble by condensation of hydroquinone with formaldehyde.

Dichlorodicyanobenzoquinone cannot be used because the cyano groups
are split off by hydrolysis. Nevertheless it shows the effect of the
substituents on the potential. Electron-withdrawing groups raise the
redox potential. On the other hand, chlorine-substituted quinones,
for example tetrachloro-p-benzoquinone, also called chloranil, are
very stable compounds. The chlorine substituents stabilize the molecule. The redox potential of chloranil is 664 mV in aqueous alcoholic
solution [4]. Tetramethyl-p-benzoquinone, also called duroquinone, is
another stable molecule, but has a somewhat lower potential owing to
the electron-releasing property of the methyl groups. The last on the
list, diphenoquinone, permits a much higher potential to be achieved.
This is due to the fact that two benzene rings are formed during reduction, a process which results in a large resonance-energy gain.
But diphenoquinones — including also chlorinated compounds — are not
sufficiently stable in aqueous electrolytes.

The mechanism of reduction of p-benzoquinone and duroquinone (and
of the oxidation of the hydroquinones) in the dissolved state is
largely known [6-8]. Fig. 2 shows the two reaction paths for the reduction of p-benzoquinone (as an example for its derivatives) according
to Vetter. The overall reaction is illustrated in the upper part of
the figure. Depending on the type of electrolyte, this reaction can
proceed in different ways. At a high proton concentration, $i.e.$, in
an acid electrolyte, the protonated quinone is involved in the charge-
transfer reaction. At a low proton concentration, $i.e.$, above pH 7,
the quinone molecule is reduced in a direct process. In any case we
have two charge transfer reactions and two protonation reactions giving radicals and also cations and anions as intermediates.

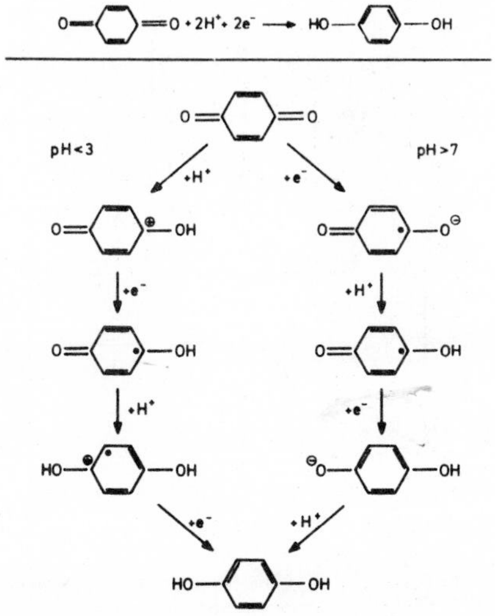

FIG. 2 *Mechanisms of the reduction of p-benzoquinone (and
derivatives) after Vetter.*

CHLORANIL IN DILUTE SULFURIC ACID

To produce a conductive electrode, the chloranil was mixed with
graphitized carbon in a weight ratio of 1:1 and ground in a ball mill
for 16 hours. Most experiments were carried out with graphite powder

(Stackpole PC 62H), which is fairly resistant to oxidation [9]. In addition, some investigations were performed with electrically conductive active carbon (Norit BRX), which has a large surface area. All measurements were made in a half-cell arrangement. The reference electrode, an autogenous hydrogen electrode, was used in the same electrolyte and at the same temperature as the test electrode and the counter electrode (graphite). The test electrode consisted of two graphite felt disks one cm^2 in surface area, held together by two tantalum gauze strips in a Plexiglas holder. The powdered chloranil-graphite mixture was put between the disks [10]. The electrolyte consisted of 2N H_2SO_4.

In order to investigate the reversibility of the reaction cyclic voltammetry was applied using a triangular voltage sweep rate of 20 mV/min. Fig. 3 shows the cyclic potentiodynamic current-voltage characteristic. If the voltage decreases, the quinone is reduced to hydroquinone, and if the voltage increases, the hydroquinone is oxidized to quinone. The two curves are symmetrical with respect to the open-circuit potential, and the open-circuit potential in the amount of 668 mV is nearly the same as the redox potential of 664 mV measured in 50 percent aqueous alcoholic solution [4]. The oxidation peak and the reduction peak, are both very narrow, suggesting that polarization is low. In addition the areas enclosed by the two curves are identical, indicating that the reaction is completely reversible. The amount of charge transferred is consistent with that calculated from the quinone content; this shows that the quinone is utilized 100 percent. The cyclic potentiodynamic curve can be reproduced periodically.

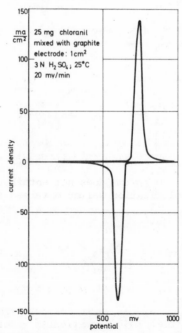

FIG. 3 *Cyclic potentiodynamic curve of a chloranil electrode in dilute sulfuric acid.*

From the narrow peaks one can expect a hard discharge curve. The galvanostatic discharge curves (Fig. 4) at medium current densities

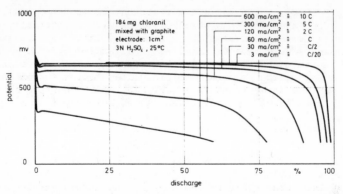

FIG. 4 *Discharge curves of a chloranil cathode at different rates in dilute sulfuric acid.*

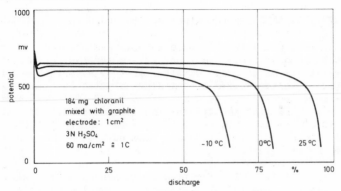

FIG. 5 *Discharge curves of a chloranil cathode at different temperatures in dilute sulfuric acid.*

are indeed horizontal, at least up to a discharge depth of 90 percent, and the discharge polarization is as low as 20 to 60 mV, *vs*. the rest potential. It is further noted from the Figure that the quinone cathode can be loaded with exceptionally high currents. The practical capacity decreases with increasing current density, but even at a discharge rate of 2 C it is still as high as 85 percent of the theoretical value. Even a load of 600 mA/cm^2 does not entail voltage breakdown. When considering the polarization values obtained for the high current densities, it should be taken into account that the potentials have not been corrected with respect to the IR drop between the test electrode and the Luggin capillary.

In contrast to manganese dioxide, the chloranil cathode can be heavily loaded even at low temperatures. At the high discharge rate of 1 C, polarization increases only slightly with decreasing temperature. As expected, however, the capacity becomes smaller (Fig. 5).

Polarization during charging is as low as during discharging. Fig. 6 shows a section of a galvanostatically recorded potential-time diagram. Discharging and recharging are carried out periodically. Although a current of C/2 is used, polarization does not increase with time. The decrease in electrode capacity was insignificant. Cycling was stopped after 100 cycles.

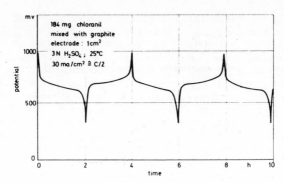

FIG. 6 *Recording chart of the discharging and recharging cycles of a chloranil cathode.*

CHLORANIL IN AMMONIUM CHLORIDE AND ZINC CHLORIDE SOLUTIONS

The electrode potentials in aqueous ammonium chloride and zinc chloride solutions were measured against a saturated calomel electrode (SCE) as reference electrode. We also recorded a cyclic potentiodynamic current-voltage curve in ammonium chloride solution (Fig. 7). The single peak observed in 3N H_2SO_4 has changed into two current peaks, both for charging and discharging. The two current maximums are about 70 mV apart. This is due to the two charge transfer reactions, whose potentials are different from each other because the semiquinone may be stabilized by the ammonium ions. It is known that in aprotic solvents, such as acetonitrile, the two reaction steps are as far as 0.7 V apart [7]. The cyclic potentiodynamic curve shows that the quinone is utilized 100 percent also in ammonium chloride solution.

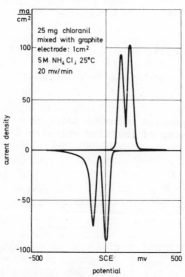

FIG. 7 *Cyclic potentiodynamic curve of a chloranil electrode in ammonium chloride solution.*

On the basis of the potentiodynamic curve, we would expect again a hard discharge curve. In fact the polarization at medium current

densities during the galvanostatic discharge is very small up to a
capacity of 80 percent of the theoretical value (Fig. 8). In accordance with the two current maxima in the potentiodynamic curve, we
find a step after half the capacity of the quinone is reached. The
second plateau is only 60 mV lower than the first one. At the beginning of the second step a small minimum occurs which is due to a
crystallization overpotential of the hydroquinone phase. Polarization
is low even at high current densities, for example 2 C, so that the
performance of the quinone cathode is almost as satisfactory in ammonium chloride electrolytes as in sulfuric acid. However, the open-
circuit potential is about 330 mV lower at pH = 0 than it is in sulfuric acid, and totals only 340 mV (vs. NHE). (The potential was
measured after about 20 percent of the quinone had been discharged.)
It is consequently almost 100 mV less than the value expected from the
theoretical pH dependence. (Since the pH is about 4, the redox potential should be only 240 mV less than that at pH = 0.) This may be due
to the effect of the ammonium ion.

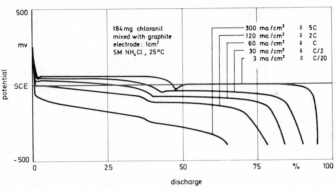

FIG. 8 *Discharge curves of a chloranil cathode at different rates
in ammonium chloride solution.*

In contrast to this, results in zinc chloride solution show the
expected behavior. The open-circuit potential is 400 mV (vs. NHE) at
pH 4.8 so that the redox potential corresponding to the theoretical
pH dependence is reached.

As can be seen from Fig. 9, polarization during discharge is somewhat higher than in ammonium chloride solution. This is probably due
to the lower conductivity of the zinc chloride solution. Nevertheless

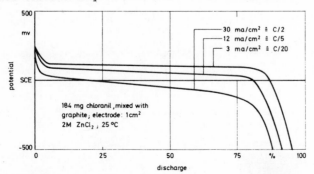

FIG. 9 *Discharge curves of a chloranil cathode at different rates
in zinc chloride solution.*

the chloranil electrode can also be used in zinc chloride solution; it can also be recharged completely. The step in the discharge curve in ammonium chloride solution is not observed in zinc chloride solution.

REGENERATION OF CHLORANIL USING HYDROGEN PEROXIDE

For these and other investigations, which will be described in a later publication [11], a porous gold electrode was used. This electrode can easily be produced for testing purposes by compacting a mixture of gold powder, chloranil/graphite mixture and sodium sulfate.

The discharge curve (Fig. 10) corresponds to the curve shown in Fig. 9. After treating the electrode for five minutes with 4M H_2O_2 solution, about 50 percent of the original capacity was regained.

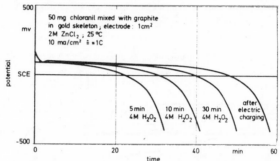

FIG. 10 *Discharge curves of a chloranil cathode at different rates in zinc chloride solution after regeneration with hydrogen peroxide.*

Nearly complete recovery of the capacity was achieved after treatment for 30 minutes. The recovery time is even shorter in dilute sulfuric acid.

On the basis of these results, the chloranil electrode can be considered as an indirect hydrogen peroxide cathode. Although operating according to a redox mechanism, the reoxidation is so fast that polarization should be negligible. After development of a suitable structure, this hydrogen peroxide electrode should be capable of being loaded with high currents.

REGENERATION OF QUINONES BY AIR

Normally, hydroquinone is not oxidized by oxygen, but if in contact with a catalytically active material, it should be possible to oxidize hydroquinone to quinone with air or oxygen because the oxidation potential of oxygen is higher than that of benzoquinones. As active carbon is known to be capable of activating oxygen, we mixed chloranil — and in another investigation, duroquinone — with Norit BRX active carbon, which is electrically conductive.

A hydrophobic electrode should be produced from such a mixture because otherwise air does not get any access to the hydroquinone. Instead, in a simple test, we used a hydrophobic fuel cell cathode as a support for the active mass. This cathode is made of BRX carbon and PTFE with a PTFE backing. When operated on air it has a rather

good performance in alkaline electrolytes whereas the current density
is low in acid electrolytes. Approximately 10 mA/cm^2 was measured in
3N H$_2$SO$_4$ at a potential of 500 mV and 1 mA/cm^2 at 700 mV.

The quinone/active carbon mixture was put on the electrolyte side
of the electrode, which was then covered with a graphite felt disk.
The discharge behavior of this electrode did not differ from that of
the other types of test electrodes. The results obtained with duroquinone in 3N H$_2$SO$_4$ are shown in Fig. 11. The discharge curve is as
hard as with chloranil; polarization is very low. After half the discharge time an open-circuit potential of 478 mV was measured; this
corresponds to the redox potential of 472 mV measured in alcoholic
solution [8]. The capacity is somewhat higher than 100 percent (related to the amount of duroquinone used). This is due to the fact
that the active carbon has a certain capacity because its surface can
be partly oxidized. Therefore, the open-circuit potential of the
completely charged electrode is about 900 mV owing to the activity of
the surface species of the oxidized active carbon with the high surface
area. Of course, during the discharge measurements air was replaced
by nitrogen.

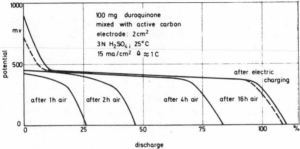

FIG. 11 *Discharge curves of a duroquinone cathode in dilute
sulfuric acid after regeneration with air.*

After having exposed the back of the electrode to air by replacing
the nitrogen — without applying any excess pressure — the discharge
measurements were repeated (in the presence of nitrogen again). The
results of Fig. 11 show that after a two-hour exposure, approximately
50 percent of the capacity was regained. After 16 hours, not only the
total capacity of the quinone but also that of the active carbon was
recovered.

Investigations performed in the same way with chloranil gave the
following results: about 15 percent recovery of the capacity after
admitting air for one hour and approximately 80 percent recovery after
16 hours.

The regeneration was governed by an electrochemical mechanism: as
the active mass was not hydrophobic, the hydroquinone was oxidized
anodically, the electrons flowing to the carbon of the supporting
electrode where oxygen was cathodically reduced. Therefore reoxidation
of duroquinone is faster than that of chloranil because its redox potential is 200 mV lower. This lower potential implies that the current density of the pure oxygen electrode operating on air is increased
by roughly a factor of 10. The course of the reoxidation reaction of
the hydroquinone by air can be pursued by the potential time curve
(Fig. 12). A potential of 480 mV is measured during a period of about

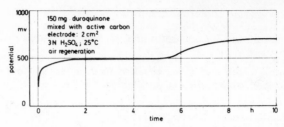

FIG. 12 *Potential-time curve of a discharged duroquinone cathode in dilute sulfuric acid during regeneration with air.*

five hours. After that time the potential increases, indicating that the regeneration of the quinone is complete.

It should be mentioned that chloranil could be regenerated with air in ammonium chloride solution, too.

APPLICATION OF QUINONE CATHODES

As we have shown in the foregoing, quinone cathodes are capable of being highly loaded in acid electrolyte (and also at medium pH). They can be regenerated with hydrogen peroxide and air. It is thus possible to develop hydrogen peroxide cathodes for continuous operation and air cathodes for intermittent duty. It seems possible to increase the recovery time for air regeneration by developing suitable electrodes, which are partly hydrophobic even in the layer containing the active mass. Besides, it should be possible to find carbon catalysts with a higher activity for the reduction of oxygen so that the reaction rate can also be improved.

Another application of quinones would be their use as active mass in secondary batteries. The theoretical energy densities of two quinone/zinc couples for pH 4 have been calculated:

duroquinone/HCl/Zn   179 Wh/kg
chloranil/HCl/Zn   167 Wh/kg

These values compare favorably with the theoretical value of 165 Wh/kg for the lead-acid battery. Moreover, the materials for the construction of the organic active mass would be relatively light.

It should be noted that quinones are suitable for cathodes in high energy secondary batteries with organic electrolytes. While their performance is at least the same as that of heavy metal halides, they are still superior to the latter because of the insolubility of certain representatives [10].

The investigations aimed at elucidating the reaction mechanism of solid quinone electrodes are being continued. This research has to be supplemented by development work towards practical electrodes.

ACKNOWLEDGMENT

The assistance of Frau G. Klempert in the experimental work is gratefully acknowledged.

REFERENCES

[1]   W. M. Latimer, *Oxidation Potentials*, Prentice-Hall, Inc., Englewood Cliffs, N. J. (1952).

[2]   K. Schwabe, *Polarographie und Chemische Konstitution Organischer Verbindungen*, Akademie-Verlag, Berlin (1957).

[3]   M. v. Stackelberg, "Elektrochemische Potentiale Organischer Stoffe" in *Methoden der Organischen Chemie*, by E. Müller (Ed.), Vol. III2, Georg Thieme Verlag, Stuttgart (1955).

[4]   W. M. Clark, *Oxidation-Reduction Potentials of Organic Systems*, Williams & Wilkens Co., Baltimore (1960).

[5]   B. A. Gruber, E. A. McEltrill, and D. L. Williams, *Research on Organic Depolarizers*, AD 454913, Clearinghouse, Department of Commerce, Washington, D.C. (1965).

[6]   K. J. Vetter, *Z. Naturforschung* 7a, 328 (1952); 8a, 823 (1953).

[7]   L. Jeftic and G. Manning, *J. Electroanal. Chem. and Interfacial Electrochem.*, 26, 195 (1970).

[8]   J. K. Dohrmann, and K. J. Vetter, *Berichte Bunsengesellschaft Physik. Chemie*, 73, 1068 (1969).

[9]   H. Binder, A. Köhling, K. Richter, and G. Sandstede, *Electrochimica Acta*, 9, 255 (1964).

[10] H. Alt, H. Binder, A. Köhling, and G. Sandstede, paper presented at the meeting of the Comité International de Thermodynamique et de Cinétiques Electrochimiques, Prague, Sept. 1970, Extended Abstracts, p. 319.

[11] H. Alt, H. Binder, G. Klempert, A. Köhling, and G. Sandstede, paper presented at the meeting of the Comité International de Thermodynamique et de Cinétique Electrochimiques, Strasbourg, April 1971.

SOLID ELECTROLYTES FOR BATTERIES

W. Baukal

Battelle-Institut e.V., Frankfurt am Main, Germany

ABSTRACT

New types of solid electrolytes have been discovered during the last few years and applications in new battery systems have been developed. The properties of the most important solid electrolytes are reviewed and a tentative classification is presented. Most of the early battery systems contained silver halides as electrolyte. Modern batteries include the compounds Ag_3SI, $RbAg_4I_5$, LiI, CaF_2 and $\beta-Al_2O_3$. The review covers the state of development of these batteries, the problems encountered in practice with the individual systems, and their possible applications.

I. INTRODUCTION

It is well known that various compounds exhibit electrical conductivity in the solid state which is due to high ion mobility in the crystal lattice. The conductivity of some of these solid ionic conductors at ambient temperature is as high as that in concentrated aqueous solutions.

Solid oxide electrolytes with oxygen ion conductivity and their application in high-temperature fuel cells have been extensively investigated during the last decade [1,2]. This fuel cell research has focused new attention on solid electrolytes where charge transfer is effected by silver, sodium, lithium, and fluorine ions, with a view to their utilization for new types of batteries.

So far there have been two ranges of application for solid electrolyte batteries: 1) rechargeable batteries with high energy density and power density, and 2) all-solid batteries with low energy and power density.

Recently a battery has been proposed that combines the advantages of both types [3].

High energy density can be achieved by using light-weight materials as battery electrodes: alkali or alkaline metals as anodes, and sulfur or fluorides as cathodes. These electrode materials require, of course, an electrolyte free of protons.

At the present stage of development high power density can only be obtained by raising the operating temperature of the battery to levels of 250° to 500°C. During operation, the Joule heat generated in the electrolyte maintains thermal equilibrium of the battery.

All-solid batteries offer the advantages of sturdiness and long storage life, and the possibility of minimization.

345

Recent reviews of solid electrolytes and their application in batteries have been presented by Foley [4], Hull [5,6], Heyne [7], and Raleigh [8].

II. STRUCTURE AND CONDUCTIVITY OF SOLID ELECTROLYTES

Fig. 1 shows the conductivity of the most important solid electrolytes. Their structure and the mobile ionic species are summarized in Table I. The first column presents a tentative classification of these substances:

1. NORMAL SOLID ELECTROLYTES

The behavior of class A electrolytes may be described by the classical theory of ion migration in ionic crystals. At sufficiently high temperatures, ionic defects such as vacant ion sites and ions at normally unoccupied lattice sites (interstitials) are in thermal equilibrium with the lattice. Usually one particular defect pair (Schottky or Frenkel type) dominates over others [9,10]. The concentration of the defects and their mobility determine the electrical, *i.e.*, ionic conductivity.

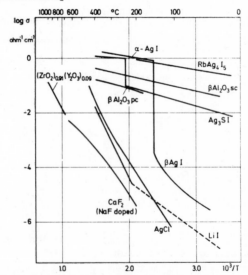

FIG. 1 *Electrical conductivity of solid electrolytes.*

The concentration of these thermally generated defects decreases with decreasing temperature until the influence of aliovalent impurities which are — intentionally or unintentionally — present in the crystal becomes significant. In this new temperature region, which is called impurity controlled or "extrinsic", one of the two defects which constitute the defect pair is preponderant. It behaves like a mobile charge carrier dissolved in the lattice. The transition temperature from the intrinsic (high temperature) to the extrinsic region is shifted to higher values by higher dopant concentrations.

In the range of dilute solutions, the charge carrier concentration, and consequently the ionic conductivity, is directly proportional to

TABLE I

Structure of Solid Electrolytes

Tentative Classifi- cation	Electro- lyte	Struc- ture	Lattice Type	Defects	Mobile Species
A	AgCl	cub.	NaCl	cations Frenkel disordered	Ag^+
A	β-AgI	hex.	wurtzite	cations Frenkel disordered	Ag^+
A	LiI	cub.	NaCl	Schottky disordered	Li^+
A	CaF_2	cub.	fluorite	anions Frenkel disordered	F^-
A	ZrO_2 (cubic stab.)	cub.	fluorite	anions Frenkel disordered	$O^=$
B	α-AgI	cub.	anions bcc	cations "fused"	Ag^+
B	β-Ag_3SI	cub.	anions prim. cub. ordered	cations "fused"	Ag^+
B	$RbAg_4I_5$	cub.		cations "fused"	Ag^+
C	"β-Al_2O_3"	hex.	spinel block	foreign cations mobile in cleavage planes	Na^+

the doping concentration. Among the solid electrolytes listed in Table I are the following examples for doping with aliovalent impurities in the dilute-solution range:

1) Doping of AgCl with Cd^{++} makes the Ag^+ vacancies preponderant over the Ag^+ interstitial ions and causes silver ion conductivity [11,12].

2) Doping of LiI with Mg^{++} creates Li^+ vacancies which predominate over the I^- vacancies and caused lithium ion conductivity [13]. The values of Fig. 1 were taken from [14].

3) An example for the alternative preponderance of one defect is found in CaF_2: doping with Na^+ gives F^--vacancies, doping with Y^{+++} leads to F^--interstitials, both of them causing high fluorine ion conductivity [15,16]. Simultaneous doping with identical concentrations of Na^+ and Y^{+++} would compensate these fluoride defects. The values of Fig. 1 correspond to the solubility limit of NaF in CaF_2 [15].

The conductivity values of AgI below 146°C correspond to a β-modification with wurtzite structure [17] and show the normal thermally induced conductivity behavior [18]. The effect of doping on the conductivity has not been systematically investigated to date. The existence of another low-temperature γ-modification with zinc blende structure is still being discussed [18,19,19a]. The high-temperature α-form of AgI is listed under class B.

At temperatures below the extrinsic conduction range, the mobile charge carriers and the immobile doping ions, which are always oppositely charged, form complexes by electrostatic attraction. This is why the concentration of mobile defects and the conductivity decrease more strongly with decreasing temperature than in the extrinsic range. The behavior of heavily doped zirconia $(ZrO_2)_{0.9}(Y_2O_3)_{0.1}$ has been explained on the basis of this effect [20]. The Arrhenius plot of the oxygen ion conductivity shows a downward bend at temperatures of around 800°C; it was assumed that complexes $[V_O^{\cdot\cdot} \times Y_{Zr}^{'}]$ form below that temperature.

However, cubic stabilized zirconia always contains at least 10 mole percent of lower valent doping cations. At these concentrations the charge carriers no longer follow the ideal laws of dilute solutions. Besides complex formation, we rather have to expect strong interactions, long-range ordering in the lattice ("superstructures") or cooperative phenomena.

The conductivity of doped zirconia, for example, decreases strongly with increasing dopant concentration [21]. The maximum conductivity is at or in the vicinity of the lower stability limit of the cubic mixed crystal [22]. In doped thoria, which is isostructural but has no lower stability limit, a distinct maximum of ion conductivity can be observed [23]; it is found at concentrations of about 3 percent vacancies in the anion sublattice, corresponding to about 12 percent doping cations in the cation sublattice, *i.e.*, to the formula $(ThO_2)_{0.935}(Y_2O_3)_{0.065}$. The falling conductivity in the heavy doping range was predicted [24] using a computer-simulation technique. Several models were taken as a basis; *e.g.*, an existing anion vacancy prevents other vacancies in the neighborhood from participating in the conduction process.

The features of highly doped solid electrolytes may also be expected in calcium fluoride doped with yttrium fluoride. CaF_2 can dissolve YF_3 up to concentrations of about 50 mole% [25]; however, solid electrolytes of this type [3] have not yet been investigated.

2. HIGHLY CONDUCTIVE SOLID ELECTROLYTES

The characteristic feature of class B type solid electrolytes is the "internal melting." The solid state of these ionic crystals is determined by the lattice coherence of only one of the ionic species.

Taking α-AgI as an example, only the iodine anions are fixed in a bcc lattice. The cations, however, do not form a fixed lattice, but are randomly distributed over a large number of interstitial positions. The unit cell of α-AgI contains 2 anions and 2 cations, the latter being distributed over 42 possible sites [17]. They are virtually in a liquid state inside a rigid skeleton of anions. For this reason the high ionic conductivity of α-AgI [26] is not due to the formation of defects, either by doping or by raising the temperature, but is an intrinsic property of the crystal structure.

The same phenomenon is observed in other compounds summarized in Table I under class B; they are all silver ion conductors. Fig. 1 shows that the activation energy of silver ion migration is very low compared with that involved in solid electrolytes of class A.

Among silver-free compounds, only α-Li_2SO_4 is known to exhibit also an abnormally high conductivity (1 $ohm^{-1}cm^{-1}$ at 600°C). This is presumably due to an internally fused Li^+ sublattice [27]. The compound has not yet been investigated as solid electrolyte for batteries; molten lithium and solid chlorides (*e.g.*, nickel chloride) could form the active electrode materials; the reaction product would be fused lithium chloride.

The low temperature modification of Ag_3SI contains one formula unit per structural unit [28]: the I^- ion occupies one corner of a primitive cubic array, the $S^=$ ion occupies the center. Three Ag^+ ions are distributed randomly over 12 possible sites. On transition to the α-form, the two anion sites are no longer distinguishable, so that $S^=$ and I^- together form a bcc-lattice. At the same time, the number of possible cation sites increases from 12 to 42 — the same number as in α-AgI — allowing the three silver ions to fluctuate more easily within the rigid anion lattice.

To date, $RbAg_4I_5$ is the compound observed to have the highest ionic conductivity at room temperature. The cations are again in a virtually molten state [29,30]: one structural unit contains 56 lattice positions which may be occupied by silver ions, but only 16 Ag^+ are available. The conductivity data of Fig. 1 were taken from [31].

Recently, Owens [32] described highly conductive double salts formed from AgI and several tetraalkylammonium iodides. Their conductivity and the activation energy of conduction is of the same order as that of $RbAg_4I_5$. The structure has not yet been identified.

3. ION-EXCHANGING SOLID ELECTROLYTES

In Table I the solid electrolyte β-alumina has been grouped in class C. It has the empirical formula $Na_2O \cdot 11\ Al_2O_3$ and a hexagonal structure of the spinel-block type [33]. In the direction of the c-axis, the unit cell contains two blocks of aluminum and oxygen ions which are arranged in a way similar to the spinel structure. A plane perpendicular to the c-axis separates these blocks; it contains all the sodium ions. They are mobile in the two dimensions of the plane. The conductivity data of Fig. 1 have been taken from the papers by Weber and Kummer [34], and by Miles and Jones [35].

Two effects result from this structure: first, polycrystalline materials exhibit a lower conductivity than a properly oriented single crystal. Second, $β-Al_2O_3$ resembles an ion-exchanging substance; for instance, the Na^+ ions can be rapidly and almost completely replaced by Ag^+ or Tl^+ ions from nitrate melts. The equilibrium relations between solid and melt were established for Ag^+, Tl^+, K^+, Rb^+, Li^+ and Cs^+ ions [36]. In this sequence, the tendency towards incorporation in the solid electrolyte decreases. Electrolytes of class A or B have not been observed to exhibit ion-exchange properties.

The binary phase diagram $Na_2O-Al_2O_3$ contains another sodium polyaluminate with Na^+ conducting properties; it is the β"-phase with the approximate empirical formula $Na_2O \cdot 6\ Al_2O_3$ and three spinel blocks in one unit cell [37,38]. This phase is capable of taking up appreciable amounts of MgO into solid solution. The phase stability of $β"-Al_2O_3$ even seems to increase with increasing MgO content [39]. In the polycrystalline state, this MgO-stabilized β"-phase has a resistivity of only 5 Ωcm at 300°C, compared with 20 Ωcm for the MgO-free β-phase. In the ternary system $Na_2O-MgO-Al_2O_3$, two other compounds were discovered with structures corresponding to the β- and β"-phases. They were named $β"'-$ and $β""-Al_2O_3$ [39].

III. BATTERIES WITH SILVER HALIDES AS SOLID ELECTROLYTE

1. PRIMARY CELLS

The early solid electrolyte cells were of the primary type and contained, in most cases, silver halides as electrolyte. Selected examples are listed in Table II.

TABLE II

Batteries with Silver Halides as Solid Electrolyte

No.	Cell	Op. Temp.	Electrolyte			OCV V	Short Circuit Current
			Type	Cond. at Op.Temp. $\Omega^{-1}cm^{-1}$	Thickness		
I	Ag/AgI/I,C	amb.	β-AgI	10^{-6}	several microns	0.7	200 $\mu A/cm^2$
II	Ag/AgI/Ag$_2$S/ S,C	200°C	α-AgI	1	25 μ	0.2	180 mA/ cm^2
III	Ag/AgI/I$_2$ (CsI$_4$)	300°C	α-AgI	1-2		0.67	1-20 mA
IV	Ag/AgCl/Cl$_2$ (KICl$_4$),C	amb.	AgCl	10^{-6} to 10^{-8}	several microns	1.0	15-30 $\mu A/cm^2$
V	Ag/AgI/Pt	amb.	β-AgI	10^{-6}	0.25 mm	0.5-0.25	100-600 $\mu A/cm^2$

Cell I (see Table II) is the battery investigated by Lehovec and Broder [40], who used a mixture of pressed iodine and carbon as cathode. Short circuit currents were, however, only a few hundred $\mu A/cm^2$. Therefore, the same authors used the highly conductive α-form of AgI as solid electrolyte (cell II) and obtained an increase in short circuit current by about three orders of magnitude. Fused sulfur was contained in a carbon pellet and served as cathode. The silver sulfide layer formed chemically on contact of the cathode with the AgI layer and electrochemically during discharge.

Cells III and IV are gas-activated cells. The cathodes consisted of polyhalide compounds which dissociated under formation of the respective halogen. CsI$_3$ and CsI$_4$ served as a source of iodine vapor at elevated temperature [41], in cell IV chlorine was liberated by KICl$_4$ [42].

An extensive description of these and other primary cells may be taken from the reviews by Foley [4] and Hull [5,6].

2. CONCENTRATION CELLS

The only successful attempts to make a silver halide cell rechargeable was undertaken by Mrgudich [43]. He investigated cells of type V (see Table II), using sputtered Ag and Pt films as electrodes and a pressed AgI pellet as electrolyte. This device has to be regarded as a concentration cell because the driving force of cell V is the difference in silver concentration between the Ag anode and the Pt cathode.

Mrgudich observed flash currents of 100 to 600 $\mu A/cm^2$, steady discharge currents of 0.1 to 0.5 $\mu A/cm^2$, and rechargeability during repeated cycles.

This system has been further developed to an all-thin film device with electrolyte thicknesses of 5 to 10 μm [6,44]. No data on drain currents or short-circuit currents have been reported to date.

Concentration cells similar to type V have been recently investigated by Kennedy, *et al.*, [45] for the application in coulometric timers and memories. A certain amount of silver is plated onto a gold electrode and stripped off during operation. As soon as the gold electrode is silver free, there is a rapid voltage increase which may be used as a signal. A stripping cell with a thin AgBr film as solid electrolyte was reported to have a timing accuracy of a few tenths of a percent [46]. Silver bromide behaves in a similar fashion as AgCl in its electrolyte properties. By using special evaporation techniques, it can be made free of pinholes in thicknesses of 10 μm and above [46].

IV.   MODERN SOLID ELECTROLYTE BATTERIES

1.   GENERAL

New types of solid electrolyte batteries have been proposed during the last four years. They are listed in Table III. Except for the sodium-sulfur battery, they are all-solid systems.

TABLE III

Modern Solid Electrolyte Batteries
(SCC short-circuit current, DISC discharge current)

No.	Cell	Op. Temp.	Electrolyte Type	Cond. at Op.Temp. $\Omega^{-1}cm^{-1}$	Thickness	OCV V	Current
VI	$Ag/Ag_3SI/$ I,C	amb.	β-Ag_3SI	$7\cdot10^{-3}$	1.5 mm	0.68	1 mA/cm^2 DISC
							10 mA/cm^2 SCC
VII	$Ag/RbAg_4I_5/$ I,RbI$_3$,C	amb.	$RbAg_4I_5$	0.25	0.5 mm	0.66	900 mA/cm^2 SCC
	$Ag/RbAg_4I_5/$ I,TBAI,C	amb.	$RbAg_4I_5$	0.25	1.5 mm	0.56	1 mA/cm^2 DISC
							15 mA/cm^2 SCC
VIII	$Li/LiI/AgI$, Ag	amb.	LiI	10^{-7}	15 μm	2.1	120 $\mu A/cm^2$ SCC
IX	$Ca/CaF_2/NiF_2$	400-500°C	doped CaF_2	$2\cdot10^{-3}$	10 μm	2.8	
X	Na/β-$Al_2O_3/$ S,C	300°C	β-Al_2O_3	0.3	1 mm	2.1	100 mA/cm^2 DISC

Gaseous or fluid materials taking part in an electrode reaction guarantee an easy transport of reactants and reaction products and an intimate contact át the site of the electrochemical reaction (the three-phase boundary). If an electrode contains only solids, however, the three-phase boundary must be extended to form the bulk of a composite. This composite has to contain the active material, an auxiliary electrolyte and an auxiliary electronic conductor. The two conducting phases must form continuous conduction paths throughout the composite. These paths have to span the distance from the solid-electrolyte layer to a current-collecting layer.

Raleigh [8,47] discussed the important role of geometry in electrodes of this type; a merely geometrical mismatch of different components, e.g., active material and electrolyte, may limit the performance of the battery.

A simplified model of the three-phase structure has recently been proposed by the author of this review [3]. It has been shown that the reaction site and the progression of the electrode reaction depend on the type of conductivity involved in the active material. If this material is an electrolytic conductor itself, it might be that the whole electrode becomes rechargeable and that the composite maintains a stable geometric structure.

The cathodes of cells VI, VII and VIII (see Table III) are formed by three-phase mixtures. No attempts have been made to date to achieve a defined structure within these electrodes. Instead, components of low particle size are mixed mechanically and pressed together. Some of the silver anodes in cells VI to VIII have also been incorporated in electrode mixtures. This was because it had turned out that amalgamation of silver was sufficient for polarization-free conductivity measurements, but not for battery discharge. Three-phase mixtures have also been proposed for the electrodes of cell IX [3].

A problem encountered with metal anodes combined with a cation-conducting solid electrolyte is that the anode leaves behind cavities on discharge. As the metal has to deposit in an empty space on recharge, there is the risk of dendrite formation. Cell IX, on the other hand, combines a metal with an anion-conducting electrolyte, so that the recharge reactions find predetermined sites in both of the electrodes.

2. Ag_3SI-BATTERIES

Takahashi and Yamamoto [48] proposed Ag_3SI as an electrolyte in a primary cell operating at ambient temperature:

$Ag/\beta-Ag_3SI/I,C$.

Polarization losses at the electrodes have been kept small by amalgamation of silver and selection of a suitable carbon black. Steady discharge currents of up to one mA/cm^2 were measured; short-circuit currents were even one order of magnitude higher. However, long storage life of this battery is doubtful. The electrolyte becomes a partial n-type electronic conductor in chemical equilibrium with silver [49], and this might lead to self-discharge [6]. However, the electrolyte seems to be chemically unstable in contact with iodine [50].

3. $RbAg_4I_5$-BATTERIES

Various authors [51,52,53] are studying at present the cell

$Ag/RbAg_4I_5/I,QI_3,C.$

A mixture of iodine with either RbI_3 [51] or organic polyiodides in
the form of tetrabutylammonium polyiodide [53] is used as active material in the cathode. The open circuit voltage of 0.66 V corresponds
to the decomposition voltage of the electrolyte.

Oxley [54] reported on the rate-determining steps of the electrode
reactions: in the anode, silver has to diffuse over internal surfaces
to the electrolyte; in the cathode a simple concentration polarization
is built up as iodine has to diffuse out of the polyiodide to the electrolyte interface.

$RbAg_4I_5$ is thermodynamically unstable at temperatures below 27°C.
A disproportionation reaction to Rb_2AgI_3 and AgI occurs when catalytic
traces of water vapor are present [52]. This may be overcome by sealing the cells. However, if RbI_3 is the active component in the cathode
mix, the desired highly conductive reaction product, $RbAg_4I_5$, cannot
be formed at low temperatures. A new possibility was opened up by the
discovery of electrolytically conductive organic double salts [32]
which are stable down to -55°C. The corresponding tetraalkylammonium
polyiodide may be used as active cathode material [52,53]. The rechargeability of this type of battery is under investigation [53,55].
As a primary system, the cell has applications in the ordnance field
[56].

Besides the application of cell VII in batteries, its use for timers
and capacitors is being studied [54].

Another type of cathode for the $RbAg_4I_5$ battery was proposed by
Takahashi [57]. He investigated Se and Te, which form Ag_2Se and Ag_2Te
on discharge and yield rather low cell voltages (0.22 and 0.27 V, respectively). However, the low vapor pressure of Te and Se — low in
comparison with the vapor pressure of iodine — allows for elevated operation temperatures. This might lead to high current densities; at
150°C no polarization was reported at a discharge current density of
10 mA/cm^2 and only slight polarization at 20 mA/cm^2.

4.   LiI-BATTERIES

An all-solid battery with lithium iodide as solid electrolyte

$Li/LiI/AgI,Ag$

$Li/LiI/CuI,Cu$

has been designed as a thin-film device and is under study by Liang
and his coworkers [58,59]. Although the conductivity of LiI is very
low at room temperature (about $10^{-7}\Omega^{-1}cm^{-1}$), the internal electrolyte
resistance is kept tolerable by using an electrolyte layer only 15 µm
thick. This allows for short-circuit currents of 120 $\mu A/cm^2$. The
anode and cathode layers are also in the thickness range of 10 to 30
µm. The ion transport number in the LiI electrolyte is close to unity
[60].

The development started with AgI as active cathode material, but
this turned out to be unstable because of interdiffusion of Li and Ag.
This problem can be overcome by replacing AgI by CuI [61]. In this
case a LiI-CuI mixture is pressed onto a Cu foil and, subsequently,
layers of pure LiI and of Li are evaporated on top.

The battery is expected to have applications in the ordnance field.
Moreover, Greatbatch [62] reported on the miniaturization of the battery for an implantable device for heart pacemakers. He uses a pro-
prietary iodide complex for the cathode and predicts a storage life of
more than ten years.

5. CaF$_2$-BATTERY

An all-solid battery with doped calcium fluoride as solid electrolyte

Ca/CaF$_2$/NiF$_2$,Ni

has been proposed recently [3]. Alternative electrode materials are
Mg for the anode and FeF$_2$ or CrF$_2$ for the cathode. The thicknesses of
electrolyte and electrode layers are planned to be in the 10 to 20 µm
range. If the formation of blocking layers in the electrodes can be
avoided by making the active materials electrolytically conductive and
by developing a proper structure for the electrode mixes, the battery
will be rechargeable. It is intended to stack several cells one on top
of the other and to connect them electrically in series by means of a
copper foil acting as current collector. The stacks will be electrically connected in parallel. Above a power range of a few kW the operating temperature of 400° to 500°C will be self-sustaining during
operation. The system will have to be sealed against the atmosphere.

Possible improvements of electrolyte conductivity may allow for
lower operating temperatures. Promising electrolyte compositions seem
to include high YF$_3$ concentrations, the ternary system CaF$_2$-CaYF$_4$-NaF,
and possibly the use of LiF or YbF$_3$ as dopants.

Of high theoretical and practical interest are the transport num-
bers in doped CaF$_2$. According to Wagner [63], pure CaF$_2$ might become
a partial n-type conductor when equilibrated with metallic calcium.
First experiments have been carried out by Hinze and Patterson [64],
however, their results are not yet conclusive. According to Kröger's
theory [65,66], doping of the electrolyte will influence the onset of
electronic conductivity at low chemical fluorine potentials. Monova-
lent fluorides such as NaF will shift the limit to lower fluorine potentials; trivalent fluorides such as YF$_3$ will shift it in the undesired direction towards higher fluorine potentials:

$$n = K_1 \cdot p_{F_2}^{-1/2} \cdot [Na'_{Ca}]^{-1}$$

$$n = K_2 \cdot p_{F_2}^{-1/2} \cdot [Y^{\cdot}_{Ca}]^{+1}$$

(n: electron concentration; p(F$_2$): theoretical fluorine partial pres-
sure; [Na$^{\cdot}_{Ca}$] and [Y$^{\cdot}_{Ca}$]: concentrations of doping cations Na$^+$ and Y^{+++},
respectively; K$_1$ and K$_2$: constants). This theory does not predict the
behavior in the heavy doping range, e.g., at YF$_3$ concentrations of
several mole percent.

Another problem in the development of the battery might be interdiffusion of the components (fluorides at the interface electrode/
electrolyte, metals at the interface electrode/current collector).
This could be solved by lowering the battery temperature to medium
levels during idling periods.

The low equivalent weight and the low layer thicknesses of the materials make energy densities and power densities possible which may
approach or even exceed those of the sodium-sulfur battery. Table IV

TABLE IV

Theoretical Energy Density of Batteries

System	Reaction Products	Temp. °C	1 kg divided into			OCV V	En.dens. Wh/kg
			Anode g	Cathode g	Charge Ah		
Li/LiCl, KCl/Cl$_2$	nil, LiCl	475-650	164	836	635	3.5	2220
Li/LiI, KI/S	nil, Li$_2$S$_2$	280-400	178	822	688	2.25	1550
Ca/CaF$_2$/ NiF$_2$	CaF$_2$, Ni	400-500	293	707	393	2.78	1090
Na/β-Al$_2$O$_3$ /S	nil, Na$_2$S$_3$	275-300	324	676	379	2.10	795
Li/LiI/AgI	nil, LiI	amb.	28.7	971	111	2.10	286
Ag/RbAg$_4$I$_5$/ RbI$_3$ + 2 I	nil, RbAg$_4$I$_5$	amb.	375	625	93.3	0.66	61.5

shows the theoretical energy densities of several batteries. For comparison, the lithium-sulfur battery [67] and the lithium-chlorine
battery [68] are included as high-energy systems (fused electrolytes),
and the LiI and RbAg$_4$I$_5$ batteries as low-energy systems. The values
have been calculated by dividing one kilogram into weights of anode
and cathode such that both electrodes generate the same amount of
charge. The product of this charge and the open circuit voltage is
the maximum theoretical energy density. As the assumptions were the
same for all the systems included, a comparison is reasonable. The
experimental data suggest that approximately one quarter of the maximum value can be realized. If this is confirmed by the development of
the CaF$_2$ battery, the CaF$_2$ system will have applications in the traction field.

6.   SODIUM-SULFUR BATTERY

In this type of battery, the sodium-ion-conducting electrolyte
β-Al$_2$O$_3$ separates a liquid anode of fused sodium and a liquid cathode
made up of fused sulfur, sodium polysulfides and graphite:

Na/"β-Al$_2$O$_3$"/Na$_2$S$_x$,C   .

The operating temperature is about 300°C. The open circuit voltage
depends on the state of discharge and ranges from 2.1 to 1.8 V. On
discharge of the cathode, x must not become smaller than 3, because
polysulfides containing more sodium are solid. Charge and discharge
current densities amount to 100 mA/cm^2 at an overall polarization of
about 1 V [34]. An energy density of 330 Wh/kg and a power density of
400 W/kg have been anticipated [34,69], so that an application of the
battery in the traction field seems feasible [69]. It is the only
solid electrolyte system whose high recharge performance has been verified experimentally.

Current problems involved in the development of the sodium-sulfur
battery include the sealing of the sodium container, the development
of sulfur-resistant, light-weight casing materials, and the chemical
and phase stability of the solid electrolyte [70].

It has been reported that sodium is deposited along the grain boundaries of the solid electrolyte [35]. This may be due either to a simple chemical reaction with impurities which have accumulated at the
grain boundaries, or to a partial electronic conductivity in the bulk
of the crystals by equilibration with metallic sodium. The latter
mechanism seems less probable as it had been shown on Ag-exchanged
β-Al$_2$O$_3$ that the electronic transport number is very low [71]. Extremely pure materials, special methods of preparation, and MgO as
stabilizing agent may solve this problem.

The graphite matrix in the cathode compartment must have a suitable
structure in order to minimize the polarization losses occurring during
recharge. These losses are encountered as soon as elemental sulfur is
formed from Na$_2$S$_5$ [70,72]. At this stage, a change in the reaction
mechanism seems to take place, as the polysulfides between Na$_2$S$_3$ and
Na$_2$S$_5$ — which are present at deeper discharge states — differ only
slightly in their properties [73].

REFERENCES

[1] H. Binder, et al., Electrochim. Acta, 8, 781 (1963).

[2] W. Baukal, Chem.-Ing.-Tech., 41, 791 (1969).

[3] W. Baukal, Paper Presented at the 138th Meeting of The Electro-
     chemical Society, Atlantic City, October 1970.

[4] R. T. Foley, J. Electrochem. Soc., 116, 13C (1969).

[5] M. N. Hull, Proc. 22nd Annual Power Sources Conference, PSC Publ.
     Comm., Red Bank, N. J. (1968) p. 106.

[6] M. N. Hull, Energy Conversion, 10, 215 (1970).

[7] L. Heyne, Electrochim. Acta, 15, 1251 (1970).

[8] D. O. Raleigh, to be printed in Adv. Electroanal. Chem.

[9] C. Wagner and W. Schottky, Z. Phys. Chem., B11, 163 (1930).

[10] J. Frenkel, Z. Phys., 35, 652 (1926).

[11] J. P. Gracey and R. J. Friauf, J. Phys. Chem. Solids, 30, 421
     (1969).

[12] H. C. Abbink and D. S. Martin, Jr., J. Phys. Chem. Solids, 27, 205
     (1966).

[13] Y. Haven, *Rec. Trav. Chim. Pays-Bas,* 69, 1471 (1950).

[14] D. C. Ginnings and T. E. Phipps, *J. Amer. Chem. Soc.,* 52, 1340 (1930).

[15] R. W. Ure, Jr., *J. Chem. Phys.,* 26, 1363 (1957).

[16] F. K. Fong, *in Progress in Solid-State Chemistry,* Vol. 3, by H. Reiss (Ed.), Pergamon Press, Oxford (1967) p. 135.

[17] L. W. Strock, *Z. Phys. Chem.,* B25, 441 (1934); B31, 132 (1935).

[18] K. H. Lieser, *Z. Phys. Chem. NF,* 9, 216, 302 (1956).

[19] G. Burley and H. E. Kissinger, *J. Res. Nat. Bur. Stand.,* 64A, 403 (1960).

[19a] T. Takahashi, K. Kuwabara and O. Yamamoto, *J. Electrochem. Soc.,* 116, 357 (1969).

[20] J. E. Bauerle and J. Hrizo, *J. Phys. Chem. Solids,* 30, 565 (1969).

[21] D. W. Strickler and W. G. Carlson, *J. Amer. Ceram. Soc.,* 47, 122 (1964).

[22] W. Baukal, *these Proceedings,* p. 247.

[23] B. C. H. Steele and C. B. Alcock, *Trans. Met. Soc. AIME,* 233, 1359 (1965).

[24] W. W. Barker and O. Knop, The British Ceramic Society, Basic Science Section, Meeting on "Mass Transport in Non-Metallic Solids", London, December 1969.

[25] H. Hahn, W. Seemann and H. L. Kohn, *Z. Anorg. Allg. Chem.,* 369, 48 (1969).

[26] C. Tubandt and E. Lorenz, *Z. Phys. Chem.,* 87, 513 (1914).

[27] A. Kvist and A. Lundén, *Z. Naturforsch.,* 20A, 235 (1965).

[28] B. Reuter and K. Hardel, *Naturwiss.,* 48, 161 (1961).

[29] S. Geller, *Science,* 157, 310 (1967).

[30] J. N. Bradley and P. D. Greene, *Trans. Farad. Soc.* 63, 424, 2516 (1967).

[31] B. B. Owens and G. R. Argue, *J. Electrochem. Soc.,* 117, 898 (1970).

[32] B. B. Owens, *J. Electrochem. Soc.,* 117, 123C (1970).

[33] W. L. Bragg, *et al., Z. Kristallogr.,* 77, 255 (1931).

[34] N. Weber and J. T. Kummer, *Proc. 21st Annual Power Sources Conference,* PSC Publ. Comm., Red Bank, N. J. (1967) p. 37.

[35] L. J. Miles and I. W. Jones, *Power Sources 1970,* Brighton.

[36] Y.-F. Y. Yao and J. T. Kummer, *J. Inorg. Nucl. Chem.,* 29, 2453 (1967).

[37] R. C. DeVries and W. L. Roth, *J. Amer. Ceram. Soc.,* 52, 364 (1969).

[38] M. Bettman and C. R. Peters, *J. Phys. Chem.,* 73, 1774 (1969).

[39] N. Weber and A. F. Venero, *Ford Motor Co.; Scient. Res. Staff, Techn. Rept. No. SR 69-86 and SR 69-102.* Papers presented at the 76th Meeting of the Am. Ceram. Soc., Philadelphia, May 1970.

[40] K. Lehovec and J. Broder, *J. Electrochem. Soc.,* 101, 208 (1954).

[41] J. L. Weininger, *J. Electrochem. Soc.*, 106, 475 (1959).

[42] D. M. Smyth, *J. Electrochem. Soc.*, 106, 635 (1959).

[43] J. N. Mrgudich, *et al.*, *IEEE Trans. Aerospace Electron. Systems* AES-1, 290 (1965).

[44] P. Vourous and J. I. Masters, *J. Electrochem. Soc.*, 116, 880 (1969).

[45] J. H. Kennedy, F. Chen and J. Willis, *J. Electrochem. Soc.*, 117, 263 (1970).

[46] J. H. Kennedy and F. Chen, Paper presented at the 138th Meeting of The Electrochemical Society, Atlantic City, October 1970.

[47] D. O. Raleigh, Paper presented at the 138th Meeting of The Electrochemical Society, Atlantic City, October 1970.

[48] T. Takahashi and O. Yamamoto, *Electrochim. Acta*, 11, 911 (1966).

[49] B. Reuter and K. Hardel, *Ber. Bunsenges. Phys. Chem.*, 70, 82 (1966).

[50] B. B. Owens and G. R. Argue, Paper presented at the 153rd Meeting of the ACS, Miami Beach, April 1967.

[51] G. R. Argue, B. B. Owens and I. J. Groce, *Proc. 22nd Annual Power Sources Conference*, PSC Publ. Comm., Red Bank, N. J. (1968), p. 103.

[52] J. E. Oxley and B. B. Owens, *Power Sources 1970*, Brighton.

[53] M. DeRossi, G. Pistoia and B. Scrosati, *J. Electrochem. Soc.*, 116, 1642 (1969).

[54] J. E. Oxley, Paper presented at the 138th Meeting of The Electrochemical Society, Atlantic City, October 1970.

[55] B. Scrosati, Paper presented at the 138th Meeting of The Electrochemical Society, Atlantic City, October 1970.

[56] D. L. Warburton and P. Arbesman, Paper presented at the 138th Meeting of The Electrochemical Society, Atlantic City, October 1970.

[57] T. Takahashi and O. Yamamoto, *J. Electrochem. Soc.*, 117, 1 (1970).

[58] C. C. Liang and P. Bro, *J. Electrochem. Soc.*, 116, 1322 (1969).

[59] C. C. Liang, J. Epstein and G. H. Boyle, *J. Electrochem. Soc.*, 116, 1452 (1969).

[60] C. C. Liang, *Trans. Farad. Soc.*, 65, 3369 (1969).

[61] C. C. Liang, Paper presented at the 138th Meeting of The Electrochemical Society, Atlantic City, October 1970.

[62] W. Greatbatch, Paper presented at the 138th Meeting of The Electrochemical Society, Atlantic City, October 1970.

[63] C. Wagner, *J. Electrochem. Soc.*, 115, 933 (1968).

[64] J. W. Hinze and J. W. Patterson, *Special Report*, Iowa State University, May 1970.

[65] F. A. Kröger, *The Chemistry of Imperfect Crystals*, Interscience, New York 1964.

[66] H. Schmalzried, *Z. Phys. Chem. NF*, 38, 87 (1963).

[67] N. P. Yao, *et al.*, Paper presented at the 138th Meeting of The Electrochemical Society, Atlantic City, October 1970.

[68] D. A. J. Swinkels, *J. Electrochem. Soc.*, 113, 6 (1966).

[69] J. L. Sudworth and I. Dugdale, *Power Sources 2 (1968)* by D. H. Collins (Ed.), p. 547, Pergamon Press, Oxford (1970).

[70] J. L. Sudworth and M. D. Hames, *Power Sources 1970*, Brighton.

[71] M. S. Whittingham and R. A. Huggins, Paper presented at the 138th Meeting of The Electrochemical Society, Atlantic City, October 1970.

[72] S. M. Selis, *Electrochimica Acta,* 15, 1285 (1970).

[73] N. K. Gupta and R. P. Tischer, Paper presented at the 137th Meeting of The Electrochemical Society, Los Angeles, March 1970.

HIGH POWER ELECTROCHEMICAL SYSTEMS*

E. J. Cairns, R. K. Steunenberg and H. Shimotake

Argonne National Laboratory, Argonne, Illinois

There is a growing need for high-specific-power, high-specific-energy battery systems suitable for use in spacecraft, communications, propulsion of civilian and military vehicles, off-peak energy storage at central power stations, and for other similarly demanding applications. Most of these applications require electrically rechargeable batteries having a specific power of at least 200 W/kg and a specific energy of 200 W-hr/kg. These criteria can probably be met only by electrochemical cells that operate at temperatures above about 200°C, and contain alkali metal anodes and chalcogen or halogen cathodes.

The sodium/sulfur cell is the most actively investigated of all the high-temperature cells [1-5]. It makes use of a solid electrolyte called beta alumina ($Na_2O \cdot 11\ Al_2O_3$), which conducts sodium ions between a liquid sodium anode and a liquid sulfur cathode. The products of the cell reaction are sodium polysulfides, which accummulate in the sulfur compartment. These cells have open circuit voltages of about two volts, and generally operate at current densities of 0.2-0.5 A/cm^2 at 1.75 to 1.5 V. Rapid recharge and long cycle lives have been difficult to achieve because of the tendency of the electrolyte to crack, presumably due to sodium deposition within the electrolyte [3,4]. The projected specific energy capability of the sodium/sulfur battery is above 200 W-hr/kg [1,2].

The lithium/chlorine cell [6-11] operates at 650°C and makes use of a molten lithium chloride electrolyte which is also the product of the cell reaction. The chlorine cathode is porous carbon, and can support very high current densities if high-purity chlorine is used in order to avoid diffusion limitations caused by the accumulation of impurities in the pores of the carbon. Power densities as high as 40 W/cm^2 can be maintained for short periods of time. Some work has been done on electrical recharge of lithium/chlorine cells, but no information is available on cycle life. The problems requiring further work center mainly on corrosion and seals.

A cell closely related to the lithium/chlorine cell is one that makes use of a lithium-aluminum alloy anode (in which lithium is the active material), a LiCl-KCl eutectic electrolyte, and a high specific-area carbon cathode which stores chlorine by adsorption [12-14]. Operating at 475°C, several hundred cycles have been obtained from rather large, sealed cells (25 × 15 × 1 cm). With some improvements in the capacity of the cathode these cells are expected to be capable of storing 150 W-hr/kg.

Lithium/chalcogen cells [15-21] make use of liquid lithium anodes, molten salt electrolytes containing lithium halides, and liquid chalcogens (Te, Se, or S) as the cathode. The current density-voltage

Work performed under the auspices of the United States Atomic Energy Commission.

curves for a number of these cells are shown in Fig. 1. Typical open
circuit voltages range from 1.75 to 2.3 V; short-circuit current densities are 8-13 A/cm^2, corresponding to peak power densities of 3.5 to
8 W/cm^2. Because these cells are still in the early stages of development, only rather short cycle lives have been demonstrated. Lithium/
selenium cells have experienced failure because of selenium transport
to the lithium electrode. The principal mode of failure of lithium/
sulfur cells has been loss of sulfur by evaporation from unsealed cells.
The projected specific energy and specific power for lithium/selenium
and lithium/sulfur cells are above 200 W-hr/kg and above 200 W/kg, respectively.

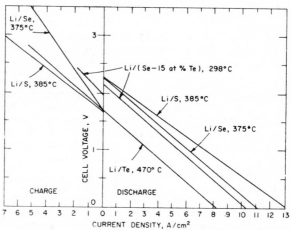

FIG. 1 *Current-voltage curves for lithium/chalcogen cells.*

The characteristics of a number of high-temperature cells are summarized in Table I. The values in the table were taken from various
references. The performance data for a given type of cell are not necessarily all from the same cell; therefore, all of the characteristics
should not be assumed to have been demonstrated together for a single
cell design. The intent of the table is to present an impression of
the performance capabilities which can be achieved, and to point out
some of the problems requiring attention. The projected specific power
vs. specific energy relationships which have been presented in the
literature for various cells are shown in Fig. 2. Note that all of
the high-temperature cells show projected characteristics which are
superior to those for any conventional cells. It is this promise of
superior performance that whets the appetite of electrochemists working with high-temperature cells.

It can be concluded from this brief review* that both sodium/sulfur
and lithium/chalcogen cells show promise for high specific energy and
high specific power applications. Further work on recharge is needed
for lithium/chlorine cells, and improved specific energy is needed for
lithium (aluminum)/chlorine (carbon) cells. Both sodium/sulfur and
lithium/chalcogen cells have similar materials and seals problems, and
some problems associated with electrolytes. The laboratory results

More detailed reviews of high-temperature cells can be found in
references [22] and [23].

TABLE I

Characteristics of High-Temperature Cells

Cell	Temp °C	E_{oc} Volts	Pmax W/cm^2	q A-hr/cm^2	Cycle Life	Life hr.	Problems
Na/Na$_2$O·11Al$_2$O$_3$/S	300	2.1	1.0	0.35	300	2000	Na penetration
Li/LiCl/Cl$_2$	650	3.5	40.0	-	-	1	Corrosion
Li-Al/LiCl-KCl/Cl-C	475	3.2	0.5	0.07	500	1500	Low capacity
Li/LiF-LiCl-LiI/Se	375	2.2	2.3	0.4	15	60	Se migration
Li/LiBr-RbBr/S	375	2.3	4.0	0.5	50	750	S loss

are encouraging, but a great deal of research and development work remains to be done before any high-temperature cell becomes a practical energy-storage device.

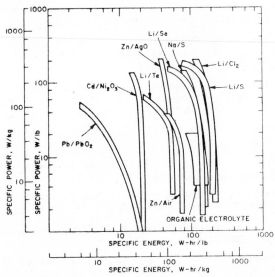

FIG. 2 *Projected specific power vs. specific energy relationships.*

REFERENCES

[1] J. T. Kummer and N. Weber, SAE Automotive Engineering Congress, Detroit, Mich., Jan. 1967, paper No. 670179.

[2] N. Weber and J. T. Kummer, in *Proc. 21st Ann. Power Sources Conf.*, PSC Publ. Comm., Red Bank, N. J. (1967), p.37.

[3] J. L. Sudworth and M. D. Hames, presented at The Internat. Power Sources Symp., Brighton, Sept. 1970, preprint No. 13.

[4] L. J. Miles and I. Wynn Jones, *ibid*, preprint No. 14.

[5] S. Hattori, Yuasa Battery Co., Ltd., private communication (1970).

[6] D. A. J. Swinkels, *J. Electrochem. Soc.*, 113, 6 (1966).

[7]   D. A. J. Swinkels, *IEEE Spectrum,* 5, 71 (May, 1968).

[8]   D. A. J. Swinkels, *J. Electrochem. Soc.,* 114, 812 (1967).

[9]   E. H. Hietbrink, J. J. Petrarts, D. A. J. Swinkels and G. M. Craig, *Technical Report to Air Force Aero Propulsion Lab TR-67-89,* General Motors Co. (Aug. 1967).

[10] D. A. J. Swinkels, *Electrochem. Tech.,* 5, 396 (1967).

[11] T. G. Bradley, *General Motors Research Laboratory Report No. GMR-795,* (Aug. 1968).

[12] R. A. Rightmire and A. L. Jones, *in Proc. 21st Ann. Power Sources Conf.,* PSC Publ. Comm., Red Bank, N. J. (1967), p. 42.

[13] E. J. Dowgiallo, D. H. Bomkamp, R. A. Rightmire and J. W. Sprague, presented at SAE International Automotive Engrs. Congress, Detroit, Mich., Jan. 1969, paper No. 690207.

[14] J. L. Benak, presented to the Washington D.C. Section of The Electrochemical Society, Dec. 1969.

[15] E. J. Cairns, C. E. Crouthamel, A. K. Fischer, M. S. Foster, J. C. Hesson, C. E. Johnson, H. Shimotake and A. D. Tevebaugh, Argonne National Laboratory, Report, ANL-7316 (1967).

[16] H. Shimotake, G. L. Rogers and E. J. Cairns, presented at Electrochem. Soc. Meeting, Chicago, Oct. 1967; see also *Extended Abstracts of The Battery Div.,* 12, 42 (1967).

[17] H. Shimotake and E. J. Cairns, presented at Comité International de Thermodynamique et de Cinetique Electrochimiques Meeting, Detroit, Sept. 1968; *Extended Abstr.,* p. 254.

[18] H. Shimotake and E. J. Cairns, presented at The Electrochem. Soc. Meeting, New York, May, 1969.

[19] N. P. Yao, L. A. Herédy and R. C. Saunders, presented at Electrochem. Soc. Meeting, Atlantic City, October 1970, paper No. 60.

[20] H. Shimotake, G. L. Rogers and E. J. Cairns, *Ind. Eng. Chem., Process Design and Development Quarterly,* 8, 51 (1969).

[21] H. Shimotake, A. A. Chilenskas, R. K. Steunenberg and E. J. Cairns, *in Proc. 5th IECEC,* Amer. Nuclear Soc., 1970.

[22] E. J. Cairns, R. K. Steunenberg and H. Shimotake, *in Kirk-Othmer Encyclopedia of Chemical Technology Supplement,* Interscience, in press.

[23] E. J. Cairns and H. Shimotake, *Science,* 164, 1347 (1969).

SUMMARY OF DISCUSSION AND COMMENTS: SESSION CHAIRMAN VII

T. R. Beck

Boeing Scientific Research Laboratories, Seattle, Washington

It appears appropriate to have a session on primary and secondary batteries in a fuel cell symposium because the two are technologically closely related. The relationship exists in operation as the same principles of electrochemical engineering, *i.e.*, thermodynamics, kinetics, mass transport and current distribution, apply to both. The major difference is that batteries operate by batch or cyclic processes whereas fuel cells are by definition continuous reactors. In end use both will compete in the same economic, social and political field.

With a continuation of the long-term trend towards an all-electric economy it is probable that batteries and fuel cells will in the future play a more prominant role in transportation and portable equipment. The electrochemical storage and conversion systems available today are but crude predecessors of what must come. An increasing sophistication in understanding and utilizing materials, and use of electro-chemical engineering systems approach in battery research and development, would appear necessary to achieve the goals of high-energy, high-energy-density, long-life electrochemical systems.

H. Tannenberger summarized the technical similarities and differences of existing battery and fuel cell systems and their fields of application. Possible fields of application of metal-air batteries, high-energy batteries and hydrocarbon fuel cells were delimited on the basis of published data. Discussion related to use of these devices in automobiles. Tannenberger pointed out that existing U.S. electric energy capacity would have to be increased by 50% if rechargeable batteries replaced the internal combustion engine in all automobiles.

J. Giner discussed the aluminum-chlorine battery with molten salt electrolyte. Energy density is high but corrosion problems are severe.

K. D. Beccu described construction and operational characteristics of metal-air systems. These are hybrid systems which can continuously cathodically reduce oxygen from the air using a porous electrode and batchwise oxidize metal anodes. Historically it is interesting that development of the air cell contributed to fuel-cell technology and in turn the intensive fuel cell development in the past decade may contribute to metal-air systems. The major problems with high-energy-density metal-air systems at present, however, are connected with the metal electrode. Polarization and limited utilization of active metal are the main problems. Only mechanically rechargeable systems are currently available. Electrochemical recharging, if possible, often results in dendritic deposits or changes in electrode dimensions. In the discussion, J. Huff commented that he has a program in which a zinc electrode has had 100 cycles at 50% depth of discharge. He projects a power density of 30 watts/lb and an energy density of 30 watt-hrs/lb for a rechargeable zinc-air system. K. Kordesch commented that there is a good possibility of a throwaway aluminum-air battery with 85% efficiency using KOH electrolyte.

G. Sandstede reported on progress on quinone/hydroquinone cathodes that can be recharged reversibly by electrochemical means or by air or H_2O_2 when pasted on a porous hydrophobic carbon electrode. Two quinones, chloranil and duroquinone, found to be sufficiently insoluble in aqueous electrolytes and completely stable, show promise as cathode materials for high-energy-density secondary batteries.

W. Baukal reviewed new solid electrolytes for batteries and mechanisms of conduction. Two ranges of applications for solid electrolyte batteries are elevated temperature with high energy and power density and at room temperature with low energy and power density. During the discussion the applications of low-temperature solid-electrolyte batteries were questioned. It was suggested that they could be incorporated in integrated micro-electronics circuits and used in implantable pacemakers. Problems are self-discharge and poor cycle life due to dendrites.

E. J. Cairns reviewed high-power electrochemical systems operating above 200°C and containing alkali-metal anodes and chalcogen or halogen cathodes. Although laboratory results are encouraging, much more R&D must be done before the potential 200 w/kg and 200 w-hr/kg systems become commercial realities.

DEVELOPMENTAL GOALS AND PROSPECTS

Chairman

K. V. Kordesch

FUEL CELLS FOR PRACTICAL ENERGY CONVERSION SYSTEMS

W. T. Reid

Battelle Memorial Institute, Columbus, Ohio

ABSTRACT

Future demands for energy will be so great that every effort must be made to convert the energy in fuels into electricity with the greatest possible efficiency. Fuel cells are unique, offering the only energy-conversion system not Carnot-cycle limited, so that conversion efficiency can be maximized and thermal pollution can be minimized. However, costs are out of line by about three orders of magnitude. Nevertheless, further attempts to develop lower-cost fuel cell systems for the generation of electricity in congested areas are well justified. Air pollution from automotive vehicles likewise will provide a strong incentive to devise a practical fuel-cell system for automobiles. The potential market here is very large, but until an inexpensive and reliable electrode system is developed, and perhaps a fuel as yet unknown, that market cannot be realized. Further work on fuel cells will be worthwhile indeed if it leads even partially to these two tremendous applications.

INTRODUCTION

Of all the energy-conversion systems being considered to provide man's future demands for electricity, fuel cells stand in the most unique position. They alone offer both radically higher overall conversion efficiency to minimize thermal pollution, and the capability of generating electricity within congested areas completely without air-polluting emissions. Further, automotive vehicles using fuel cells would not contaminate the local environment, and there would be no limit to the number of vehicles permitted in an urban area.

This Utopia would be here today if practical fuel-cell systems were available at an acceptable cost. Unfortunately, significant improvements in fuel-cell technology have not kept pace with the demand for reasonably priced energy-conversion systems. As a result, fuel cells have found their place only in aerospace applications such as the Gemini and the Apollo missions, where capital costs are secondary to performance. Earthbound applications have been disappointing, mainly because fuel cells as they are known today are too costly by about three orders of magnitude.

Yet, hopefully, this situation may not always exist. The tremendous capabilities of fuel-cell systems, related both to high conversion efficiency and to air-pollution control, will continue to stimulate fuel cell enthusiasts. And although the visions of these researchers will not be evident to tight-fisted business men, a nucleus of scientists and engineers will keep up their search for improved fuel cell systems, just as this seminar is evaluating new developments in electrocatalysis.

369

It is difficult to foresee when practical fuel cell systems will be available. What is strikingly apparent is the rapidly growing need for such systems. Two such needs are discussed here: substitution of fuel cells to generate a share of the electricity now being supplied by large central-station power plants, and fuel cell power for automotive vehicles in place of internal-combustion engines. These presently are the richest goals for fuel cells. Although many lesser opportunities on a much smaller scale will turn up for fuel cells, it is in these two tremendously large applications that the greatest impact can be made.

ELECTRICAL POWER GENERATION

PROJECTIONS OF FUTURE DEMAND

Energy consumption in the United States is growing at an essentially constant yearly percentage rate. Except for a minor reversal of the usual trends between 1930 and 1935 — the Great Depression years — total energy consumption has doubled about every 25 years since 1860 [1], or at an annual rate of increase of 2.8 percent. Many projections of future energy consumption have been made, but a recent one by the U. S. Bureau of Mines [2] is particularly detailed. It suggests a continuation of the present trend in growth at least through 2000 A.D.

Fig. 1 shows these data graphically, with the Bureau of Mines projections listed for low, medium, and high values. The other projection based on expected population growth with an increase in total energy

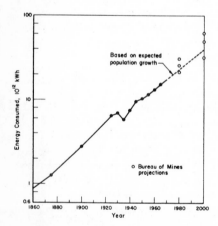

FIG. 1 *Energy consumption in the United States.*

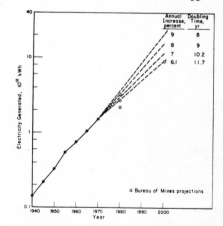

FIG. 2 *Electricity generation in central-station power plants in the United States.*

consumption of 1.05 percent annually per capita, shown as the broken line, falls somewhat below the medium rate expected by the Bureau of Mines.

Projections like these are seldom optimistic. Because long-range estimates of energy consumption usually turn out to be too small, providing 40×10^{12} kWh equivalent energy by the turn of the century can be considered as a minimum goal.

Fig. 2 shows the yearly generation of electricity in the U. S. by the public utilities since 1940, with projections beyond 1970 based on

the same Bureau of Mines estimates. These call for an annual increase
of 6.1 percent to 2000 A.D., for a doubling time of 11.7 years, a most
conservative estimate. The broken lines show extrapolations beyond
1970 at annual percentage gains more in line with recent trends. An
eight percent annual increase, for a doubling time of nine years, is
considered most likely. At this rate, central-station power plants
will be generating some 15×10^{12} kWh annually at the turn of the century, or ten times their output in 1970.

It is evident from these long-range projections that there will be
a much swifter increase in the rate of demand for electricity than for
other forms of energy. Fig. 3 illustrates how electricity will be given

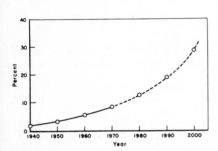

FIG. 3 *Percentage of total
energy consumed in United States
used as electricity generated
by public utilities.*

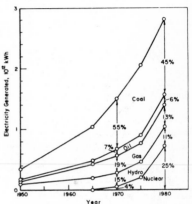

FIG. 4 *Source of energy
for generation of electricity
(NAPCA).*

an increasingly larger share of the total demand based on an annual
increase of only seven percent beyond 1970. If the annual rate of
increase were eight percent, as noted above, then electricity from
central-station power plants would be providing 38 percent of the
total demand for energy in 2000 A.D.

These predictions of future demand for electricity are limited to
central-station power plants. Other sources of electricity are less
significant; it is this public-utility-generated electricity that will
provide the largest potential market for fuel cells outside the automotive field.

SOURCES OF ENERGY FOR ELECTRICITY GENERATION

The fossil fuels coal, oil, and natural gas have been the major
sources of energy for electricity generation. In past years, hydro-
electric generation produced as much as a fifth of the total output,
but the percentage is decreasing as fewer hydro sites remain to be
developed. Nuclear power is still a largely unknown factor.

Fig. 4 shows trends in sources of energy for generating electricity
[3]. Although coal will supply a smaller fraction of the total energy
input as nuclear generation increases, solid fuels will still provide
a larger total energy input in 1980 than in 1970. Oil will also decrease percentagewise but increase in total consumption, but the exact
figures are in doubt because of the indefinite status of present air-
pollution-control measures. Some of these rely on low-sulfur oil which

is more readily available on the East Coast than low-sulfur coal. How far this trend will go is uncertain, but the general unavailability of large amounts of low-sulfur residual fuel oil suggests only a limited replacement of coal with fuel oil over the entire nation. Natural gas is presently in short supply, with no indication that large quantities of low-cost gas will be available in the years ahead. Hence the projections on natural gas may need to be revised downward.

Coal is the only major fuel available in nearly unlimited quantities. Although data on reserves differ, it is usually conceded that at least 1×10^{12} tons of coal may be recovered from proven reserves, although an unknown fraction of these reserves cannot be produced within today's mining costs. Another important factor is the location of these coal reserves; around half the total subbituminous coal and lignite is in the mid-western states. Almost all of these reserves of low-rank coal have less than one percent sulfur, but their distance from thermal stations, most of which are located east of the Mississippi, explains why these coals have not been widely exploited. That situation is gradually changing.

This tremendous reserve of energy in the form of coal can take care of our needs for generating electricity for many hundreds of years. Hence any consideration of fuel cells in the years ahead must be based on an energy source originating from coal. This energy source probably will be a fuel gas, either for a high-temperature fuel cell as proposed by Westinghouse, where producer gas would be suitable to provide the oxygen gradient between the electrodes, or as hydrogen for direct-electrochemical oxidation.

The ultimate system, of course, would be the carbon fuel cell. Unfortunately, the likelihood is poor of devising a practical carbon-oxidation cell where a large fraction of the carbon could be oxidized electrochemically. Yet the return would be so great that any approach to such a development would be worth exploring.

LARGE-SCALE FUEL CELL SYSTEMS

Many studies have been made of high-temperature fuel cell systems for generating large blocks of electrical energy. One of the most extensive of these was made by the Central Electricity Generating Board in England during 1962. Unpublished, the report on that evaluation concluded that molten carbonate fuel cells in which unreformed hydrocarbons would be the fuel offered the most promise economically, although the technical feasibility of such a system remained to be proved. Fused-salt cells using reformed hydrocarbons had little chance of economic success, according to the CEGB assessment.

An excellent study of fuel cell systems for integrated central stations is that of Hart and Womack [4], who considered a system in which propane was the fuel, reformed to H_2 and CO, with a thermal efficiency of 75 percent. This reformed gas was then to be oxidized electrochemically in molten-salt fuel cell battery modules each providing 125 kW at 62.5 volts, with waste-heat boilers to recover thermal energy. Their final conclusion was that the overall generation efficiency would be 46.4 percent. With the most optimistic cost estimates, the initial investment would be $120 per kW of capability; if a five-year life were assumed for the fuel cells compared with 20 years for a gas-turbine plant, for example, the equivalent duty cost would be $225 per kW. Hence, these authors concluded that there is no foreseeable likelihood of achieving an economically attractive fuel cell central-

station power plant. Their low-cost estimates for such items as a
DC-to-AC inverter are not in line, and thus the actual capital investment of this system may be even greater than proposed; so there is no
question but that very large fuel cell installations are economically
unattractive.

Where, then, do fuel cells have any appeal for generating large
blocks of electricity? Since large central-station plants are not
attractive, how can fuel cells still provide a responsible part of that
15×10^{12} kWh to be generated in 2000 A.D.? A possible solution is the
installation of many fuel cell "substations" in energy-consuming areas,
each to be supplied with a "fuel" through a distribution network and
each furnishing electricity to a small number of customers.

It must be recalled that transmitting and distributing electricity
is the most costly way of moving energy from one point to another,
whereas moving a gas by pipeline is one of the cheapest. According to
an early Senate study [5], network transmission of electricity costs
from 16 to 44 times more than moving the same mount of energy as natural gas by pipeline. Even allowing for a conversion efficiency of gas
into electricity of 40 percent, energy movement by electricity still
is six to 18 times more costly. No such data are available on the distribution of electricity or natural gas in urban areas, but if the pro-
portionate costs are comparable, a strong incentive remains for generating electricity as near as possible to the ultimate consumer.

Rather than generating electricity from natural gas in each home,
as is the intent of the TARGET program, a more flexible arrangement
would be to provide a large number of 1-MW fuel-cell substations in
residential areas, each substation to serve not more than 50 households.
Since these 50 houses normally might occupy a square area roughly 700
feet on a side, the maximum distance of the substation would be 500 feet.
Over such short distances, electricity could be transmitted as direct
current at some nominally low voltage so that copper or aluminum water
pipes could serve as the conductors with minimum insulation problems.
At peak demand of 20 kW, each house would be drawing 400 amperes at 50
volts, indicating the necessity of furnishing a large conductor between
the substation and the home. Providing direct current appliances would
pose no difficulties for most household applications if the market were
sufficiently large.

A major advantage of the 1-MW substation is that it would diversify
the maximum demand. It is quite unlikely that all 50 residences would
demand peak power at the same instant, so that the maximum load would
be spread over a longer time interval than would be the case with individual home installations. Also, surplus capacity would be built into
the fuel cell battery with the redundant cells serving as stand-by
units. Thus, if any cells failed to function properly, they could be
switched out of the circuit automatically and one of the stand-by units
substituted. When enough cells had failed, the unattended substation
would signal the need for attention and repairs could be made.

Hydrogen would provide the ideal "fuel" for such a system, but
other readily oxidizable fluid fuels could be used. Produced in large
quantities, hydrogen might cost no more than 30 cents per thousand
cubic feet delivered by pipeline from conversion plants using coal,
located outside the city limits to minimize air-pollution problems.
At this rate, fuel costs would be about 0.6 cent per kWh at an efficiency of 65 percent, or 0.78 cent for an efficiency of 50 percent.
Fuel cell capital costs are indeterminate; an acceptable goal might
be $250 per kW.

Obviously this concept requires detailed cost analysis to demonstrate whether it can compete successfully with central-station-generated electricity priced to the consumer presently at an average 2.2 cents per kWh. It has the advantage over individual domestic fuel cell installations of permitting a relatively sophisticated design at a 1-MW size, of higher reliability, and of requiring no attention whatsoever by the householder.

Since fuel cells probably will not be competitive in very large installations simply to replace conventional central-station power plants burning coal, some alternative scheme such as this will be necessary if fuel cells are ever to take over a share of the rapidly increasing market for electricity. The substation idea may provide fuel cells with that opportunity.

ELECTRIC AUTOMOBILES

By now everyone is aware of the major problem with electric automobiles — providing them with enough energy to achieve a reasonable range and speed. Exotic battery systems have not been brought to the point where they can be applied successfully to automotive vehicles.

High-temperature, alkali-metal secondary batteries being studied for electric automobiles are still in the developmental stage. Batteries based on lithium-sulfur and sodium-sulfur are being investigated by NAPCA with the basic research schedule to terminate in August 1972. If proof of principle is demonstrated by then, 2-, 5-, and finally 20-kW batteries will be constructed, with a target date of 1978. If metal-air batteries are not proved by 1974, with a 20-kW battery to follow, the metal-air battery program will be abandoned by NAPCA in favor of a 2-kW alkali-metal battery if that program is successful [6]. High-temperature batteries have been under development by others, notably the lithium-chlorine battery by General Motors and the sodium-sulfur battery by Ford, both outstanding batteries in the laboratory, but not demonstrated as yet for over-the-road performance.

Hence, with few exceptions, most electric automobiles have used lead-acid batteries with an energy density up to 15 watthours per pound. In an early analysis [7], it was shown that an electrical Volkswagen with a range of 150 miles at 35 miles per hour would require an energy input of 37.3 kWh. Table I shows the characteristics of four types of conventional batteries able to provide this amount of energy. It is evident that existing batteries are both too costly and too heavy, and it is this reason, of course, why most electric automobiles are limited to moderate speeds and short ranges, typically 40 miles at 35 miles per hour. Experience has confirmed that electric automobiles driven on city streets at speeds not exceeding 35 miles per hour consume about 0.25 kWh per ton mile.

Fuel cells were suggested quite early for energy converters on electric vehicles. Table II gives the characteristics of a fuel cell battery to give that Volkswagen its 150-mile range. There is nothing insurmountable here, and the weight of 700 pounds for the fuel cell battery and the fuel system is far more attractive than the minimum equivalent lead-acid battery weighing more than three times as much. Table III lists the characteristics of the cell and the battery; again they present nothing spectacular and they are certainly well within the present state of the technical art.

TABLE I

Storage Batteries to Provide 37.3 Kilowatt-Hours

Battery Type	Weight, lb	Volume, ft^3	Cost Battery	Per Mile
S-L-I	2270	16.7	$ 900	$0.035
Heavy Duty Lead-Acid	3130	16.7	2,940	0.015
Alkaline Nickel-Iron	2940	20.0	5,900	0.018
Nickel-Cadmium	2940	21.7	12,400	0.040

TABLE II

Fuel-Cell Battery for Electric Automobile

Based on 15-HP motor and 37.3 kWh total demand:

Weight of 11.2-kW fuel-cell battery - 450 lb
Volume of 11.2-kW fuel-cell battery - 5.6 ft^3

Hydrogen required, per kWh   - 18 ft^3 (70% eff.)
Hydrogen required, for 150-mile trip - 670 ft^3
Oxygen required, for 150-mile trip   - 335 ft^3

Volume hydrogen and oxygen at 3000 psi   - 5 ft^3
Weight of gases and resin-fiber glass tank - 250 lb

Total weight, fuel cell and fuel system - 700 lb

TABLE III

Characteristics of Fuel Cell for Electric Automobile

Electrode:

Current Density - 200 amps/ft^2
Cell Potential - 0.6 volt
Maximum Continuous Output - 100 watts/ft^2

Battery:

Power Density - 0.025 kW/lb
Volume Density - 2 kW/ft^3

Fig. 5 shows the General Motors "Electrovan," displayed in 1967 to demonstrate that fuel cells could indeed power an automotive vehicle [8]. With 32 kW of H_2-O_2 fuel cells built by Union Carbide, using cryogenic H_2 and O_2, the vehicle had a top speed of 70 miles per hour and a range of 100 to 150 miles. The sophisticated power train, which weighed 3,930 pounds compared with 870 pounds for a conventional van with an IC engine, included solid state logic and trigger circuits to control frequency and amplitude of power pulses fed to a 125-horsepower, specially designed, three-phase electric motor. Energy losses from the fuel cells required an elaborate plumbing system to circulate electrolyte

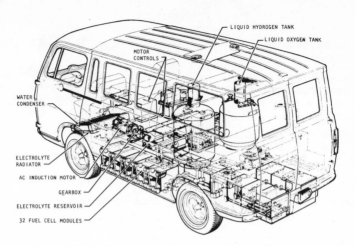

LIQUID HYDROGEN TANK
LIQUID OXYGEN TANK
MOTOR CONTROLS
WATER CONDENSER
ELECTROLYTE RADIATOR
AC INDUCTION MOTOR
GEARBOX
ELECTROLYTE RESERVOIR
32 FUEL CELL MODULES

FIG. 5 *Fuel cell powered electrovan.*

through heat exchangers, with 44 percent of the weight of the fuel-cell power plant going to these auxiliaries. The Electrovan consumed about 1 kWh per mile with a gross vehicle weight of 7,100 pounds, or 0.28 kWh per ton mile.

Problems faced in running the Electrovan were many, but the investigators concluded that "...the rate of progress in this field and the strong advantages of fuel cells are sufficient incentives to maintain this effort." Despite this optimistic viewpoint, no further work seems to have been undertaken by the automotive industry to devise an improved fuel-cell-powered automobile.

The reason, of course, is cost. Although no price was quoted for the fuel cells in the Electrovan, it is generally assumed that the 1-kW fuel cells cost approximately $10,000 each, or $320,000 for the installation on this vehicle. Although perhaps one fifth this amount might be recoverable platinum, the costs are so high as to be completely unrealistic. Since the fuel cell energy converter and the electric motor and its controls must compete with an IC engine, transmission, and drive train costing less than $1,000, it is evident that the cost of the fuel cells cannot exceed about $20 per kW. This seemingly impossible goal indicates the challenge involved for fuel cell researchers.

With air pollution from automotive vehicles causing more and more revision of old concepts of vehicle propulsion systems, alternative nonpolluting systems are being investigated in depth even though their economics may be generally unfavorable. With sales of 10 million passenger cars annually, the market is very big indeed. The opportunity here for fuel cells cannot be overestimated, but whether fuel cells ever power any of these vehicles will depend almost certainly on the development of an electrode system and a compatible, readily available fuel that can be sold at a reasonable price. That is the goal that must be sought by fuel cell technologists if the automotive industry is ever to convert to an electrical energy-conversion system.

REFERENCES

[1]   W. T. Reid, "The Energy Explosion," *J. Inst. Fuel*, 43, (349), 43
      (February 1970).

[2]   W. E. Morrison and C. L. Readling, "An Energy Model for the United
      States, Featuring Energy Balances for the Years 1947 to 1965 and
      Projections and Forecasts to the Years 1980 and 2000," *U.S. Bur.
      Mines, Inf. Circ. 8384* (1968).

[3]   "National Air Pollution Control Adminstration," *Chem. & Eng. News*,
      p. 16 (June 15, 1970).

[4]   A. B. Hart and G. J. Womack, *Fuel Cells, Theory and Operation*,
      Chapman and Hall, London (1967).

[5]   *Report of the National Fuels and Energy Study Group on an Assess-
      ment of Available Information on Energy in the United States*,
      Committee on Interior and Insular Affairs, United States Senate,
      87th Congress, Document No. 159, Government Printing Office (1962)
      p. 168.

[6]   J. J. Brogen, "Almost Pollution-Free Powerplants are Scheduled
      for Early 1975," *J. Auto. Eng.*, SAE, 78, (12), 40 (1970).

[7]   W. T. Reid, "Energy Sources for Electrically Powered Automobiles,"
      Battelle Tech. Rev., 14, (4), 9 (April 1965).

[8]   C. Marks, E. A. Rishavy and F. A. Wyczalek, "Electrovan — A Fuel
      Cell Powered Vehicle," *SAE Paper 670176* (1967).

REQUIREMENTS OF AN ELECTROCHEMICAL POWER SOURCE FOR A DUAL-MODE URBAN TRANSPORTATION SYSTEM

T. R. Beck

Boeing Scientific Research Laboratories, Seattle, Washington

ABSTRACT

A great deal of discussion has been carried on about replacing the internal combustion engine in automobiles by batteries or fuel cells, and there are certainly some promising electrochemical systems under development. A possible alternative to the present automobile, less demanding on an electrochemical power source is described here. This is a dual-mode system utilizing a small vehicle operated on a secondary electrochemical power source on city streets and on electric power supplied through a trolley while on an electrified rail system. Power and energy requirements for the dual-mode vehicle are within the present state of the art of lead-acid batteries, although a higher energy density electrochemical system would be desirable.

PRESENTATION

The present ground transportation system in the United States depends very heavily on internal-combustion-engine automobiles. Urban and suburban developments in the United States in the past half century have been built around the automobile as the dominant mode of transportation. The automobile has many advantages such as a very high two-dimensional mobility on the network of highways, roads and streets, and it permits users to travel between any two points unrestricted by schedules and without waiting at terminals.

The internal combustion automobile is not an unmixed blessing though as it has a number of disadvantages which are becoming more clear with the increasing population of automobiles. Some of these are summarized in Table I. Chemical and noise pollution problems are of course

TABLE I

Contributions to Obsolescence of the Internal Combustion Engine Automobile

- Pollution problem - chemical, noise
- Excessive land requirement >1/3 of city area
- Low efficiency - 10% (vehicle), <1% (person)
- Exhaustion of low-cost fuel

widely discussed now. Streets, freeways and parking lots require over one third of the land area of a modern city such as Los Angeles. The thermal efficiency of an internal combustion engine is about 10 percent and a 4,000-pound automobile used to transport a 150 pound person gives

an overall thermal efficiency of less than one percent. We are also
exhausting our low-cost petroleum fields in the contiguous United States
and adjacent continental shelves [1]. Increasing use of imported petroleum and use of liquid fuel derived from oil shale or coal will
surely increase the price.

Some alternatives to the internal combustion engine automobile are
shown in Table II. We have heard in this conference about development
of the hydrogen/air fuel cell [2] and the sodium/sulfur battery [3] as
replacements for the internal combustion engine. These or other electrochemical power sources may perhaps some day serve this purpose.
Trains and buses are becoming obsolete for urban transportation except
in older high-density cities. A third alternative, not well known, is
a dual-mode system. Studies of a dual-mode system have been made by
the Boeing Company [4] and other organizations [5]. Accelerated interest in rapid transit in recent years makes it appropriate to discuss
the concept now.

TABLE II

Alternatives to Internal Combustion Engine Automobiles

- Electric cars, same performance
- Train, bus
- Dual-mode system

A dual-mode system would consist of electrically powered vehicles
capable of operating on city streets, and an electrically powered system capable of guiding the vehicles over longer distances. Typical of
such vehicles is the Urbmobile proposed by Weinberg [5] and illustrated
in Fig. 1. This would be a small-size vehicle operated by battery power

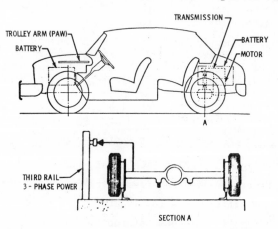

FIG. 1 *Illustration of a dual-mode vehicle [4].*

and an electric motor on city streets. It would have pneumatic tires
for city streets and flanged wheels for guidance on an electrified rail
system as illustrated in Fig. 2. Adjacent pairs of tracks would be provided for two-way traffic. On-off ramps would be located at distances
of, say, two to five miles. The radius of operation of vehicles off
the system is indicated by the dashed circles.

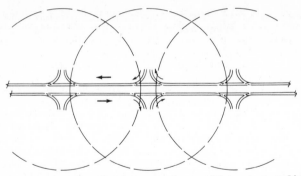

FIG. 2 *System track, station spacing, off-system radius.*

In operation the system would work as follows. A vehicle, either privately owned or leased from the system, would be parked over night in a patron's carport. Its battery would be trickle charged overnight, and the vehicle would be ready to use in the morning. Heaters could be programmed to heat the vehicle prior to use. The patron would drive the vehicle under battery power to the nearest on-ramp of the system. Arriving at an on-ramp, he would instruct the system, via an IBM card or other suitable means, of his destination. At this point the system would take over and the patron could relax and read his paper. Vehicles on the track would travel at a constant rate of perhaps 60 miles per hour. Sensors located at appropriate intervals along the track would inform the computer controlling the system of the whereabouts and spacing of all vehicles on the system. At the arrival of an appropriate space between vehicles the system would automatically accelerate the patron's vehicle up to speed on the on-ramp and feed it into this space. The system would automatically exit his vehicle at the desired off-ramp. The patron could then drive his vehicle to a parking lot at his place of work, which for system-owned vehicles could be close packed, thus saving space.

An alternative in a high-density metropolitan area would be for the patron to leave the vehicle on a side track in an unloading area and for the system to automatically return the vehicle to the active tracks for use in delivering packages and mail during working hours. Vehicles could again be made available in a boarding area when required for the homeward commute.

The main features of the dual-mode system are summarized in Table III. Vehicles on the system would recharge their batteries on system

TABLE III

Main Features of Dual-Mode System

ON SYSTEM	• Electric power from system
	• Constant speed
	• Computer controlled
	• High traffic density
OFF SYSTEM	• Electrochemical power
	• Operates on city streets
	• Limited range and speed

power. The system would operate at constant speed and only admit vehicles when there was space. Computers would control the traffic density on the system and automatically admit and discharge vehicles.
This would allow a high traffic density and avoid the overload and slow-
down condition experienced on present freeways during commuting hours.
Vehicles off the system would operate on their own electrochemical
power. They would operate on the existing grid of city streets and
thereby require only a limited range and speed as compared with the
present internal-combustion-engine automobile.

A feature of this transportation system is that it embodies a concept utilized in telephone networks; that all technologically new components are compatible with the existing system. The vehicles would be
compatible with existing grid of streets and vehicles and the electrified rail system could be installed as a more efficient alternative to
more freeways. Initially it is envisaged that a dual-mode system would
be used principally for commuting in urban areas, but if the concept
were successful it could be expanded to intercity use and for use with
dual-mode buses and delivery vans.

Requirements of the electrochemical power source for the commute
vehicles are listed in Table IV. Off the system the vehicles would
require a range of only about 1/10 that of a commercial internal combustion engine automobile, or about 20 miles. As the speed on city

TABLE IV

Requirements of Electrochemical Power Source

- 1/10 distance of internal combustion engine car
- 1/4 power of internal combustion engine car
- Rechargeability

streets is a maximum of about 35 miles per hour as compared with 70 or
80 miles per hour on the present freeways, the power requirement turns
out to be only about 1/4 that of the present internal combustion engine
car of similar weight. For a vehicle of limited range, recharging
would be necessary while on the system and while parked in a user's
carport. For a 2,000-pound vehicle traveling city streets, these requirements translate to about 10 kilowatts of power and about 3 kilowatt-hours of energy. These requirements are within the present state
of the art of lead/acid batteries, in that about 350 pounds of lead/
acid battery could do the job. It is anticipated, however, that if such
a system were to come into general usage that there would be a large
incentive to develop higher-energy-density batteries.

Would such a dual-mode system overcome the objections to the internal combustion engine automobile noted in Table I? In terms of
pollution, the chemical effluents would be eliminated and the noise
should be much reduced on city streets. System noise might be comparable to existing freeway noise. In terms of land requirement the
track system operating at high traffic density and constant speed
could move many more vehicles per hour per unit area than our existing
freeway system. As already mentioned, parking space could be reduced
by close packing of system-owned vehicles. In terms of efficiency,
the system would be operated from thermal power plants with an efficiency of 40 percent. Considering the transmission and distribution

losses and losses in the motor and controls of the vehicle, the overall efficiency for vehicles on the system still ought to be nearly
twice that of the internal-combustion-engine automobiles. Special-
purpose vehicles for commuting could also be made lighter in weight
than the existing all-purpose internal-combustion-engine car. While
off the system, operating on batteries, the overall efficiency may be
somewhat better than the internal combustion engine for a battery discharge efficiency of about 50 percent. As the vehicles would spend
more than half of their mileage on the system, there would be a net
gain in overall thermal efficiency over internal-combustion-engine
automobiles. In respect to exhaustion of low-cost liquid fuel, the
system would be powered by predominantly thermal power either from
coal-fired plants or from nuclear-powered plants.

Adoption of this dual-mode concept is consistent with the trend to
an all-electric economy pointed out by W. T. Reid [6] in this conference. The purpose of describing the dual-mode concept here is to make
those working in electrochemical power sources aware of a possible new
power source requirement to which they might direct efforts in the
future.

REFERENCES

[1]   M. King Hubbart, "Energy Resources," in *Resources and Man*,
      National Academy of Sciences, National Research Council,
      W. H. Freeman and Co., San Francisco (1969).

[2]   K. V. Kordesch, these proceedings, p. 157.

[3]   H. Tannenberger,(unpublished presentation).

[4]   T. R. Beck, Boeing internal report, August 1964.

[5]   M. I. Weinberg, Society of Automotive Engineers, meeting preprint,
      May 20-24, 1968.

[6]   W. T. Reid, these proceedings, p. 369.

COMMENT ON THE PAPER OF T. BECK BY J. O'M. BOCKRIS

Dr. Beck's concept of "battery-third rail" has also been developed
at SOHIO. Would it not apply mainly to major highways? Most of the
work in the towns would be the province of battery-driven cars. The
gasoline consumed in towns represents about two thirds of the energy
consumed in the driving of cars.

This seems a reasonable point at which to recall the advantage of
municipally owned public cars. The charge plate, magnetic taped,
could not be withdrawn unless the user had plugged it in for recharge
at a lamppost charging point. Study of this system at the University
of Pennsylvania has shown a saving of traffic space because the commuter
cars can be used by shoppers during the day. Private cars are reserved
for turnpike driving, later electrified and programmed.

SOME BASIC ASPECTS OF ELECTROCATALYSIS

J. O'M. Bockris, J. McHardy and R. Sen

University of Pennsylvania, Philadelphia, Pennsylvania

ABSTRACT

Despite Ostwald's recognition in 1894 of the intrinsic advantages of electrochemical energy conversion over thermal methods, practical difficulties and a series of misunderstandings delayed significant advances until the last decade. The first treatment of electrocatalysis was implicit in a 1935 paper by Horiuti and Polanyi, but no explicit discussion appeared until the 1960's. No effect of the electrode material is anticipated by thermodynamics for simple electron transfer (nonbonding) reactions, but a dependence of rate upon the work function Φ can arise from double-layer concentration effects. Analogous effects in the space-charge region of a semiconductor electrode can also give rise to a reaction rate dependent on Φ. For bonding reactions, large differences in rate observed among metal electrodes can be rationalized from a simple model that predicts a linear variation of activation energy with the adsorption energy of a reactant or product. For reactions involving two adsorbed species, e.g., the oxidation of ethylene, the problem of competitive adsorption must be considered. Current ideas about electrocatalysis by semiconductors are discussed only briefly because of their tentative nature. The quantum mechanics of electron transfer reactions at interfaces is still in dispute 40 years after Gurney first formulated it. For bonding reactions typified by the hydrogen evolution reaction, two main approaches now prevail. In the Gurney-based approach, thermal activation of an H^+-OH_2 bond satisfies the condition necessary for radiationless electron transfer, and the process is treated by tunneling theory. In the approach developed by Levich, a lattice model is used for the water structure, and the harmonic motion of H_2O dipoles in this lattice provides electrostatic activation energy for the reaction. The Levich model is treated by the adiabatic perturbation method of Platzman and Frank. Both models predict the same type of dependence for the reaction rate upon hydrogen adsorption energy as did the empirical model discussed earlier. Few of the factors governing electrocatalysis are adequately understood. Areas that require research range from the chemical and physical properties of catalyst surfaces, to double-layer structure at solid electrodes, and the quantum theory of charge-transfer processes. Additional prospects for accelerating electrode reactions include photo-activation and the use of nonaqueous electrolytes.

I. INTRODUCTION

The thermodynamic advantages of the electrochemical method over the thermal method of energy conversion have been recognized for almost a century [47], and it is considerably longer since the first fuel cell was discovered [29]. In spite of such early beginnings, practical difficulties and a series of misunderstandings [12] resulted in the virtual abandoning of electrochemical energy conversion in favor of the intrinsically less efficient thermal method. More than just efficiency was sacrificed, however, and when environmental factors are taken into consideration, the thermal method ceases to be even an acceptable alternative. A pollution-free technology will not be possible until electrochemical energy conversion is commercially feasible, and in this context the problems of electrocatalysis are paramount. The role of electrocatalysis in the performance of a fuel cell has been discussed at length elsewhere [13] and only a brief treatment is warranted here.

A rigorous analysis of electrode kinetics in a porous electrode gives rise to equations that can be solved only numerically and by a computer [5]. Fortunately, most of the important parameters can be appreciated by considering planar electrodes, and little is lost for purposes of the present article if one of the electrodes is assumed to be ideally nonpolarizable. The terminal voltage of the fuel cell under load is then

$$V = IR_1 = E^0 - \Sigma\eta \quad , \tag{1}$$

where I is the current, R_1 is the load resistance, E^0 is the thermodynamic cell EMF, and $\Sigma\eta$ is the net overpotential.

$$\Sigma\eta = \eta_{act} + \eta_{diff} + \eta_{ohm} \quad . \tag{2}$$

Each overpotential is a function of the current, and can be expressed in terms of kinetic parameters. Eq. (1) then becomes

$$IR = E^0 - \left(\frac{RT}{\alpha F} \ln \frac{I}{Ai_0}\right) + \left(\frac{RT}{nF} \ln\left[1 - \frac{I}{Ai_L}\right]\right) - IR_{int} \tag{3}$$

where α is the transfer coefficient, A the electrode area, i_0 the exchange current density, n the number of electrons passed per act of the overall reaction, i_L the limiting current density, and R_{int} is the cell resistance. A slightly more general treatment in which both electrodes were polarizable would introduce extra activation and diffusion terms into Eq. (3) resulting from processes at the second electrode. From a knowledge of E^0, the α's, the i_0's and the i_L's, it is then possible to calculate the curve of current $vs.$ terminal voltage, and such curves typically exhibit the three regions illustrated in Fig. 1. The efficiency at any current, $\rho(I)$ is related to the terminal voltage by the equation

$$\rho(I) = \frac{nVF}{\Delta H} \tag{4}$$

where ΔH is the enthalpy change for the overall reaction. The power output of a cell, being the product of current and voltage, follows a curve such as that in Fig. 2.

Clearly, all three sources of overpotential should be minimized, but under most operating conditions it is the activation overpotential that is usually the largest offender. Differentiation of Eq. (1) with respect to $\ln i_0$ leads to the relations:

$$\left(\frac{\partial\rho}{\partial\ln i_0}\right)_I = \frac{nRT}{\alpha\Delta H} \tag{5a}$$

and

$$\left(\frac{\partial P}{\partial \ln i_0}\right)_I = \frac{IRT}{\alpha F} \quad , \tag{5b}$$

i.e., both efficiency and power at a given current are linear functions of $\ln i_0$, and hence depend directly on the activity of the electrocatalyst. The purpose of this paper is to review present knowledge about electrocatalysis, especially as it applies to reactions of significance to electrochemical energy conversion.

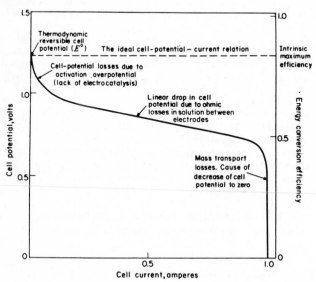

FIG. 1 *Typical cell potential and efficiency vs. current relation of a fuel cell showing losses due to three types of overpotential [13].*

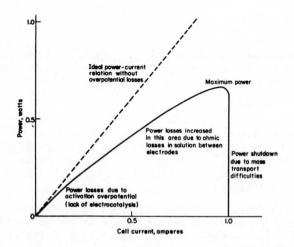

FIG. 2 *Typical power vs. current density relation of a fuel cell showing losses due to three types of overpotential [13].*

II. THE PRESENT THEORY OF ELECTROCATALYSIS

1. BACKGROUND

Unfortunately, several authors have employed the term 'electrocatalysis' to refer to the accelerative effect of the interfacial electric field upon electron transfer reactions. In the sense originally intended [30] and also employed here, electrocatalysis refers to the catalytic influence of the electrode material upon an electrode reaction under standard conditions. The choice of a reference potential for comparing electrocatalysts has been the subject of much discussion [24,48,51], but the most common one is the thermodynamic reversible potential. The exchange current density, i_0 thus emerges as the primary criterion of activity for an electrocatalyst. Another parameter of importance in practical applications is the transfer coefficient, α, because of the way it influences η_{act}, and hence ρ and P (Eqs. 3 and 5).

The first treatment of electrocatalysis was implicit in a paper by Horiuti and Polanyi [32]. Butler [16] introduced a surface bonding term into the quantum mechanical model of Gurney [31] and effectively laid the groundwork for present theories of electrocatalysis, but explicit treatments did not appear until the 1960's. The evolution of quantum mechanical treatments of bonding-type electrode reactions is discussed in Section 4 below.

2. INFLUENCE OF THE ELECTRODE MATERIAL ON NONBONDING REACTIONS

A. Metals

A series of equations will be first assembled to facilitate subsequent discussion.

In an isolated uncharged phase, the Fermi level is equivalent to the chemical potential of a single electron in that phase:

$$E_{f,0} = \frac{\mu_e}{N} \, ,$$
(6)

where N is Avogadro's number. If $E_{f,0}$ is referred to the energy of a free electron at infinity, it can be related to the thermionic work function Φ by the equation:

$$\Phi = -E_{f,0} + e\Delta X_0 \, ,$$
(7)

where e is the electronic unit of charge and ΔX_0 is the surface dipole p.d. for the phase *in vacuo*. The chemical potential of electrons in the phase is therefore:

$$\mu_e = -N \Phi + F\Delta X_0 \, .$$
(8)

In a charged phase, *e.g.*, an electrode in contact with a solution, the chemical potential μ_e must be replaced by the electrochemical potential $\bar{\mu}_e$, *i.e.*, the Fermi level is equivalent to the <u>electrochemical</u> potential of a single electron:

$$E_f = \frac{\bar{\mu}_e}{N} \, .$$
(9)

By introducing the concept of the inner electrostatic potential in the metal ϕ^M, $\bar{\mu}_e$ may be formally separated into chemical and electrostatic components:

$$\bar{\mu}_e = \mu_e - F\phi^M \quad , \tag{10}$$

the negative sign resulting from the negative charge of the electron.

No effect of the electrode material is anticipated for simple electron transfer (nonbonding) reactions for the following reason. Consider the redox reaction:

$$A^Z + e \rightleftharpoons A^{Z-1} \quad . \tag{11}$$

At equilibrium, $i.e.$, at the thermodynamic reversible potential, the electrochemical potentials of reactants and products respectively are equal:

$$\bar{\mu}_Z + \bar{\mu}_e = \bar{\mu}_{Z-1} \quad . \tag{12}$$

Neither $\bar{\mu}_Z$ nor $\bar{\mu}_{Z-1}$ can depend on the electrode; therefore $\bar{\mu}_e$, which is the molar equivalent of the Fermi level, (Eq. (9)) must be the same for all electrode materials:

$$\mu_e - F\phi^M = \text{constant} \quad . \tag{13}$$

At first sight, this conclusion is paradoxical because the value of ϕ^M, since it must compensate for μ_e, will vary from one electrode material to another, and might be expected to produce variations in the cell E.M.F. — a thermodynamic impossibility. The paradox is resolved by considering the metal-metal junction between working and reference electrodes. The effect of ϕ^M on the p.d. at that junction will be equal and opposite to its effect on the p.d. at the metal-solution interface. The net contribution of ϕ^M to the cell EMF is thus zero.

The preceding discussion concerned an electrode only at the reversible potential, but the conclusions are equally valid at any given potential. If comparisons were made at a potential η volts below the reversible potential, the inner potential of the electrode would be lowered by η volts and the new (molar) Fermi level would be:

$$\bar{\mu}'_e = \text{constant} + F\eta \quad , \tag{14}$$

which would still be the same for all electrode materials.

Experimental results for the Fe^{3+}/Fe^{2+} redox couple on a series of electrodes confirmed that the activation energy of the reaction was independent of the substrate, but also indicated a dependence of $\log i_0$ upon Φ - Fig. 3 [10]. Such a dependence was predicted by Parsons [48] because of double-layer effects. Parsons followed an elaborate course of reasoning that involved the potential of zero charge (p.z.c.). The following argument is more direct. The redox reaction is assumed to occur at the outer Helmholtz plane (O.H.P.), and in general the electric potential there (ϕ^H) will lie somewhere in between the inner potential of the metal (ϕ^M) and the inner potential of the bulk of the solution (ϕ^B). The difference in electric potential between the O.H.P.

and the solution bulk causes different concentrations of ions at the two locations.

In the manner of Bockris, Devanathan and Muller [6], we define three "integral capacities" per unit area:

$$K_H = \frac{Q}{\phi^M - \phi^H} \; ; \; K_G + \frac{Q}{\phi^H - \phi^B} \; ; \; K = \frac{Q}{\phi^M - \phi^B} \quad , \tag{15}$$

where H signifies the Helmholtz layer, G the Guoy layer and Q a charge. K is the integral capacity of the overall double layer. From Eq. (15) we may obtain the relation:

$$(\phi^M - \phi^H) = \frac{K}{K_H} (\phi^M - \phi^B) \quad , \tag{16}$$

substituting for ϕ^M from Eq. (10):

$$\left(\frac{\mu_e - \bar{\mu}_e}{F} - \phi H\right) = \frac{K}{K_H} \left(\frac{\bar{\mu}_e - \mu_e}{F} - \phi^B\right) ,$$

or $\quad (\mu_e - \bar{\mu}_e)\left(1 - \frac{K}{K_H}\right) = F\phi^H - \frac{K}{K_H} F\phi^B \quad . \tag{17}$

Except close to the potential of zero charge, the capacitance of the Guoy layer is very large, so to a good approximation $\frac{K}{K_H} \simeq 1$, and Eq. (17) becomes

$$(\mu_e - \bar{\mu}_e)\left(1 - \frac{K}{K_H}\right) = F(\phi^H - \phi^B) \quad . \tag{18}$$

The exchange current density is given by

$$i_0 = \vec{k} \, a_Z^H \exp \frac{-\beta(\phi^M - \phi^H)F}{RT} = \overleftarrow{k} \, a_{Z-1}^H \exp \frac{(1-\beta)(\phi^M - \phi^H)F}{RT} \quad , \tag{19}$$

where the ϕ's are now the values at the reversible potential. The activities of the ions at the OHP are related to those in the bulk of the solution by the equations:

$$\left(\frac{a^H}{a^B}\right)_Z = \exp\left[-\frac{ZF}{RT}(\phi^H - \phi^B)\right] \quad , \tag{20a}$$

and

$$\left(\frac{a^H}{a^B}\right)_{Z-1} = \exp\left[-\frac{(Z-1)F}{RT}(\phi^H - \phi^B)\right] \quad , \tag{20b}$$

which are readily derived by equating the appropriate electrochemical potentials. The exponential terms in Eq. (19) may be written in terms of the Galvani potential difference $(\phi^M - \phi^B)$ by the transformation:

$$(\phi^M - \phi^H) = (\phi^M - \phi^B) - (\phi^H - \phi^B) \quad .$$

The exponential dependences of a_Z^H and a_{Z-1}^H upon $(\phi^H - \phi^B)$ (Eq. (20)) result in new expressions for i_0:

$$i_0 = \vec{k}\, a_Z^B \exp\left[\frac{F}{RT}\, (\phi^H - \phi^B)(\beta-Z)\right] \exp\left[\frac{-\beta F}{RT}\, (\phi^M - \phi^B)\right] \qquad (21a)$$

and

$$i_0 = \overleftarrow{k}\, a_{Z-1}^B \exp\left[\frac{-F}{RT}\, (\phi^H - \phi^B)(1-\beta+Z-1)\right] \exp\left[\frac{(1-\beta)F}{RT}\, (\phi^M - \phi^B)\right]. \qquad (21b)$$

The coefficient of $(\phi^H - \phi^B)$ in each case reduces to $\frac{-F}{RT}\,(Z-\beta)$. Note that the value of $(\phi^M - \phi^B)$, while different for different electrodes, is compensated for by differences in $\vec{k}$ and $\overleftarrow{k}$ to satisfy the condition of a fixed Fermi level (Eq. (13)). Thus,

$$\ln i_0 = \frac{-F(Z-\beta)}{RT}\, (\phi^H - \phi^B) + \text{constant},$$

or, from Eq. (8) and Eq. (18)

$$\ln i_0 = \frac{-F(Z-\beta)}{RT} \cdot \frac{1}{F}\, (-N\Phi + F\Delta X_0 - \bar{\mu}_e)\left(1 - \frac{K}{K_H}\right) + \text{constant}$$

so that

$$\frac{d\ln i_0}{d\Phi} = \frac{(Z-\beta)}{kT}\left(1 - \frac{K}{K_H}\right) \qquad (22)$$

assuming that K/K_H and ΔX_0 are independent of Φ. For Fe^{3+}/Fe^{2+}, $Z = 3$ and at 25°C, $kT = 0.026$ eV. The experimental slope of the plot of $\log i_0$ *vs.* Φ was about 2.5 eV^{-1} for both 0.01 M and 0.1 M solution (Fig. 3). Substituting these values in Eq. (22) yields a value for K/K_H of about 0.94, which is not at all unreasonable.

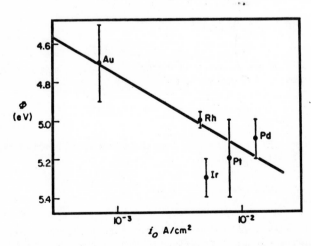

FIG. 3 *The exchange current densities on noble metals vs. their thermionic work functions, in 0.1 M Fe^{3+}/Fe^{2+} (ammonium sulphate) solution [10].*

Bockris, Mannan and Damjanovic [10], following Parsons [48], derived an empirical value for $\dfrac{d\log i_0}{dV_{p.z.c.}}$ from the data and obtained reasonable agreement with a value calculated from published data on mercury.

B.   Semiconductors

The limited supply of carriers at an energy close to the Fermi level in a semiconductor usually results in a smaller exchange current density for a given reaction than that observed on metal electrodes. The kinetics of the reaction are influenced by the concentrations of electrons and holes in the same way as they are by the concentrations of ions in the solution. Factors that influence the concentrations of electrons and holes thus influence i_0, but the maximum value of i_0 that results from such changes cannot exceed the value observed on a metal electrode, since there the kinetics no longer depend on the concentration of carriers.

The space-charge region of a semiconductor is closely analogous to the Guoy layer in an electrolyte. Unlike the Guoy layer, however, the space-charge region often absorbs a major part of the Galvani p.d. Following arguments similar to those in the last section, one obtains a series of relations, *e.g.*, for electrons:

$$(\mu_e^M - \bar{\mu}_e)\left(1 - \frac{K}{K_{SC}}\right) = F\phi^S - \frac{K}{K_{SC}} F\phi^B \quad , \tag{23}$$

and

$$i = \vec{k}\, a_e^S \exp \frac{-\beta(\phi^S - \phi^B)F}{RT} \quad , \tag{24}$$

where M signifies the bulk of the semiconductor, S the surface, and where K_{SC} is the integral space-charge capacity. For a typical semiconductor electrode, K/K_{SC} is close to unity, thus

$$(\mu_e^M - \bar{\mu}_e)\left(1 - \frac{K}{K_{SC}}\right) = F(\phi^S - \phi^B) \quad . \tag{25}$$

Equating electrochemical potentials for electrons in the bulk and at the surface respectively, we obtain

$$a_e^S = a_e^M \exp\left[\frac{-F}{RT}(\phi^M - \phi^S)\right] \quad , \tag{26}$$

so that, making the same assumptions as before, and neglecting double-layer effects in the solution,

$$\frac{d\ln i_0}{d\phi} = \frac{\beta - 1}{kT}\left(1 - \frac{K}{K_{SC}}\right) \quad . \tag{27}$$

A similar relationship is obtained if one considers holes, because for the cathodic formulation selected here, i_0 is inversely proportional to the activity of holes in the surface. From Eq. (21) one can see that the dependence of i_0 upon ϕ increases as $\dfrac{K}{K_{SC}}$ decreases, *i.e.*, as

the portion of the Galvani p.d. across the space-charge region decreases.
One way this can occur is through the introduction of surface states.
Kuznetsov [35] arrived at qualitatively similar conclusions by the ap-
plication of Fermi Dirac statistics to the problem.

3.   INFLUENCE OF THE ELECTRODE MATERIAL
ON BONDING REACTIONS

A.   Metals

Whatever model is selected to account for a charge-transfer step,
the energies of the initial and final states can each be represented
by a surface in "reaction coordinate" space. The surfaces approximate
paraboloids and intersect each other in a U-shaped curve. The most
probable reaction path is *via* the minimum in this curve since it corresponds to the smallest activation energy, and the essential features of
the system may be isolated by considering only the plane (the reaction
plane) defined by this point and the minima of the two paraboloids. As
an example we will consider the cathodic discharge step of a solvated
proton to give a chemisorbed hydrogen atom. A diagram of the reaction
plane is shown for two substrates A and B in Fig. 4. The adsorption

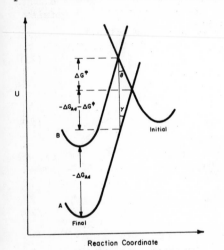

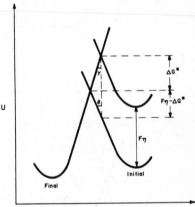

FIG. 4 *Potential energy
curves for proton discharge
on two substrates — A and B.*

FIG. 5 *Potential energy
curves for proton discharge at
two electrode potentials differing by η volts.*

energy of hydrogen is greater by ΔG_{Ad} on substrate A than on substrate
B, and the activation energy for the discharge step is lower on substrate A by an amount $\Delta G^{\ddagger}$. Approximating the intersecting curves by
straight lines, the quantities ΔG_{Ad} and $\Delta G^{\ddagger}$ are related by the equation:

$$\Delta G^{\ddagger} \tan \theta = (-\Delta G_{Ad} - \Delta G^{\ddagger}) \tan \gamma$$

since $\Delta G^{\ddagger}$ and ΔG_{Ad} have opposite signs, or

$$\Delta G^{\ddagger} = -\beta \Delta G_{Ad} \tag{28}$$

where

$$\beta = \frac{\tan \gamma}{\tan \theta + \tan \gamma} \quad .$$

Neglecting possible differences in entropy of adsorption, Eq. (28) may be written in terms of enthalpies of adsorption, thus

$$\frac{i_{0,A}}{i_{0,B}} = \exp\left[\frac{\beta}{RT} (\Delta H_B - \Delta H_A)\right] \quad . \tag{29}$$

By considering Fig. 5, it can readily be shown that the symmetry factor appearing in the expression relating current to overpotential for the discharge step is then equal to $(1-\beta)$. The two initial state curves differ in energy by an amount, $F\eta$, corresponding to a difference in the electrode potentials of η volts. In this case, the change in activation energy ΔG^* is related to η by the equation:

$$\Delta G^* \tan \gamma = (F\eta - \Delta G^*) \tan \gamma \quad ,$$

i.e.,

$$\Delta G^* = (1 - \beta)F\eta \quad , \tag{30}$$

Note also, that if a linearized approximation of the potential energy curves is used (Fig. 6), the energy $\Delta\varepsilon$ is related to the energy ΔE_0 by

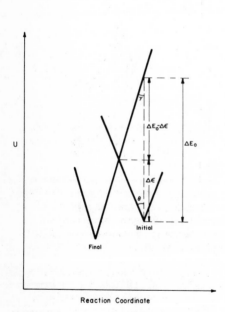

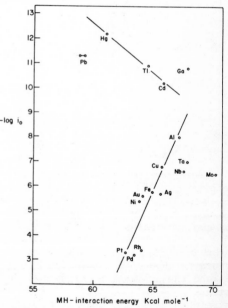

FIG. 6 *Linearized potential energy curves for proton discharge.*

FIG. 7 *The exchange current densities for H.E.R. on a series of metals vs. the calculated M-H bond enery showing two types of dependence [17,43].*

$$\Delta\varepsilon \tan \Theta = (\Delta E_0 - \Delta\varepsilon) \tan \gamma \quad ,$$

or

$$\frac{\Delta\varepsilon}{\Delta E_0} = \beta \quad . \tag{31}$$

The significance of Eq. (31) will emerge in Section 4, below.

Returning to Eq. (29), the rate of the discharge step is clearly favored by a large (negative) adsorption energy. If on the other hand, the rate-determining step involved desorption, *e.g.*,

$$H^+ + H(ads) + e \rightarrow H_2 \quad ,$$

a high rate would be favored by a small adsorption energy.

Both types of dependence have been observed experimentally, see Fig. 7 (updated from Conway and Bockris [17] with the data of Matthews [43]. A linear increase of log i_0 with bond energy on a series of Pt-Au and Pd-Au alloys has also been observed, Fig. 8 [18].

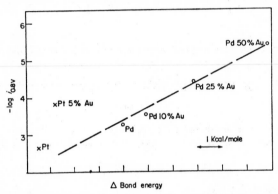

FIG. 8 *The current density at 0.8 V vs. H.E. for oxygen reduction on a series of metals and alloys vs. the calculated relative M-O bond energy [19].*

More complex bonding reactions are exemplified by the oxidation of ethylene on a series of metals and alloys studied by Kuhn, Wroblowa and Bockris [34]. When the activity of the electrocatalysts was plotted against their heat of sublimation L, the points delineated a volcano-shaped curve, Fig. 9. L is a parameter related to the bond strength of the metal and hence, *via* the Pauling [49] relation, to the adsorption bond strength, *e.g.*, for oxygen:

$$D_{M-O} = 1/2(D_{M-M} + D_{O-O}) + 23.06(X_M - X_O)^2 \tag{32}$$

where D is a dissociation energy, X is an electronegativity, and the subscripts refer to the metal, oxygen or bonds between them.

The overall reaction:

$$C_2H_4 + 4 H_2O \rightarrow 2 CO_2 + 12 H^+ + 12 e$$

was found to proceed by the reaction path

$$C_2H_4 \rightleftarrows C_2H_4(ads) \tag{33}$$

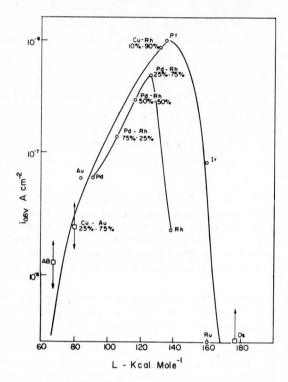

FIG. 9 *The current density at 0.6 V vs. H.E. for ethylene oxida-*
tion on a series of metals and alloys vs. their latent heat of sub-
limation [34].

$$H_2O \rightleftarrows OH(ads) + H^+ + e \quad , \tag{34}$$

$$C_2H_4(ads) + OH(ads) \rightarrow CH_2{-}O{-}CH_3(ads) \quad , \tag{35}$$

and on all substrates except Pt, step (35) was rate determining. The
rate of the reaction is thus

$$i = \text{constant} \cdot \Theta_{Eth} \cdot \Theta_{OH} \cdot \exp\left(-\frac{\Delta G^{\ddagger}}{RT}\right), \tag{36}$$

and is dependent upon the adsorption of two species. The rate-determin-
ing step involves the breaking of an M–O bond, and $\Delta G^{\ddagger}$ should <u>increase</u>
with the bond strengths. However, this effect is offset by the fact
that OH adsorption apparently follows Temkin [52] conditions, whereby
the adsorption energy (and hence $\Delta G^{\ddagger}$) tends to decrease as the coverage
increases. The primary control of rate thus falls upon the Θ terms in
Eq. (36). Both Θ_{Eth} and Θ_{OH} tend to increase with L, but at some
point a competition for adsorption sites begins; OH radicals tend to
win the competition and Θ_{Eth} falls off.

The maximum in the volcano curve occurs at a value of L, corresponding to platinum, where the rate of step (35) is so great that step
(34) is actually rate determining.

B. Semiconductors

Virtually all catalysts of commercial importance are semiconductors. Noble metals are too expensive, and other metals are too unstable for most applications. It is logical then, that the search for new electro-catalysts should now enter the field of semiconductors. Findings are too sparse yet for the formulation of solid theoretical treatments, but some approaches do look hopeful.

Paramagnetic susceptibility (frequently discussed in terms of the apparent number of unpaired d-electrons per atom) has long been recog-nized as a feature of many good catalysts and electrocatalysts. Goldstein and Tseung [28] have proposed a model based on an idea of Evans [22] to account for the importance of paramagnetism for catalysis of the oxygen reduction reaction. The paramagnetic electrons in the electrode are supposed to couple with the paramagnetic electrons in an oxygen molecule and thus facilitate side-on adsorption. In this orientation, easy split-ting of the O-O bond is possible by a mechanism involving a chain of hydrogen bonds — the so-called "pseudo-splitting" mechanism. Arising from the model, a series of promising ferromagnetic electrocatalysts based on $LaCoO_3$ has been developed for oxygen reduction in alkaline solution [45,53].

Davtyan, Misyuk [20] and coworkers have discussed an apparent corre-lation between the hole conductivity of an oxide semiconductor and its activity for oxygen reduction. They propose an explanation in terms of a "hole complex" in which a hole, instead of being localized on a single atom, is distributed over a small group of atoms. The atoms of the hole complex in effect share an unattached bond and thereby possess enhanced chemical activity. Hole complexes at the surface of the electrode are proposed as reaction intermediates in the reduction of oxygen, so that the rate of the reaction will increase with hole concentration in the semiconductor.

Davtyan's model can perhaps be regarded as a classical forerunner to the general scheme proposed by Gerischer [26] of molecular orbital formation between reactants and substrate. Overlap between symmetri-cally compatible wave functions of the semiconductor and an adsorbing species can create new wave functions which may be more favorable for electron exchange with the conduction or valence bands. The formation of a stable bond, however, depends on how the new levels are populated; stability requires that some of the antibonding orbitals remain vacant. The existence of states in the semiconductor (including surface states) that are occupied by unpaired electrons will clearly result in fewer populated orbitals, and this fact constitutes another possible explana-tion for the correlation between paramagnetism and catalysis.

A recently proposed radical new mechanism of chemical catalysis by semiconductors [36] may also find application in explaining electro-catalysis. The idea behind the mechanism is that energy generated at the surface by the annihilation of electron-hole pairs is utilized in activating the reaction being catalyzed. The supply of electron-hole pairs is maintained either by thermal generation in the bulk followed by diffusion to the surface, or by the influence of light. Gerischer [26] has discussed the specific case of photo-activated electrocatalysis in somewhat similar terms.

4.   MECHANISM OF INTERFACIAL CHARGE TRANSFER

The actual mechanism of the charge-transfer process at an electrode surface is still not very clear. The process essentially consists of transferring an electron from the metal to an ion in the solution, or *vice versa*. Several models have been proposed to explain such a process, but none of them can describe the process completely. The models will be discussed in two groups.

A.   The Gurney-Based Model

The first model was proposed by Gurney [31] and later made more sophisticated by other authors [25,11]. It is well known that a large number of charge-transfer processes are accompanied by the emission or absorption of radiation. Since such a radiative process is very rare during an electrode reaction*, the model basically assumes that the charge-transfer process occurring at the electrode is radiationless, imposing the constraint that the energies of the electron in the donor and the acceptor must be equal. For the proton-discharge reaction, the model visualizes the donor electron as being in the Fermi level of the metal, and the acceptor, which is the H_3O^+ ion, as being in its ground state at the OHP. The model then assumes that, due to the vibrations of the H^+-OH_2 bond, the energy levels of the acceptor state can change. When the H_3O^+ species reaches a certain vibrationally excited state, the energy level of the electron state in H_3O^+ becomes equal to the Fermi level of the metal, and a radiationless electron transfer takes place. The rate of the reaction is proportional to the product of the probability of the H_3O^+ species existing in the desired vibrationally excited state and the probability of the electron transfer's taking place by the tunneling mechanism.

The role of overpotential in changing the energy of the initial state (see above) is interpreted on this model in terms of its effect on the minimum H-OH_2 vibrational energy needed to satisfy the tunneling condition. An increase in cathodic overpotential raises the Fermi level and the energy of the initial state so that less energy is required of the H^+-OH_2 bond and a larger fraction of the solvated ions contain electron states suitable for radiationless electron transfer from the metal.

The reaction coordinate can be identified with the spatial location of the hydrogen nucleus between the electrode and the water molecule to which it is initially bonded. The "reaction plane" defined in Section 3 is thus the plane perpendicular to the electrode that includes the hydrogen nucleus. The initial state of the system under consideration is a solvated proton close to the electrode surface and an electron in the Fermi level of the metal. The energy of the initial state may thus be written as $-E_f + E_L + E_R$ where E_f = Fermi energy, E_L = proton solvation energy and E_R = repulsive energy between proton and the metal. E_L and E_R are functions of the distance between H^+ and H_2O, and hence give rise to the initial state energy curve. The final state is the hydrogen atom adsorbed on the metal surface, and the energy is given by $E_I + E_{Ad}$ where E_I = ionization energy of the hydrogen atom and E_{Ad} = adsorption energy of the atom on the electrode surface. E_{Ad} is a function of the distance between the metal and hydrogen atom and

*For a discussion of the exceptions, see Marcus [42].

gives rise to the final state curve. The complete potential energy
profile for the reaction is then as shown in Fig. 10. The vertical

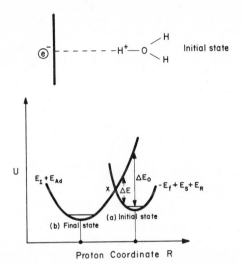

FIG. 10 *The Gurney-type model for proton discharge. The potential
energy surfaces are functions of a single variable — R, the proton
coordinate — and hence can be represented in two dimensions.*

transition ΔE_0 corresponds to the process

$$e(M) + (H_3O^+)_{dl} \rightarrow (M\text{---}H\text{---}OH_2)_{dl} \quad ,$$

i.e., taking an electron from the Fermi level of the metal to an H_3O^+
ion in its ground rotation-vibration state, with no change in the proton discharge coordinate. Thus, such a vertical transition for an H_3O^+
ion not in its ground state will be accompanied by some energy change
ΔE. Stretching of the $H^+\text{-}OH_2$ bond causes ΔE to decrease. Eventually
at the intersection point x in Fig. 10, $\Delta E = 0$. Thus, it is at this
point that radiationless electron transfer from the metal to H_3O^+ becomes possible. The resultant H atom interacts with the water molecules and the metal according to curve (b). At the intersection point

$$-E_f + E_R - E_{Ad} = E_I - E_L \tag{37}$$

So this is the condition for radiationless electron tunneling. Thus,
we find that for a given value of the Fermi level, the vibrational
level to which an electron can transfer depends on the interaction
energy of the hydrogen atom and the metal.

Quantifying this picture, the rate of the hydrogen-evolution reaction may be written as:

$$\vec{i} = e \int n(E_f) P_T \, P(\Delta \varepsilon) \, n_{H_3O^+} \quad , \tag{38}$$

where $n(E_f)$ is the number of electrons with energy E_f, the Fermi energy.

$$n(E_f) = \frac{4\pi m (kT)^2}{h^3} \tag{39}$$

P_T is the tunneling probability. For a rectangular barrier, it is given as

$$P_T = \exp - \frac{4\pi l}{h} [2m(E_x - E_f)]^{1/2} \quad , \tag{40}$$

where E_x is the height of the barrier and l the width of the barrier.

$P(\Delta\varepsilon)$ is the probability that the H^+-H_2O bond is adequately stretched to permit electron tunneling. The Boltzmann expression is used to find this probability

$$P(\Delta\varepsilon) = \exp\left(- \frac{\Delta\varepsilon}{kT}\right) . \tag{41}$$

$n_{H_3O^+}$ is the number of hydrated protons per cm^2 at the OHP which can be expressed according to Bockris and Reddy [12], as:

$$n_{H_3O^+} = 2rN \, C_{H_3O^+} \quad , \tag{42}$$

where r = radius of the water molecule and C = concentration in moles ml^{-1}. So now, selecting the reversible potential in order to define E_f,

$$i_0 = e_0 \frac{4\pi m(kT)^2}{h^3} \exp\left\{- \frac{4\pi l}{h}[2m(E_x - E_f)]^{1/2}\right\} P(\Delta\varepsilon) 2rN \, C_{H_3O^+} . \tag{43}$$

It was shown in Section 3 that an increase in the molar adsorption energy of hydrogen on the electrode by ΔG_{Ad} resulted in a decrease in the molar activation energy by $\beta\Delta G_{Ad}$. The factor β was shown to be equal to the ratio $\frac{\Delta\varepsilon}{\Delta E_0}$, which can now be interpreted as the ratio of the activation energy when H^+-OH_2 bond stretching occurs to the hypothetical activation energy required in the absence of bond stretching. Clearly, an increase in atomic adsorption energy results in a decrease of equal magnitude in ΔE_0. Since $\frac{\Delta\varepsilon}{\Delta E_0} = \beta$ we may write Eq. (41) in the form

$$P(\Delta\varepsilon) = \exp -\left(\frac{\beta\Delta E_0}{kT}\right) , \tag{44}$$

comparing rates on two substrates A and B,

$$i_{0,A} = e \, n(E_f) P_T \, n_{H_3O^+} \exp\left(- \frac{\beta\Delta E_0^A}{kT}\right) , \tag{45a}$$

$$i_{0,B} = e \, n(E_f) P_T \, n_{H_3O^+} \exp\left(- \frac{\beta\Delta E_0^B}{kT}\right) , \tag{45b}$$

so that

$$\frac{i_{0,A}}{i_{0,B}} = \exp\left[\frac{\beta}{kT} (\Delta E_0^B - \Delta E_0^A)\right] . \tag{46}$$

Since $\Delta\Delta E_0$ is equal to the difference in atomic adsorption energy and since this can reasonably be assumed to arise primarily from a difference in the enthalpy of adsorption, Eq. (46) in molar units reduces to:

$$\frac{i_{0,A}}{i_{0,B}} = \exp\left(\beta \frac{\Delta H_B - \Delta H_A}{RT}\right) \;,$$ (47)

confirming the result derived empirically in Section 3. The full expression for i_0 on the Gurney-based model is:

$$i_0 = 2e_0\left[\frac{4\pi m (kT)^2}{h^3}\right] \exp\left\{-\frac{4\pi l}{h}\left[2m(E_x - E_f)\right]^{1/2}\right\}$$

$$\left\{\exp\left[-\frac{\beta\Delta E_0}{kT}\right]\right\} NrC_{H_3O^+} \;.$$ (48)

Both the potential dependence and the dependence on substrate are contained in the term $\left(\frac{\beta\Delta E_0}{kT}\right)$.

B.   The Levich Model

Models belonging to the other group are based on an original suggestion of Libby [40] and developed for nonbonding reactions by Marcus [41] and for both bonding and nonbonding reactions by Levich [37,38]. These models also accept the point that the transfer should be radiationless in nature. However, they criticize the Gurney-based model on the ground that the H^+-OH_2 bond should not be characterized by a simple Boltzmann distribution, but should be treated as a quantum system. Thermal energy cannot then activate the H^+-OH_2 bond to a higher vibrational level because for this bond $h\omega \gg kT$, thus they suggest another source of activation, namely, "solvent fluctuation."

As the solvent dipoles around the central ion fluctuate, the energy of the ion also changes, and for some orientation of the dipoles, the ion gets enough energy so that a radiationless charge transfer becomes possible. A very crude order of magnitude calculation [39] shows that for a reaction with an activation energy of the order of 0.5 eV, all the water molecules within a radius of 20 Å have to fluctuate. The actual process of transfer according to this model, again taking the hydrogen-evolution reaction as an example, is as follows: The potential energy surfaces of the initial and final states are paraboloids shown in Fig. 11, where R is the proton coordinate, and [q] denotes the generalized solvent coordinate. When the solvent fluctuates (*i.e.*, its coordinate [q] changes), the potential energy of the system in the initial state changes as well, and the system moves continuously along the energy surface $U(qR)$, preserving the constant value $R = R_{0,i}$ for the proton coordinate; *i.e.*, oscillations along the bond H^+-OH_2 remain unexcited. At some point, the solvent reaches a coordinate where the energy of the proton becomes equal to the energy of the molecule M-H and the system performs a quantum transition; *i.e.*, the proton crosses over from the normal state in the molecule H^+-OH_2 into the normal state of molecule M-H. This transition takes place through a barrier, which changes continuously during the transition. Then, the solvent, moving continuously along the coordinate [q], leaves the fluctuation state, and the molecule M-H together with the solvent which surrounds it comes to the state with minimal energy $U(0,R_{0,a})$. The transition probability is calculated using adiabatic perturbation method of Platzmann and Frank [50].

The main question that arises from this sort of model is whether it can explain the dependence of the rate on the nature of the substrate. The answer, however, is yes. That such a dependence will exist becomes

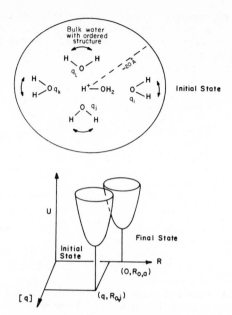

FIG. 11 *The Levich-type model for proton discharge. The potential
energy surfaces are functions of two variables — R, and q, the solvent
polarization coordinate, — and hence are represented in three dimensions.*

obvious when the final expression for the activation energy is examined.
This expression is [27]

$$\Delta\epsilon = \frac{(E_{SR} + \Delta J)^2}{4E_{SR}} \tag{49}$$

where ΔJ is the reaction heat and E_{SR} is the solvent repolarization
energy. The reaction heat, by definition, is dependent on the energy
of the final state, which in this case is the adsorption energy of
hydrogen on the metal. Thus, we should get a rate dependent on the
electrode metal; and although the Levich model for activation is radically different from that of the Gurney-based model, each predicts a
similar type of dependence upon adsorption energy as that of the empirical treatment discussed in Section 3.

It may be noted, in passing, that reasons for accepting the Levich
model for electron transfer do not at present seem pressing. The fact
that Levich has been able to assume approximately 10^3 water molecules
fluctuating together is because he has assumed an ordered lattice
structure of water. He uses the techniques of treating fluctuations
in ionic crystals to calculate the fluctuations in water. This certainly is a very crude approximation and is open to criticism. Moreover it is not at all apparent how this approach can be utilized for
other slightly complicated charge-transfer reactions, like oxygen reduction in acid solution on a bare electrode, where the rate determining step seems to be [19]

$$O_2 + H^+ + e(M) \rightarrow HO_{2\,ads}$$

The objection raised against the use of Boltzmann statistics in the Gurney-based model may be theoretically correct, but the same approximation has been used consistently in chemical kinetics with extremely good results. There does not seem to be any reason why it cannot be applied to electrochemical kinetics.

III. FUTURE DIRECTIONS IN ELECTROCATALYSIS

1. THE NEED FOR FURTHER RESEARCH

Considering that the subject is only 10 to 15 years old, electro-catalysis is remarkably well understood. Unfortunately, fuel cells must compete with devices founded on a century of experience, and in this perspective our understanding of. electrocatalysis is rudimentary at best.

The following paragraphs indicate some of the areas requiring urgent attention.

2. CRITERIA FOR NEW ELECTROCATALYSTS

Stability requirements for an electrocatalyst are very stringent, and only rarely does a material classified as "chemically stable" turn out to withstand satisfactorily the rigors of pH, potential, and re-active gases encountered in a fuel cell. Many inorganic materials purported to be chemically inert have been examined as possible cata-lysts for oxygen reduction by various workers. Conversations with these people indicate similar findings: whatever the bulk composition of the catalyst, the surface rapidly becomes covered with an oxide. The best hope for new catalysts, therefore, is to start with a material that is thermodynamically stable under the conditions of service, and to modify its properties by doping. Clues as to what kinds of modifi-cation should be sought are gradually appearing, *e.g.*, from Tseung's work on $LaCoO_3$ [53], from Davtyan's work on oxide semiconductors [20], and from our own work on Pt-doped Na_xWO_3, reported elsewhere in these proceedings [44].

3. THE CATALYST SURFACE

The abrupt termination of a crystal lattice creates a region — "the surface" — with properties that are bound to differ drastically from those of the bulk material. The bulk properties are *propter hoc* unreliable foundations for any theory of catalysis, and many of our present difficulties in formulating theories of catalysis arise for just this reason. Experimental techniques have been developed over the past few years that in principle are capable of providing all the information needed to characterize a surface. Ellipsometry [7] and internal reflection [4] have demonstrated great potential for *in situ* examination of electrode surfaces (*e.g.*, [7,4]), but have yet to be ex-ploited in the realm of electrocatalysis.

Powerful techniques, eminently well suited for examining surfaces are the various types of electron spectroscopy that have recently been perfected. Primary electron spectroscopy ("ESCA") can yield detailed information on the chemical environment of each element in the surface. Secondary electron spectroscopy ("Auger") can yield chemical informa-tion from a surface layer less than 10 Å thick. Low-energy electron diffraction ("LEED") can yield precise structural information about

the surface atoms. Application of these and other surface techniques
(such as ion probe and electron probe microanalysis) to electrode materials cannot fail to help elucidate the mechanism of electrocatalytic
action.

4. CHEMIBONDING IN THE PRESENCE OF ADSORBED WATER

In predicting probable trends in electrocatalysis for a situation
in which the rate-determining step is known, the bond strengths assumed
cannot be those obtained from gas-phase adsorption measurements (although they may be expected to follow them). The presence of adsorbed
water and ions reduces the bonding energy considerably, but probably
not by the same fraction for all bonds. Thus, Breiter [15] has shown
that the heat of adsorption of hydrogen varies with coverage in three
different ways for platinum, iridium and rhodium electrodes.

5. DOUBLE-LAYER STRUCTURE

As indicated above, some aspects of electrocatalysis concern only
the double-layer structure. Certainly, the most unresearched of all
the aspects discussed here is the double-layer structure at solid electrodes. The only body of knowledge here is from radiotracer studies
[2] and some determinations of the potential of zero charge [1].

The essential purpose for studying the double layer is to obtain
the electric field gradient in the interface. However, this can only
be done if the differential adsorption situation (cation and anion as
a function of distance) is known. Recent developments that show promise for measuring differential interfacial tension at solid-metal-
solution interfaces have been reported by Beck [3], and by Fredlein,
Damjanovic and Bockris [23]. Thereafter, from the Gibbs equation:

$$\Gamma_i = - \left(\frac{\partial \gamma}{\partial \mu_i} \right)_{\Gamma_j, \eta} \quad , \tag{50}$$

the individual ionic surface excesses can be found.

6. RATE-DETERMINING STEP

Establishing the rate-determining step of a reaction is essential
to being able to think about catalysis. Research in the technique of
rapid determinations are necessary, for example, a rationalization of
the potential-sweep patterns.

Experience hitherto suggests that the rate-determining step for
organic oxidation reaction is early in the sequence [9,14,8] and this
greatly reduces the alternative possibilities [21].

7. MECHANISM OF CHARGE TRANSFER

The quantum mechanical model for the charge transfer at an interface is being intensively studied by a large group of Russian theoretical physicists under Levich. However, there are inconsistencies
between prediction and fact in the present formulation, which does beg
development by quantum chemists.

8. MODEL STUDIES

It is necessary to choose a number of simple reactions, and of course bond-forming ones. For example, O_2 reduction and C_2H_4 oxidation seem suitable test reactions.

It is also necessary to make temperature-dependent studies to obtain preexponential as well as heat-of-activation terms (the former often vary catalytically and exert an influence comparable with the latter).

Such model reactions should be used in studies for the establishment of the fundamental data. For example, a rational variation of substrate and bonding, and its effect on enthalpy and entropy of activation could be correlated with spectroscopic investigations of surface structure.

9. PHOTO ACTIVATION

Requirements of stability and economy elimiate almost all materials but semiconductors as candidates for fuel cell electrodes. By definition, the Fermi level of a semiconductor lies in the forbidden energy band, with the result that no electrons or holes are available with the optimum energy for charge-transfer reactions. The situation can be improved 1) by having surface states present in the energy gap; 2) by doping and hence changing the work function (see Section II 2B); or 3) by inducing charge transfer to occur directly to or from the conduction and valence bands. Gerischer [26] has discussed ways in which the last process can occur under the influence of light.

Two effects of light are possible; it can create electron-hole pairs in the semiconductor, or it can excite adsorbed molecules to a reactive state. If electron-hole pairs are created, the relative concentration of minority carriers is likely to be increased, and an electrode reaction that requires them will be stimulated. The effect is catalytically significant if the minority carrier reaction produces a highly reactive radical. For example, the electrooxidation rate of alcohols on ZnO (an n-type semiconductor) proceeds at negligible rates in the dark, but accounts for significant currents under UV illumination. Holes created by the light react with adsorbed alcohol molecules to produce reactive intermediates that can then oxidize further by injecting electrons into the valence band:

$$R-CH_2-OH + p^+ \rightarrow R-CH-OH^* + H^+$$

$$R-CH-OH^* \rightarrow R-CHO + e^- + H^+ \quad .$$

Corresponding effects of light have been observed [46] for the reduction of $S_2O_8^=$ and H_2O_2 on p-type semiconductors.

If light is absorbed by molecules on the surface of the electrode (notably dyes), molecules can be raised to an excited state, where they can inject electrons (or holes) into the semiconductor and thereby be oxidized (or reduced) themselves. The effect can be utilized to oxidize (or reduce) other species in the solution if they can react to regenerate the dyes. Gerischer [26] reports that quantum yields approaching 100 percent have been achieved in this way.

10. NONAQUEOUS ELECTROLYTES

The importance of electrolyte properties in the mechanism of charge-transfer reactions (Section II, 4) suggests that the rates of such reactions would differ if the electrolyte were changed. Both quantum mechanical models indicate that a solvent which interacted weakly with the dissolved ions would favor adsorptive reaction steps, and the Levich model indicates that such steps would also be favored by solvents with large dipole moments. To some extent these requirements may be mutually incompatible, but some optimum situation can probably be found.

A further reason for contemplating nonaqueous solutions is suggested by the work of Hoytink [33], which demonstrated remarkably large rate constants for a series of organic reductions on mercury in dimethyl formamide (DMF) solution. The rate constants were all consistent with the predictions of the Marcus [41] nonbonding theory, meaning that electrocatalysis was not involved. The fact that large reaction rates are possible without the need for adsorption of intermediate species indicates that it should be worth examining the possibility of finding solvents in which electrode reactions proceed by nonbonding mechanisms. At the very least, a nonbonding mechanism would eliminate problems associated with optimizing coverages and orientations of reaction intermediates.

GLOSSARY OF SYMBOLS

ELECTROCHEMICAL QUANTITIES

E^0	Thermodynamic reversible potential in volts
I	Total current in amps
i	Current density in amps cm^{-2}
i_0	Exchange current density
i_L	Limiting current density
K	Integral capacity in microfarads cm^{-2}
n	Number of electrons passed per act of the cell reaction
P	Power in watts
Q	Charge in coulombs
R_{int}	Cell resistance in ohms
R_l	Load resistance in ohms
V	Potential difference in volts
α	Transfer coefficient
β	Symmetry factor
Γ_i	Gibbs Surface Excess of species (i)
η	Overpotential in volts
θ_i	Fractional surface coverage by adsorbed species (i)
ρ	Efficiency of energy conversion
ϕ	Electrostatic (inner) potential in volts
ΔX_0	Surface dipole p.d. *in vacuo*, in volts

ENERGY — MOLAR UNITS (kcals mole^{-1})

D	Bond dissociation energy
G	Gibbs free energy
H	Enthalpy
L	Latent heat of sublimation
μ	Chemical potential
$\bar{\mu}$	Electrochemical potential

ENERGY — ATOMIC UNITS (eV)

E   General energy term
ΔJ Enthalpy of reaction
$\Delta \varepsilon$ Activation energy
ϕ Thermionic work function

MISCELLANEOUS

A   Electrode area
a   Reactant activity
N   Avogadro's number
q   Solvent polarization coordinate
R   Proton coordinate
X   Electronegativity
γ Surface energy in ergs cm^{-2}

SUPERSCRIPTS

B   Bulk electrolyte value
H   Outer Helmholtz Plane value
M   Bulk electrode value
S   Electrode surface value
$\ddagger$,* Activation energy
Z   Ionic charge

SUBSCRIPTS

Ad   Adsorption
dl   Double layer
e   Electron
f   Fermi level
G   Gouy layer
H   Helmholtz layer
I   Ionization
M   Metal
O   Oxygen
p.z.c.   Potential of zero charge
R   Repulsion
S   Solvation
SC   Space charge
SR   Solvent repolarization
z   Species of ionic charge Z

REFERENCES

[1]   S. D. Argade, Ph.D. Thesis, University of Pennsylvania (1968).

[2]   N. A. Balashova and V. E. Kazarinov, *Usp. Khim.*, **34**, 1721 (1965).

[3]   T. A. Beck, *J. Phys. Chem.*, **73**, 466 (1969).

[4]   K. H. Beckmann and N. J. Harrick, in *Optical Properties of Dielectric Films*, N. N. Axelrod (Ed.), Electrochemical Soc., Inc., New York (1968) p. 123.

[5]   J. O'M. Bockris and B. D. Cahan, *J. Chem. Phys.*, **50**, 1307 (1969).

[6]   J. O'M. Bockris, M.A.V. Devanathan and K. Muller, *Proc. Roy. Soc.*,
      (London), A274, 55 (1963).

[7]   J. O'M. Bockris, M. A. Genshaw and V. Brusic, *Disc. Faraday Soc.*,
      in press.

[8]   J. O'M. Bockris, E. Gileadi and G. E. Stoner, *J. Phys. Chem.*,
      73, 427 (1969).

[9]   J. O'M. Bockris, J. W. Johnson and H. Wroblowa, *Electrochim. Acta*,
      9, 636 (1964).

[10] J. O'M. Bockris, R. J. Mannan and A. Damjanovic, *J. Chem. Phys.*,
      48, 1898 (1968).

[11] J. O'M. Bockris and D. B. Matthews, *Proc. Roy. Soc.*, *(London)*,
      A292, 479 (1966).

[12] J. O'M. Bockris and A.K.N. Reddy, *Modern Electrochemistry*,
      Vol. 1, Plenum Press, New York (1970).

[13] J. O'M. Bockris and S. Srinivasan, *Fuel Cells, Their Electro-
      chemistry*, McGraw-Hill Book Company, New York (1969).

[14] J. O'M. Bockris, H. Wroblowa, E. Gileadi and B. J. Piersma,
      Trans. Faraday Soc., 61, 2531 (1965).

[15] M. W. Breiter, *Ann. N. Y. Acad. Sci.*, 101, 709 (1963).

[16] J.A.V. Butler, *Proc. Roy. Soc.*, *(London)*, A157, 423 (1936).

[17] B. E. Conway and J. O'M. Bockris, *J. Chem. Phys.*, 26, 532 (1967).

[18] A. Damjanovic and V. Brusic, *Electrochim. Acta*, 12, 615 (1967).

[19] A. Damjanovic and V. Brusic, *Electrochim. Acta*, 12, 1171 (1967).

[20] O. K. Davtyan, E. G. Misyuk, *et al.*, *Zh. Fiz. Khim.*, 43, 321
      (1969); 43, 1467 (1969); 43, 1475 (1969).

[21] A. R. Despic, *Bull. Acad. Serbe Sci. et Arts, Cl. Sci. Math.
      Nat., Sci. Nat.*, 12 (1969).

[22] U. R. Evans, *Electrochim. Acta*, 14, 197 (1969).

[23] R. A. Fredlein, A. Damjanovic and J. O'M. Bockris, *Surface Sci.*,
      in press.

[24] A. N. Frumkin, *J. Electroanal. Chem.*, 9, 173 (1965).

[25] H. Gerischer, *Z. Physik Chem.*, *(Frankfurt)*, 26, 223 (1960).

[26] H. Gerischer, *Surface Sci.*, 18, 97 (1969).

[27] E. D. German, *et al.*, *Soviet Electrochem.*, 6, 342 (1970).

[28] J. R. Goldstein and A.C.C. Tseung, *Nature*, 222, 869 (1969).

[29] W. R. Grove, *Phil. Mag.*, 14, 127 (1839).

[30] W. T. Grubb, *Nature*, 198, 883 (1963).

[31] R. W. Gurney, *Proc. Roy. Soc.*, *(London)*, A134, 137 (1931).

[32] J. Horiuti and M. Polanyi, *Acta Physicochim.*, *URSS*, 2, 505 (1935).

[33] G. J. Hoytink, *Surface Sci.*, 18, 1 (1969).

[34] A. T. Kuhn, H. Wroblowa and J. O'M. Bockris, *Trans. Faraday Soc.*,
      63, 1458 (1967).

[35] A. M. Kuznetsov, *Electrochim. Acta*, 13, 1293 (1968).

[36] V. J. Lee, *J. Catalysis*, <u>17</u>, 178 (1970).

[37] V. G. Levich, *Advan. Electrochem. Electrochem. Eng.*, <u>4</u>, (1966).

[38] V. G. Levich, *et al.*, *Dokl. Akad. Nauk. SSSR*, <u>179</u>, 137 (1968).

[39] V. G. Levich, <u>in</u> *Physical Chemistry, an Advanced Treatise*, by H. Eyring, D. Henderson and W. Jost (Eds.), Academic Press, New York (1970) Vol. 9B, Chapt. 12.

[40] W. F. Libby, *J. Phys. Chem.*, <u>56</u>, 863 (1952).

[41] R. A. Marcus, *J. Chem. Phys.*, <u>43</u>, 679 (1965).

[42] R. A. Marcus, *J. Chem. Phys.*, <u>43</u>, 2654 (1965).

[43] D. B. Matthews, Ph.D. Thesis, University of Pennsylvania (1965).

[44] J. McHardy and J. O'M. Bockris, these proceedings, p. 109.

[45] D. B. Meadowcroft, *Nature*, <u>226</u>, 847 (1970).

[46] R. Memming, cited by Gerischer [26].

[47] W. Ostwald, *Z. Electrochem.*, <u>1</u>, 122 (1894).

[48] R. Parsons, *Surface Sci.*, <u>2</u>, 418 (1964).

[49] L. Pauling, *The Nature of the Chemical Bond*, Cornell University Press, Ithaca, New York (1948) p. 60.

[50] R. Platzmann and T. Frank, *Zh. Fiz. Khim.*, <u>15</u>, 296 (1941).

[51] S. Srinivasan, H. Wroblowa and J. O'M. Bockris, *Advan. Catal. Relat. Subj.*, <u>17</u>, (1967).

[52] M. I. Temkin, *Zh. Fiz. Khim.*, <u>15</u>, 296 (1941).

[53] A.C.C. Tseung and H. L. Bevan, *Abstract No. 8*, Electrochem. Soc., 138th Meeting, Atlantic City, (October 1970).

SUMMARY OF DISCUSSION AND COMMENTS: SESSION CHAIRMAN VIII

K. V. Kordesch

Union Carbide Corporation, Parma, Ohio

This final session was characterized by interesting talks and argumentations from speakers of completely different backgrounds and commitments to the case of fuel cells.

J. R. Huff, planning for new power sources to provide the U. S. Army with the future energy estimate of 2 kW per man, expressed his belief in the acidic hydrogen-air fuel cell. Alkaline cells of the liquid fuel type, *e.g.*, small hydrazine cells, may fill in the gaps on the lower power scale.

W. T. Reid, raised the spectre of future energy demands in the civilian area to the tune of 15×10^{12} kWh per year in 2,000 A.D., of which 30 percent will be electrical power. He saw energy distribution in form of gas lines as the best solution, worthy of large efforts, but he seemed very skeptical about fuel cells for automobiles, and pointed out some of the hurdles to be overcome.

T. Beck, talking about a rapid transit system utilizing rail vehicles and personal cars in a combination that could use present capabilities, claimed increases in efficiency and lower costs as the main advantages, benefiting both the individual and the environment.

H. Binder, gave a survey about "Trends in Fuel Cell Development." He stressed that the fuel cell, according to a thorough study of various competitive power sources, could replace conventional energy sources even in electrotraction, at least to a certain extent. Providing that cost could be drastically reduced or a service life of ten years could be guaranteed, fuel cells would be able to compete with the otto or diesel engine. If, in addition, the influence of the individual power sources on environment is taken into consideration, *e.g.*, emission of noise or exhaust gases, fuel cells and other electrochemical systems will be superior to other systems.

Prof. J. O'M. Bockris, the last speaker, pleaded for basic research in electrochemistry to attack the obvious but still unknown riddles of electrocatalysis by using interdiscipline research people, capable of grasping the facts in more than one field.

The Discussion revolved around the theme: "Why are fuel cells not in wider use today?" A bright future was promised ten years ago, but now fuel cells are rejected by most U.S. companies as unprofitable — worthy of being investigated on Government contracts, but not deserving any larger investment of internal money. One obvious reason was agreed upon: the initial overemphasis did more harm than good. The forced success of the space programs left the impression of an expensive specialty item. The combined effect of the budget reductions in the U.S. space and military programs and the realization that no commercial application was in sight was disastrous in respect to private continuation of funding for fuel cell work. The cancellations of large technological programs at prestigious U.S. concerns had a psychological effect which was not counterbalanced by the partial shift to civilian programs (*e.g.*, by gas and utility companies) and concentration on new research goals, especially in Europe. People suspect "something must be basically wrong with fuel cells," otherwise "they would sell." To a certain extent the fault lies with the skepticism which always accompanies a new product — the uncertainty about its being able to do what it promised and the danger of financial losses if it fails to conquer the market. The high cost of catalysts (which is not generally applicable) and the expenses of single-item production (which are unavoidable, initially) are used as "explanations" but they sound more like excuses.

The real reason might be that the fuel cell would find its best application in the areas which are now covered by combustion engines, from the small aggregate for charging batteries to the automobile. And in performance and cost fuel cells are just not competitive, except in special applications. In a civilian market, where the premium is not performance under selected conditions at any price, but product value per dollar, the fight against the existing power source is "uphill." An outlawing of the combustion engine is not likely to occur unless the clean-up efforts should fail badly or the fossil fuel supply should get scarce.

On the other hand, when fuel cell technologists honestly search their consciences, they must admit that fuel cells are far from perfection; and in research also, there is plenty to do to improve the state of knowledge.

Low-cost electrodes, less sensitive to mistreatments are needed. Complexity must be removed and systems ought to be simplified. Small fuel cells, which probably have a better chance to compete against other batteries, should be developed. The value of fuel cell work for other parts of electrochemistry should not be underrated.

In connection with the last argument, Prof. Bockris' suggestion to form a "lobby" for electrochemical interests, to stress, among other things, the values of a future "hydrogen ecology," should receive serious thoughts.

However, ups and downs must be expected in any field: the flexible electrochemist is needed, for his own sake and for the good of his science. Fittingly, one session had been named: "Digressions on primary and secondary batteries," probably with the idea in mind that the existing talents could be used most efficiently in neighboring fields.

The writer is a "convinced" fuel cell man (otherwise he would not drive an automobile powered by a hydrogen-air system), but realities must be recognized and, since we cannot anticipate a fuel-cell-powered car in the near future, why not try to improve other batteries to a point where at least a well-performing "second car" is feasible? In the course of future work, some materials problems (which fuel cells have indeed) will be solved and difficulties with accessories and controls (which are common with present fuel cells) will disappear.

The problems of energy storage are becoming increasingly important, not only on the large scale (regenerative fuel cells?) but also in the field of smaller portable power sources, where the trend for "wireless" appliances continues. Galvanic elements have not experienced any recession! The seminar was adjourned with the general feeling that the hosts and participants had benefitted very much from the frank exchange of opinions, and probably every specialist in a given field of batteries had learned a lot from his colleagues.

CONCLUDING REMARKS

G. Sandstede

Battelle Institut e.V., Frankfurt am Main, Germany

Let us conclude the seminar with the questions: What is the present state of fuel cell research and what may be expected in the future? This short review will not be a summary. For an analysis of the present state of the art and future prospects for fuel cells, the reader is referred to the papers presented at this seminar, because the answer is different in each subarea. We may, however, ask what general conclusions can be drawn from this seminar.

1) In the last few years, much effort has been devoted to the development of catalysts. A great deal of mainly empirical research has been conducted to this end.

2) New theoretical concepts of the electrolytic double layer and the process of electrocatalysis have been worked out so that the electrochemical charge-transfer reaction can now be investigated in greater detail.

3) Improved electrode structures have been developed. Theoretical approaches to transport phenomena in porous electrodes have also been made.

4) We have learned a lot about the engineering aspects of fuel batteries, but we are also aware of the fact that batteries and accessories have to be tailor-made for specific applications.

After a closer look at the present situation in the field of fuel cells, we may say:

1) The effort required for future development work will be great. For this reason the work should be carried on jointly by several companies or organizations and each investigation should be directed toward a specific goal. The basis is so well grounded that a well-directed effort of this nature will certainly guarantee success.

2) The profitability of fuel cells and batteries (*e.g.*, hybrid drive) has to be viewed also in connection with the pollution hazard of conventional power sources. When considering the amount of time and effort spent on detoxifying the exhaust gases of engines, the cost involved in the development of electrochemical drives is comparatively low.

3) Future research should not be restricted to empirical methods. This fact has not been generally accepted so that the era of the motto "get results and do not ask why it works" lies more or less behind us. Intuitive empiricism has already received some theoretical support from quantum chemistry and solid state physics. But much more fundamental research in electrocatalysis is needed. Judging from electrode kinetics, however, one may conclude that the rate of acquiring knowledge in this field will depend exponentially on the "activation energy" and "overpotential." It is true, the "overpotential" of the researcher can and must be decreased by financial support, but it is only now, when we are

413

getting enough stimulation from science and impetus from society, that
the "activation energy" can decrease as well. We may therefore trust
that, in the not too distant future, the transition will be made from
the science of electrocatalysis to the technology of new fuel cells
and related devices.

AUTHOR INDEX

AUTHOR	SESSION/NUMBER	AUTHOR	SESSION/NUMBER
Adlhart, O. J.	IV/2	Reid, W. T.	VIII/2
Alt, H.	II/4, VII/4	Sandstede, G.	I/2, I/4, I/5
Asher, W. J.	IV/6		II/1, II/4
			III/1, VII/4
Batzold, J. S.	IV/6	Introduction	xiii
Baukal, W.	V/2, VII/5	Concluding Remarks	411
Beccu, K. D.	VII/3	Sen, R.	VIII/5
Beck, T. R.	VII/7, VIII/3	Shimotake, H.	VII/6
Binder H.	I/2, I/4, I/5	Steunenberg, R. K.	VII/6
	II/1, II/4, III/1		
	VII/4	Sverdrup, E. F.	V/3
Bocciarelli, C. V.	I/3	Tannenberger, H.	II/2, V/1
Bockris, J. O'M.	I/6, II/3, IV/1*	Voïnov, M.	II/2
	IV/4,* IV/6,* VI/1*	von Benda, K.	II/1
	VIII/3,* VIII/5	Warde, C. J.	V/3
Burrows, B.	VII/2	Wolfson, S. K., Jr.	VI/2, VI/3
Cairns, E. J.	IV/7, VII/6	Yao, S. J.	VI/2
Callahan, M. A.	IV/3		
Doniat, D.	IV/4		
Gilman, S.	IV/1, V/4		
Giner, J.	IV/5, VI/1, VII/2		
Glasser, A. D.	V/3	*Commentary	
Gregory, D. P.	III/4		
Gutjahr, M. A.	III/2		
Heath, C. E.	II/5		
Holleck, G.	VI/1, VII/2		
Huff, J. R.	I/1		
Köhling, A.	I/2, I/4, I/5		
	II/1, III/1, VII/4		
Kordesch, K. V.	III/3, VIII/6		
Kuhn, W. H.	III/1		
Lindner, W.	II/4, III/1		
McHardy, J.	II/3, VIII/5		
Markley, F.	VI/2		
Michuda, M.	VI/2		

415